Artificial Intelligence for Edge Computing

Mudhakar Srivatsa • Tarek Abdelzaher • Ting He
Editors

Artificial Intelligence for Edge Computing

 Springer

Editors
Mudhakar Srivatsa
IBM Thomas J. Watson Research Center
New York, NY, USA

Tarek Abdelzaher
University of Illinois at Urbana Champaign
Urbana, IL, USA

Ting He
Pennsylvania State University
University Park, PA, USA

ISBN 978-3-031-40786-4   ISBN 978-3-031-40787-1   (eBook)
https://doi.org/10.1007/978-3-031-40787-1

© The Editor(s) (if applicable) and The Author(s), under exclusive license to Springer Nature Switzerland AG 2023

This work is subject to copyright. All rights are solely and exclusively licensed by the Publisher, whether the whole or part of the material is concerned, specifically the rights of translation, reprinting, reuse of illustrations, recitation, broadcasting, reproduction on microfilms or in any other physical way, and transmission or information storage and retrieval, electronic adaptation, computer software, or by similar or dissimilar methodology now known or hereafter developed.

The use of general descriptive names, registered names, trademarks, service marks, etc. in this publication does not imply, even in the absence of a specific statement, that such names are exempt from the relevant protective laws and regulations and therefore free for general use.

The publisher, the authors, and the editors are safe to assume that the advice and information in this book are believed to be true and accurate at the date of publication. Neither the publisher nor the authors or the editors give a warranty, expressed or implied, with respect to the material contained herein or for any errors or omissions that may have been made. The publisher remains neutral with regard to jurisdictional claims in published maps and institutional affiliations.

This Springer imprint is published by the registered company Springer Nature Switzerland AG
The registered company address is: Gewerbestrasse 11, 6330 Cham, Switzerland

Paper in this product is recyclable.

Preface

It is undeniable that the recent revival of artificial intelligence (AI) has significantly changed the landscape of science in many application domains, ranging from health to defense and from conversational interfaces to autonomous cars. With terms such as "Google Home," "Alexa," and "ChatGPT" becoming household names, the pervasive societal impact of AI is clear. Advances in AI promise a revolution in our interaction with the physical world, a domain where computational intelligence has always been envisioned as a transformative force toward a better tomorrow. Depending on the application family, this domain is often referred to as *Ubiquitous Computing, Cyber-Physical Computing*, or the *Internet of Things*.[1] The underlying vision is driven by the proliferation of cheap embedded computing hardware that can be integrated easily into myriads of everyday devices from consumer electronics, such as personal wearables and smart household appliances, to city infrastructure and industrial process control systems. One common trait across these applications is that the data that the application operates on come directly (typically via sensors) from the physical world. Thus, from the perspective of communication network infrastructure, the data originate at the *network edge*. From a performance standpoint, there is an argument to be made that such data should be processed at the point of collection. Hence, a need arises for Edge AI—a genre of AI where the inference, and sometimes even the training, are performed *at the point of need*, meaning at the edge where the data originate.

This book explores the challenges arising in Edge AI contexts. Some of these challenges (such as neural network model reduction to fit resource-constrained hardware) are unique to the edge environment. They need a novel category of solutions that do not parallel more typical concerns in mainstream AI. Others are adaptations of mainstream AI challenges to the edge space. An example is overcoming the cost of data labeling. The labeling problem is pervasive, but its solution in the IoT application context is different from other contexts. Importantly,

---

[1] In the rest of this book, we collectively refer to this domain as the *Internet of Things* with the understanding that the aforementioned variations of the term are often applicable.

before explaining further what the book is, let us add a few clarifications on what the book is not. This book is not a survey of the state of the art. With thousands of publications appearing in AI every year, such a survey is doomed to be incomplete on arrival. It is also not a comprehensive coverage of all the problems in the space of Edge AI. Different applications pose different challenges, and a more comprehensive coverage should be more application specific. Instead, this book covers some of the more endemic challenges across the range of IoT/CPS applications. To offer coverage in some depth, we opt to cover mainly one or a few representative solutions for each of these endemic challenges in sufficient detail, rather than broadly touching on all relevant prior work. The underlying philosophy is one of *illustrating by example*. The solutions are curated to offer insight into a way of thinking that characterizes Edge AI research and distinguishes its solutions from their more mainstream counterparts. The book is broken down into three parts: core problems, distributed problems, and other cross-cutting issues. Below, we summarize each part.

Part 1: Core Problems

To kick-off a book on Edge AI, inspired by IoT applications, the first question to consider is arguably: what's different about the IoT domain? In the broader AI community, much thought has gone into specializing neural network architecture to various categories of applications, such as vision and natural language processing. Such specialization is driven by data properties that characterize the particular class of applications. In the IoT domain, the key data properties include multi-modal, multi-vantage sources, the time-series nature of observations, and the interface with physical processes that give rise to low-dimensional spatial and temporal data patterns. The first goal of this book is thus to discuss what these properties imply in terms of architectural choices of downstream neural networks that need to process IoT data. Chap. 1 addresses the aforementioned question and reviews a representative example of a neural network architecture, called DeepSense, designed specifically for time-series data.

With a neural network architecture selected and before elaborating on other IoT-inspired Edge AI challenges, it is important to consider the question of training neural networks for Edge AI applications. After all, training is a prerequisite for all that comes next. The key training bottleneck in AI is traditionally the cost of data labeling. Unlabeled data are widely available in the IoT space, but labeling is expensive. This leads to the central question of reducing the logistic tail of AI: how best to exploit the widely available *unlabeled data* to reduce labeling cost while improving inference quality? An inspiration for the answer comes from a key recent development in artificial intelligence – the introduction of foundation models for different categories of applications (such as ChatGPT for conversational interfaces, and CLIP for images). The hallmark of foundation models is their ability to learn in an unsupervised fashion (i.e., without an explicit need to label the data). While

Preface

many approaches were proposed for training foundation models, two of the most important ones are (1) contrastive learning, and (2) masking. Briefly, contrastive learning teaches the model a notion of semantic similarity by presenting it with similar data samples (e.g., an image and its rotation) and contrasting those with randomly chosen samples. Conversely, masking forces the model to extract high-level semantic structure from data by hiding parts of the input and asking the model to reproduce it from the remaining (unmasked) data. Pretraining a model using either approach allows it to leverage large amounts of unlabeled data. How do these two techniques transfer to the IoT domain and how effective are they at reducing the need for data labeling? Chapter 2 answers these questions based on recent work on contrastive learning and masked auto-encoding for the IoT domain that can be thought of as a precursor to the emergence of foundation models for IoT applications.

Once a model is trained, the next question is: how well will it work on new, previously unseen, *test* data? The answer depends on the ability of the model to *generalize* from observations. Models with a very large number of parameters have the power to memorize all input samples, associating each with the desired output (e.g., a class). However, faced with a new sample, such a model might fail dramatically due to its inability to generalize from memorized observations. In contrast, models with fewer parameters (that cannot remember all inputs) need to somehow group their inputs into a smaller number of categories defined by some boundaries, in essence resulting in generalization from multitudes of points to a smaller number of regions in an appropriate latent feature space. The dependence of a model's generalization power on the underlying neural network architecture, network size, the type of nonlinearities involved (e.g., the activation function), and the number and distribution of input samples has been investigated at length in many AI contexts. To give a flavor for this active research area, Chap. 3 investigates the problem for a specific class of simple neural networks that can easily fit on edge devices.

Chapter 4 addresses a twin problem to generalization—namely, *out of distribution detection*. The chapter answers the following important question: can a model, given an input data sample, decide that the sample is outside the ranges to which it has successfully generalized from training observations and as such is unlikely to yield a reliable output? An example solution is presented that relies on auto-encoders. For each intermediate layer of the neural network, an auto-encoder is designed to compress and then reproduce features at that layer. For in-distribution data, the reconstruction is accurate. When an out-of-distribution sample is encountered, it creates feature vectors that are significantly different from those seen in training. Thus, their reconstruction after compression will likely be inaccurate, triggering an out-of-distribution alert. The approach is an example of a category of out-of-distribution detection techniques that rely on a key property of lossy compression (characteristic to auto-encoders)—the fact that the quality of reconstruction (from compressed data) deteriorates when the input to be compressed is significantly outside the training distribution.

viii
Preface

With architecture, training, generalization, and out-of-distribution detection concerns touched upon in preceding chapters, the book turns its attention to what arguably constitutes the primary *inference* challenge of the Edge AI application space, namely, compressing AI models to fit the limited resources of edge devices. Many traditional AI models are designed for large-scale cloud environments with ample GPUs. The computational environment at the edge is substantially different. Specifically, it is much more resource-constrained. Fortunately, often edge applications are also more restricted to a data sub-domain. For example, a vision application used for edge security (e.g., detecting intruders) might only need to recognize a relatively small number of object categories compared to a full-fledged general-purpose vision agent. Can model compression be driven by inference quality needs of only the subset of relevant data categories? Chapter 5 describes such an approach, exemplified in the design of DeepIoT, an application-aware compression framework for neural network architecture. The chapter discusses DeepIoT and its extensions, showing that dramatic improvements in resource footprint are possible, taking advantage of the target data domain without sacrificing inference quality.

Part 2: Distributed Problems

Chapter 6 turns attention to distributed and decentralized settings. It revisits training and considers the case where a large number of devices are involved in carrying out model parameter updates. The key optimization framework for neural network model parameters is *stochastic gradient descent* (SGD). It back-propagates errors (penalties from the loss function being optimized) by changing model parameters in the direction that reduces loss. To do so in a distributed or decentralized fashion, the machines in question must exchange gradients or model parameter updates. The chapter discusses solutions that minimize the amount of bandwidth spent on such data exchange.

Instead of moving the training task to the edge devices where the training data reside as in distributed or decentralized SGD, an alternative approach is to move a compressed version of distributed data to a centralized server or server cluster, where model training can be performed. The advantage of the latter approach is that the heavy portion of machine learning computation is offloaded to the server, alleviating computation load on the edge devices. The disadvantage is the potential loss of accuracy due to the high compression ratio required to fit the limited uplink bandwidth of the edge devices. As an example of this alternative approach, Chap. reviews a family of data compression algorithms designed to support machine learning, by constructing small weighted datasets called *coresets* that can be used in place of the original dataset in training.

In a similar vein to the above, Chap. 8 describes algorithms for lightweight latent state exchange at the *inference* stage to maximize the quality of collective inference in a distributed system, without passing significant amounts of data across the participating nodes.

Preface

Another important concern in a distributed setting is the placement of AI services at the edge. The topic of placement of data services in distributed settings has received considerable attention. The placement problem becomes more involved if service access patterns change over time, for example, due to user mobility. A trade-off arises between bearing the cost of data transfer over a longer distance and migrating the computation closer to the data. As an example of this problem space, Chap. 9 formalizes a version of the service placement problem and offers a candidate solution, while considering user mobility as well as cost of service migration and data transfer.

Chapter 10 extends the above to the joint allocation of multiple types of resources when trying to run machine learning applications on edge computing platforms. In this chapter, the machine learning workloads are modeled as demands for various types of resources (storage, communication, computation), and the resource allocation algorithms are designed to optimally satisfy these demands within the limited resource capacities of edge clouds. Different problem formulations are offered that differ in the performance objective, the types of resources considered, and the forms of resource constraints, all originating from different application scenarios of interest. The chapter discusses how these differences lead to variations in problem complexity and applicable solutions.

Part 3: Cross-Cutting Thoughts

A book on Edge AI remains incomplete without addressing another key overarching problem that arises in resource-constrained environments. Namely, one of prioritizing machine attention to attend to important data first. The problem arises in limited-resource settings where resources need to be allocated most judiciously to maximize their "bang for the buck." Versions of this problem have been addressed in various (non-AI) settings in the real-time systems community, where criticality of various tasks differs for a variety of reasons, such as urgency, semantic importance, or safety implications. Thus, Chap. 11 introduces a criticality-aware data segmentation and resource allocation framework for real-time Edge AI, with a focus on machine perception pipelines. Mainstream machine inference frameworks commonly adopt a simple First-in-First-out (FIFO) policy to process data in a holistic manner, which results in what the chapter calls *algorithmic priority inversion*. In real-time computing, priority inversion is said to occur when data of lower priority are processed ahead of or together with data of higher priority. Algorithmic priority inversion refers to the fact that AI pipelines often do not prioritize data processing by criticality. A solution is described that segments input data into fine-grained regions of different criticality, and processes them in a priority-based manner. A general architecture is described with multiple alternative algorithms for data segmentation, prioritization, and resource allocation respectively for different edge scenarios. Experimental results on autonomous driving applications are presented to show that the framework is able to provide more timely responses to critical data.

Finally, we conclude the book with Chap. 12 that takes a macroscopic practical view of the AI development pipeline, describing the journey of building, training, validating, and deploying machine learning models in production. It addresses challenges that need to be solved to allow an organization to accelerate their end-to-end analytics pipelines from data capture to model deployment, while adhering to governance policies and continuously monitoring and adapting to changes.

New York, NY, USA · Mudhakar Srivatsa
Urbana, IL, USA · Tarek Abdelzaher
University Park, PA, USA · Ting He

Contents

Part I Core Problems

1 Neural Network Models for Time Series Data 3
Shuochao Yao and Tarek Abdelzaher

2 Self-Supervised Learning from Unlabeled IoT Data 27
Dongxin Liu and Tarek Abdelzaher

**3 On the Generalization Power of Overfitted Two-Layer
Neural Tangent Kernel Models** 111
Peizhong Ju, Xiaojun Lin, and Ness B. Shroff

4 Out of Distribution Detection 137
Wei-Han Lee and Mudhakar Srivatsa

5 Model Compression for Edge Computing 153
Shuochao Yao and Tarek Abdelzaher

Part II Distributed Problems

6 Communication Efficient Distributed Learning 199
Navjot Singh, Deepesh Data, and Suhas Diggavi

7 Coreset-Based Data Reduction for Machine Learning at the Edge ... 223
Hanlin Lu, Ting He, and Shiqiang Wang

8 Lightweight Collaborative Perception at the Edge 265
Ila Gokarn, Kasthuri Jayarajah, and Archan Misra

9 Dynamic Placement of Services at the Edge 297
Shiqiang Wang and Ting He

10 Joint Service Placement and Request Scheduling at the Edge 315
Ting He and Shiqiang Wang

Part III   Cross-cutting Thoughts

11   Criticality-Based Data Segmentation and Resource Allocation in Machine Inference Pipelines 335
Shengzhong Liu, Lui Sha, and Tarek Abdelzaher

12   Model Operationalization at Edge Devices 353
Shikhar Kwatra, Utpal Mangla, and Mudhakar Srivatsa

Contributors

Tarek Abdelzaher University of Illinois at Urbana Champaign, Urbana, IL, USA

Deepesh Data Department of Electrical and Computer Engineering, University of California Los Angeles, CA, USA

Suhas Diggavi Department of Electrical and Computer Engineering, University of California Los Angeles, CA, USA

Ila Gokarn Singapore Management University, Singapore, Singapore

Ting He Pennsylvania State University, University Park, PA, USA

Kasthuri Jayarajah University of Maryland, Catonsville, MD, USA

Peizhong Ju Department of ECE, The Ohio State University Columbus, OH, USA

Shikhar Kwatra AWS, Santa Clara, CA, USA

Wei-Han Lee IBM Research, Yorktown Heights, NY, USA

Xiaojun Lin Elmore Family School of Electrical and Computer Engineering, Purdue University West Lafayette, IN, USA

Dongxin Liu University of Illinois at Urbana Champaign, Urbana, IL, USA

Shengzhong Liu Shanghai Jiao Tong University, Shanghai, China

Hanlin Lu Bytedance Inc., Mountain View, CA, USA

Utpal Mangla IBM Thomas J. Watson Research Center, New York, NY, USA

Archan Misra Singapore Management University, Singapore, Singapore

Lui Sha University of Illinois at Urbana Champaign, Urbana, IL, USA

Ness B. Shroff Department of ECE and CSE, The Ohio State University Columbus, OH, USA

Navjot Singh Department of Electrical and Computer Engineering, University of California Los Angeles, CA, USA

Mudhakar Srivatsa IBM Thomas J. Watson Research Center, New York, NY, USA

Shiqiang Wang IBM T. J. Watson Research Center, Yorktown Heights, NY, USA

Shuochao Yao George Mason University, Fairfax, VA, USA

Part I
Core Problems

Chapter 1
Neural Network Models for Time Series Data

Shuochao Yao and Tarek Abdelzaher

Abstract A wide range of mobile sensing and computing applications require time-series measurements from such sensors as accelerometers, gyroscopes, and magnetometers to generate inputs for various signal estimation and classification applications (Lane et al., IEEE Commun. Mag., 2010).

1 Introduction

A wide range of mobile sensing and computing applications require time-series measurements from such sensors as accelerometers, gyroscopes, and magnetometers to generate inputs for various signal estimation and classification applications [1]. Using these sensors, mobile devices are able to infer user activities and states [2, 3] and recognize surrounding context [4, 5]. These capabilities serve diverse application areas including health and wellbeing [6–8], tracking and imaging [9, 10], mobile security [11, 12], and vehicular road sensing [13–15].

Although mobile sensing is becoming increasingly ubiquitous, key challenges remain in improving the accuracy of sensor exploitation. In this chapter, we consider the general problem of estimating signals from noisy measurements in mobile sensing applications. This problem can be categorized into two subtypes: regression and classification, depending on whether prediction results are continuous or categorical, respectively.

For regression-oriented problems, such as tracking and localization, sensor inputs are usually processed based on physical models of the phenomena involved. Sensors on mobile devices generate time-series measurements of physical quantities such as acceleration and angular velocity. From these measurements, other physical

---

S. Yao (✉)
George Mason University, Fairfax, VA, USA
e-mail: shuochao@gmu.edu

T. Abdelzaher
University of Illinois at Urbana Champaign, Urbana, IL, USA
e-mail: zaher@illinois.edu

© The Author(s), under exclusive license to Springer Nature Switzerland AG 2023
M. Srivatsa et al. (eds.), *Artificial Intelligence for Edge Computing*,
https://doi.org/10.1007/978-3-031-40787-1_1

quantities can be computed, such as displacement through double integration of acceleration over time. However, measurements of commodity sensors are noisy. The noise in measurements is non-linear [16] and correlated over time [17], which makes it hard to model. This makes it challenging to separate signal from noise, leading to estimation errors and bias.

For classification-oriented problems, such as activity and context recognition, a typical approach is to compute appropriate features derived from raw sensor data. These hand-crafted features are then fed into a classifier for training. This general workflow for classification face the challenge that designing good hand-crafted features can be time consuming; it requires extensive experiments to generalize well to diverse settings such as different sensor noise patterns and heterogeneous user behaviors [3].

In this chapter, we propose DeepSense, a unified deep learning framework that directly addresses the aforementioned customization challenges that arise in mobile sensing applications. The core of DeepSense is the integration of convolutional neural networks (CNN) and recurrent neural networks (RNN). Input sensor measurements are split into a series of data intervals along time. The frequency representation of each data intervals is fed into a CNN to learn intra-interval local interactions within each sensing modality and intra-interval global interactions among different sensor inputs, hierarchically. The intra-interval representations along time are then fed into an RNN to learn the inter-interval relationships. The whole framework can be easily customized to fit specific mobile computing (regression or classification) tasks by three simple steps, as will be described later.

For the regression-oriented mobile sensing problem, DeepSense learns the composition of physical system and noise model to yield outputs from noisy sensor data directly. The neural network acts as an approximate transfer function. The CNN part approximates the computation of sensing quantities within the time interval, and the RNN part approximates the computation of sensing quantities across time intervals. Instead of using a model-based noise analysis method that assumes a noise model with experience or observations, DeepSense can be regarded as a model-free noise analysis that learns the non-linear and correlated-over-time noises among sensor measurements.

For the classification-oriented mobile sensing problem, the neural network acts as an automatic feature extractor encoding local, global, and temporal information. The CNN part extracts local features within each sensor modality and merges the local features of different sensory modalities into global features hierarchically. The RNN part extracts temporal dependencies.

We demonstrate the effectiveness of our DeepSense framework using three representative and challenging mobile sensing problems, which illustrate the potential of solving different tasks with a single unified modeling methodology:

- *Car tracking with motion sensors:* In this task, we use dead reckoning to infer position from acceleration measurements. One of the major contributions of DeepSense is its ability to withstand nonlinear and time-dependent noise and bias. We chose the car tracking task because it involves double-integration and

1 Neural Network Models for Time Series Data

thus is particularly sensitive to error accumulation, as acceleration errors can lead to significant deviations in position estimate over time. This task thus constitutes a worst-case of sorts in terms of emphasizing the effects of noise on modelling error. Traditionally, external means are needed to reset the error when possible [13, 18, 19]. We intentionally forgo such means to demostrate the capability of DeepSense for learning accurate models of target quantities in the presence of realistic noise.

- *Heterogeneous human activity recognition:* Although human activity recognition with motion sensors is a mature problem, Stisen et al. [3] illustrated that state-of-the-art algorithms do not generalize well across users when a new user is tested who has not appeared in the training set. This classification-oriented problem therefore illustrates the capability of DeepSense to extract features that generalize better across users in mobile sensing tasks.
- *User identification with biometric motion analysis:* Biometric gait analysis can be used to identify users when they are walking [2, 20]. We extend walking to other activities, such as biking and climbing stairs, for user identification. This classification-oriented problem illustrates the capability of DeepSense to extract distinct features for different users or classes.

We evaluate these three tasks with collected data or existing datasets. We compare DeepSense to state-of-the-art algorithms that solve the respective tasks, as well as to three DeepSense variants, each presenting a simplification of the algorithm as described in Sect. 4.3. For the regression-oriented problem: car tracking with motion sensors, DeepSense provides an estimator with far smaller tracking error. This makes tracking with solely noisy on-device motion sensors practical and illustrates the capability of DeepSense to perform accurate estimation of physical quantities from noisy sensor data. For the other two classification-oriented problems, DeepSense outperforms state-of-the-art algorithms by a large margin, illustrating its capability to automatically learn robust and distinct features. DeepSense outperforms all its simpler variants in all three tasks, which shows the effectiveness of its design components. Despite a general shift towards remote cloud processing for a range of mobile applications, we argue that it is intrinsically desirable that heavy sensing tasks be carried out locally on-device, due to the usually tight latency requirements, and the prohibitively large data transmission requirement as dictated by the high sensor sampling frequency (e.g. accelerometer, gyroscope). Therefore, we also demonstrate the feasibility of implementing and deploying DeepSense on mobile devices by showing its moderate energy consumption and low overhead for all three tasks on two different types of smart devices.

In summary, the main contribution of this chapter is that *we develop a deep learning framework, DeepSense, that solves both regression-oriented and classification-oriented mobile computing tasks in a unified manner. By exploiting local interactions within each sensing modality, merging local interactions of different sensing modalities into global interactions, and extracting temporal relationships, DeepSense learns the composition of physical laws and noise model in regression-oriented problems, and automatically extracts robust and distinct features that contain local, global, and temporal relationships in classification-oriented problems. Importantly, it outperforms the state of the art, while remaining implementable on mobile devices.*

2 DeepSense Framework

Recently, deep learning [21] has become one of the most popular methodologies in AI-related tasks, such as computer vision [22], speech recognition [23], and natural language processing [24]. Lots of deep learning architectures have been proposed to exploit the relationships embedded in different types of inputs. For example, Residual nets [22] introduce shortcut connections into CNNs, which greatly reduces the difficulty of training super-deep models. However, since residual nets mainly focus on visual inputs, they lose the capability to model temporal relationships, which are of great importance in time-series sensor inputs. LRCNs [25] apply CNNs to extract features for each video frame and combine video frame sequences with LSTM [26], which exploits spatio-temporal relationships in video inputs. However, it does not consider modeling multimodal inputs. This capability is important to mobile sensing and computing tasks, because most tasks require collaboration among multiple sensors. Multimodal DBMs [27] merge multimodal inputs, such as images and text, with Deep Boltzmann Machines (DBMs). However, the work does not model temporal relationships and does not apply tailored structures, such as CNNs, to effectively and efficiently exploit local interactions within input data. To the best of our knowledge, DeepSense is the first architecture that possesses the capability for both (1) modelling temporal relationships and (2) fusing multimodal sensor inputs. It also contains specifically designed structures to exploit local interactions in sensor inputs.

There are several illuminating studies, applying deep neural network models to different mobile sensing applications. DeepEar [28] uses Deep Boltzmann Machines to improve the performance of audio sensing tasks in an environment with background noise. RBM [29] and MultiRBM [30] use Deep Boltzmann Machines and Multimodal DBMs to improve the performance of heterogeneous human activity recognition. IDNet [20] applies CNNs to the biometric gait analysis task. DeepX [31], RedEye [32], and ConvTransfer [33] reduce the energy consumption or training time of deep neural networks, based on software and hardware, respectively. However, these studies do not capture the temporal relationships in time-series sensor inputs, and, with the only exception of MultiRBM, lack the capability of fusing multimodal sensor inputs. In addition, these techniques focus on classification-oriented tasks only. To the best of our knowledge, DeepSense is the first framework that directly solves both regression-based and classification-based problems in a unified manner.

We introduce DeepSense, a unified framework for mobile applications with sensor data inputs, in this section. We separate our description into three parts. The first two parts, convolutional layers and recurrent layers, are the main building blocks for DeepSense, which are the same for all applications. The third part, the output layer, is the specific layer for two different types of applications; regression-oriented and classification-oriented.

For the rest of this chapter, all vectors are denoted by bold lower-case letters (e.g., **x** and **y**), while matrices and tensors are represented by bold upper-case letters (e.g., **X** and **Y**). For a vector **x**, the jth element is denoted by $\mathbf{x}_{[j]}$. For a tensor **X**, the tth matrix along the third axis is denoted by $\mathbf{X}_{\cdot\cdot t}$, and other slicing denotations are defined similarly. We use calligraphic letters to denote sets (e.g., $\mathcal{X}$ and $\mathcal{Y}$). For any set $\mathcal{X}$, $|\mathcal{X}|$ denotes the cardinality of $\mathcal{X}$.

For a particular application, we assume that there are K different types of input sensors $\mathcal{S} = \{S_k\}, k \in \{1, \cdots, K\}$. Take a sensor S_k as an example. It generates a series of measurements over time. The measurements can be represented by a $d^{(k)} \times n^{(k)}$ matrix **V** for measured values and $n^{(k)}$-dimensional vector **u** for time stamps, where $d^{(k)}$ is the dimension for each measurement (e.g., measurements along x, y, and z axes for motion sensors) and $n^{(k)}$ is the number of measurements. We split the input measurements **V** and **u** along time (i.e., columns for **V**) to generate a series of non-overlapping time intervals with width τ, $\mathcal{W} = \{(\mathbf{V}_t^{(k)}, \mathbf{u}_t^{(k)})\}$, where $|\mathcal{W}| = T$. Note that, τ can be different for different intervals, but here we assume a fixed time interval width for succinctness. We then apply Fourier transform to each element in $\mathcal{W}$, because the frequency domain contains better local frequency patterns that are independent of how time-series data is organized in the time domain [34]. We stack these outputs into a $d^{(k)} \times 2f \times T$ tensor $\mathbf{X}^{(k)}$, where f is the dimension of frequency domain containing f magnitude and phase pairs. The set of resulting tensors for each sensor, $\mathcal{X} = \{\mathbf{X}^{(k)}\}$, is the input of DeepSense.

As shown in Fig. 1.1, DeepSense has three major components; the convolutional layers, the recurrent layers, and the output layer, stacked from bottom to top. In the following subsections, we detail these components, respectively.

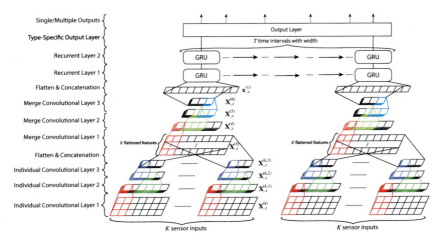

Fig. 1.1 Main architecture of the DeepSense framework

2.1 Convolutional Layers

The convolutional layers can be further separated into two parts: an individual convolutional subnet for each input sensor tensor $\mathbf{X}^{(k)}$, and a single merge convolutional subnet for the output of K individual convolutional subnets' outputs.

Since the structures of individual convolutional subnet for different sensors are the same, we focus on one individual convolutional subnet with input tensor $\mathbf{X}^{(k)}$. Recall that $\mathbf{X}^{(k)} \in \mathbb{R}^{d^{(k)} \times 2f \times T}$, where $d^{(k)}$ is the sensor measurement dimension, f is the dimension of frequency domain, and T is the number of time intervals. For each time interval t, the matrix $\mathbf{X}^{(k)}_{..t}$ will be fed into a CNN architecture (with three layers in this chapter). There are two kinds of features/relationships embedded in $\mathbf{X}^{(k)}_{..t}$ we want to extract. The relationships within the frequency domain and across sensor measurement dimension. The frequency domain usually contains lots of local patterns in some neighbouring frequencies. And the interaction among sensor measurement usually including all dimensions. Therefore, we first apply 2d filters with shape $(d^{(k)}, cov1)$ to $\mathbf{X}^{(k)}_{..t}$ to learn interaction among sensor measurement dimensions and local patterns in frequency domain, with the output $\mathbf{X}^{(k,1)}_{..t}$. Then we apply 1d filters with shape $(1, cov2)$ and $(1, cov3)$ hierarchically to learn high-level relationships, $\mathbf{X}^{(k,2)}_{..t}$ and $\mathbf{X}^{(k,3)}_{..t}$.

Then we flatten matrix $\mathbf{X}^{(k,3)}_{..t}$ into vector $\mathbf{x}^{(k,3)}_{..t}$ and concat all K vectors $\{\mathbf{x}^{(k,3)}_{..t}\}$ into a K-row matrix $\mathbf{X}^{(3)}_{..t}$, which is the input of the merge convolutional subnet. The architecture of the merge convolutional subnet is similar as the individual convolutional subnet. We first apply 2d filters with shape $(K, cov4)$ to learn the interactions among all K sensors, with output $\mathbf{X}^{(4)}_{..t}$, and then apply 1d filters with shape $(1, cov5)$ and $(1, cov6)$ hierarchically to learn high-level relationships, $\mathbf{X}^{(5)}_{..t}$ and $\mathbf{X}^{(6)}_{..t}$.

For each convolutional layer, DeepSense learns 64 filters, and uses ReLU as the activation function. In addition, batch normalization [35] is applied at each layer to reduce internal covariate shift. We do not use residual net structures [22], because we want to simplify the network architecture for mobile applications. Then we flatten the final output $\mathbf{X}^{(6)}_{..t}$ into vector $\mathbf{x}^{(f)}_{..t}$; concatenate $\mathbf{x}^{(f)}_{..t}$ and time interval width, $[\tau]$, together into $\mathbf{x}^{(c)}_t$ as inputs of recurrent layers.

2.2 Recurrent Layers

Recurrent neural networks are powerful architectures that can approximate function and learn meaningful features for sequences. Original RNNs fall short of learning long-term dependencies. Two extended models are Long Short-Term Memory (LSTM) [26] and Gated Recurrent Unit (GRU) [36]. In this chapter, we choose GRU, because GRUs show similar performance as LSTMs on various tasks [36],

1 Neural Network Models for Time Series Data

while having a more concise expression, which reduces network complexity for mobile applications.

DeepSense chooses a stacked GRU structure (with two layers in this chapter). Compared with standard (single-layer) GRUs, stacked GRUs are a more efficient way to increase model capacity [21]. Compared to bidirectional GRUs [37], which contain two time flows from start to end and from end to start, stacked GRUs can run incrementally, when there is a new time interval, resulting in faster processing of stream data. In contrast, we cannot run bidirectional GRUs until data from all time intervals are ready, which is infeasible for applications such as tracking. We apply dropout to the connections between GRU layers [38] for regularization and apply recurrent batch normalization [39] to reduce internal covariate shift among time steps. Inputs $\{\mathbf{x}_t^{(c)}\}$ for $t = 1, \cdots, T$ from previous convolutional layers are fed into stacked GRU and generate outputs $\{\mathbf{x}_t^{(r)}\}$ for $t = 1, \cdots, T$ as inputs of the final output layer.

2.3 Output Layer

The output of recurrent layer is a series of vectors $\{\mathbf{x}_t^{(r)}\}$ for $t = 1, \cdots, T$. For the regression-oriented task, since the value of each element in vector $\mathbf{x}_t^{(r)}$ is within ± 1, $\mathbf{x}_t^{(r)}$ encodes the output physical quantities at the end of time interval t. In the output layer, we want to learn a dictionary $\mathbf{W}_{out}$ with a bias term $\mathbf{b}_{out}$ to decode $\mathbf{x}_t^{(r)}$ into $\hat{\mathbf{y}}_t$, such that $\hat{\mathbf{y}}_t = \mathbf{W}_{out} \cdot \mathbf{x}_t^{(r)} + \mathbf{b}_{out}$. Therefore, the output layer is a fully connected layer on the top of each interval with sharing parameter $\mathbf{W}_{out}$ and $\mathbf{b}_{out}$.

For the classification task, $\mathbf{x}_t^{(r)}$ is the feature vector at time interval t. The output layer first needs to compose $\{\mathbf{x}_t^{(r)}\}$ into a fixed-length feature vector for further processing. Averaging features over time is one choice. More sophisticated methods can also be applied to generate the final feature, such as the attention model [24], which has illustrated its effectiveness in various learning tasks recently. The attention model can be viewed as weighted averaging of features over time, but the weights are learnt by neural networks through context. In this chapter, we still use averaging features over time to generate the final feature, $\mathbf{x}^{(r)} = (\sum_{t=1}^{T} \mathbf{x}_t^{(r)})/T$. Then we feed $\mathbf{x}^{(r)}$ into a softmax layer to generate the predicted category probability $\hat{\mathbf{y}}$.

3 Task-Specific Customization

In this section, we first describe how to trivially customize the DeepSense framework to different mobile sensing and computing tasks. Next, we instantiate the solution with three specific tasks used in our evaluation.

3.1 General Customization Process

In general, we need to customize a few parameters of the main architecture of DeepSense, shown in Sect. 2, for specific mobile sensing and computing tasks. Our general DeepSense customization process is as follows:

1. Identify the number of sensor inputs, K. Pre-process the sensor inputs into a set of tensors $\mathcal{X} = \{\mathbf{X}^{(k)}\}$ as input.
2. Identify the type of the task. Whether the application is regression or classification-oriented. Select one of the two types of output layer according to the type of task.
3. Design a customized cost function or choose the default cost function (namely, mean square error for regression-oriented tasks and cross-entropy error for classification-oriented tasks).

Therefore, if opt for the default DeepSense configuration, we need only to set the number of inputs, K, preprocess the input sensor measurements, and identify the type of task (i.e., regression-oriented versus classification-oriented).

The pre-processing is simple, as stated at the beginning of Sect. 2. We just need to align and chunk the sensor measurements, and apply Fourier transform to each sensor chunk. For each sensor, we stack these frequency domain outputs into $d^{(k)} \times 2f \times T$ tensor $\mathbf{X}^{(k)}$, where $d^{(k)}$ is the sensor measurement dimension, f is the frequency domain dimension, and T is the number of time intervals.

To identify the number of sensor inputs K, we usually set K to be the number of different sensing modalities available. If there exist two or more sensors of the same modality (e.g., two accelerometers or three microphones), we just treat them as one multi-dimensional sensor and set its measurement dimension accordingly.

For the cost function, we can design our own cost function other than the default one. We denote our DeepSense model as function $\mathcal{F}(\cdot)$, and a single training sample pair as $(\mathcal{X}, \mathbf{y})$. We can express the cost function as:

$$\mathcal{L} = \ell(\mathcal{F}(\mathcal{X}), \mathbf{y}) + \sum_j \lambda_j P_j \qquad (1.1)$$

where $\ell(\cdot)$ is the loss function, P_j is the penalty or regularization function, and λ_j controls the importance of the penalty or regularization term.

3.2 Customize Mobile Sensing Tasks

In this section, we provide three instances of customizing DeepSense for specific mobile computing applications used in our evaluation.

Car Tracking with Motion Sensors (CarTrack) In this task, we apply accelerator, gyroscope, and magnetometer to track the trajectory of a car without initial

1 Neural Network Models for Time Series Data

speed. Therefore, according to our general customization process, carTrack is a regression-oriented problem with $K = 3$ (i.e. accelerometer, gyroscope, and magnetometer). Instead of applying default mean square error loss function, we design our own cost function according to Eq. (1.1).

During the training step, the ground-truth 2D displacement of car in each time interval, $\mathbf{y}$, is obtained by GPS signal, where $\mathbf{y}_{[t]}$ denotes the 2D displacement in time interval t. Yet a problem is that GPS signal also contains noise. Training the DeepSense model to recover the displacement obtained from by GPS signal will generate sub-optimal results. We apply Kalman filter to covert displacement $\mathbf{y}_{[t]}$ into a 2D Gaussian distribution $\mathbf{Y}_{[t]}(\cdot)$ with mean value $\mathbf{y}^{(t)}$ in time interval t. Therefore, we use negative log likelihood as loss function $\ell(\cdot)$ with additional penalty terms:

$$
\mathcal{L} = -\log\left(\mathbf{Y}_{[t]}\big(\mathcal{F}(\mathcal{X})_{[t]}\big)\right)
$$
$$
+ \sum_{t=1}^{T} \lambda \cdot \max\left(0, \cos(\theta) - S_c\big(\mathcal{F}(\mathcal{X})_{[t]}, \mathbf{y}^{(t)}\big)\right)
$$

where $S_c(\cdot, \cdot)$ denotes the cosine similarity, the first term is the negative log likelihood loss function, and the second term is a penalty term controlled by parameter λ. If the angle between our predicted displacement $\mathcal{F}(\mathcal{X})_{[t]}$ and $\mathbf{y}^{(t)}$ is larger than a pre-defined margin $\theta \in [0, \pi)$, the cost function will get a penalty. We introduce the penalty, because we find that predicting a correct direction is more important during the experiment, as described in Sect. 4.4.1.

Heterogeneous Human Activity Recognition (HHAR) In this task, we perform leave-one-user-out cross-validation on human activity recognition task with accelerometer and gyroscope measurements. Therefore, according to our general customization process, HHAR is a classification-oriented problem with $K = 2$ (accelerometer and gyroscope). We use the default cross-entropy cost function as the training objective.

$$
\mathcal{L} = H(\mathbf{y}, \mathcal{F}(\mathcal{X}))
$$

where $H(\cdot, \cdot)$ is the cross entropy for two distributions.

User Identification with Motion Analysis (UserID) In this task, we perform user identification with biometric motion analysis. We classify users' identity according to accelerometer and gyroscope measurements. Similarly, according to our general customization process, UserID is a classification-oriented problem with $K = 2$ (accelerometer and gyroscope). Similarly as above, we use the default cross-entropy cost function as the training objective.

This chapter focuses on solving different mobile sensing and computing tasks in a unified framework. DeepSense is our solution. It is a framework that requires only a few steps to be customized into particular tasks. During the customization steps, we do not tailor the architecture for different tasks in order to lessen the requirement of human efforts while using the framework. However, particular changes to the architecture can bring additional performance gains to specific tasks.

One possible change is separating noise model and physical laws for regression-oriented tasks. The original DeepSense directly learns the composition of noise model and physical laws, providing the capability of automatically understanding underlying physical process from data. However, if we know exactly the physical process, we can use DeepSense as a powerful denoising component, and apply physical laws to the outputs of DeepSense.

The other possible change is removing some design components to trade accuracy for energy. In our evaluations, we show that some variants take acceptable degradation on accuracy with less energy consumption. The basic principle of removing design components is based on their functionalities. Individual convolutional subnets explore relationship within each sensor; merge convolutional subnet explores relationship among different sensors; and stacked RNN increases the model capacity for exploring relationship over time. We can choose to omit some components according to the demands of particular tasks.

At last, for a particular sensing task, if there is drastic change in the physical environment, DeepSense might need to be re-trained with new data. However, on one hand, the traditional solution with pre-defined noise model and physical laws (or hand-crafted features) would also need redesigns anyways. On the other hand, an existing trained DeepSense framework can serve as a good initialization stage for the new training process that aids in optimization and reduce generalization error [23].

4 Evaluation

In this section, we evaluate DeepSense on three mobile computing tasks. We first introduce the experimental setup for each, including datasets and baseline algorithms. We then evaluate the three tasks based on accuracy, energy, and latency. We use the abbreviations, CarTrack, HHAR, and UserID, as introduced in Sect. 3.2, to refer to the aforementioned tasks.

4.1 Data Collection and Datasets

For the CarTrack task, we collect 17,500 phone-miles worth of driving data. Namely, we collect around 500 driving hours in total using three cars fitted with 20 mobile phones in the Urbana-Champaign area. Mobile devices include Nexus 5, Nexus 4, Galaxy Nexus, and Nexus S. Each mobile device collects measures of accelerometer, gyroscope, magnetometer, and GPS. GPS measurements are collected roughly every second. Collection rates of other sensors are set to their highest frequency. After obtaining the raw sensor measurements, we first segment them into data samples. Each data sample is a zero-speed to zero-speed journey,

1 Neural Network Models for Time Series Data

where the start and termination are detected when there are at least three consecutive zero GPS speed readings. Each data sample is then separated into time intervals according to the GPS measurements. Hence, every GPS measurement is an indicator of the end of a time interval. In addition, each data sample contains one additional time interval with zero speed at the beginning. Furthermore, for each time interval, GPS latitude and longitude are converted into map coordinates, where the origin of coordinates is the position at the first time interval. Fourier transform is applied to each sensor measurement in each time interval to obtain the frequency response of the three sensing axes. The frequency responses of the accelerator, gyroscope, and magnetometer at each time interval are then composed into the tensors as DeepSense inputs. At last, for evaluation purposes, we apply a Kalman filter to coordinates obtained by the GPS signal, and generate the displacement distribution of each time interval. The results serve as ground truth for training.

For both the HHAR and UserID tasks, we use the dataset collected by Allan et al. [3]. This dataset contains readings from two motion sensors (accelerometer and gyroscope). Readings were recorded when users executed activities scripted in no specific order, while carrying smartwatches and smartphones. The dataset contains 9 users, 6 activities (biking, sitting, standing, walking, climbStair-up, and climbStair-down), and 6 types of mobile devices. For both tasks, accelerometer and gyroscope measurements are model inputs. However, for HHAR, activities are used as labels, and for UserID, users' unique IDs are used as labels. We segment raw measurements into 5-second samples. For DeepSense, each sample is further divided into time intervals of length τ, as shown in Fig. 1.1. We take $\tau = 0.25$ s. Then we calculate the frequency response of sensors for each time interval, and compose results from different time intervals into tensors as inputs.

4.2 Evaluation Platforms

Our evaluation experiments are conducted on two platforms: Nexus 5 with Qualcomm Snapdragon 800 SoC [40] and Intel Edison Compute Module [41]. We train DeepSense on Desktop with GPU. And trained DeepSense models are run solely on mobile with CPU: quad core 2.3 GHz Krait 400 CPU on Nexus 5 and dual-core 500 MHz Atom processor on Intel Edison. In this chapter, we do not exploit the additional computation power of mobile GPU and DSP units [31].

4.3 Algorithms in Comparison

We evaluate our DeepSense model and compare it with other competitive algorithms in three tasks. There are three global baselines, which are the variants of DeepSense model by removing one design component in the architecture. The other baselines are specifically designed for each single task.

DS-singleGRU: This model replaces the 2-layer stacked GRU with a single-layer GRU with larger dimension, while keeping the number of parameters. This baseline algorithm is used to verify the efficiency of increasing model capacity by staked recurrent layer.

DS-noIndvConv: In this mode, there are no individual convolutional subnets for each sensor input. Instead, we concatenate the input tensors along the first axis (i.e., the input measurement dimension). Then, for each time interval, we have a single matrix as the input to the merge convolutional subnet directly.

DS-noMergeConv: In this variant, there are no merge convolutional subnets at each time interval. Instead, we flatten the output of each individual convolutional subnet and concatenate them into a single vector as the input of the recurrent layers.

CarTrack Baseline:

- **GPS:** This is a baseline measurement that is specific to the CarTrack problem. It can be viewed as the ground truth for the task, as we do not have other means of more accurately acquiring cars' locations. In the following experiments, we use the GPS module in Qualcomm Snapdragon 800 SoC.
- **Sensor-fusion:** This is a sensor fusion based algorithm. It combines gyroscope and accelerometer measurements to obtain the pure acceleration without gravity. It uses accelerometer, gyroscope, and magnetometer to obtain absolute rotation calibration. Android phones have proprietary solutions for these two functions [42]. The algorithm then applies double integration on pure acceleration with absolute rotation calibration to obtain the displacement.
- **eNav (w/o GPS):** eNav is a map-aided car tracking algorithm [13]. This algorithm constrains the car movement path according to a digital map, and computes moving distance along the path using double integration of acceleration derived using principal component analysis that removes gravity. The original eNav uses GPS when it believes that dead-reckoning error is high. For fairness, we modified eNav to disable GPS.

HHAR Baselines:
- **HAR-RF:** This algorithm [3] selects all popular time-domain and frequency domain features from [43] and ECDF features from [44], and uses random forest as classifier.
- **HAR-SVM:** Feature selection of this model is same as the HAR-RF model. But this model uses support vector machine as classifier [3].
- **HRA-RBM:** This model is based on stacked restricted Boltzmann machines with frequency domain representations as inputs [29].
- **HRA-MultiRBM:** For each sensor input, the model processes it with a single stacked restricted Boltzmann machine. Then it uses another stacked restricted Boltzmann machine to merge the results for activity recognition [30].

UserID Baselines:

- **GaitID:** This model extracts the gait template and identifies user through template matching with support vector machine [45].

1 Neural Network Models for Time Series Data

- **IDNet:** This model first extracts the gait template, and extracts template features with convolutional neural networks. Then this model identifies user through support vector machine and integrates multiple verifications with Wald's probability ratio test [20].

4.4 Effectiveness

In this section, we will discuss the accuracy and other related performance metrics of the DeepSense model, compared with other baseline algorithms.

4.4.1 CarTrack

We use 253 zero-speed to zero-speed car driving examples to evaluate the CarTrack task. The histogram of evaluation data driving distance is illustrated in Fig. 1.2.

During the whole evaluation, we regard filtered GPS signal as ground truth. CarTrack is a regression problem. Therefore, we first evaluate all algorithms with mean absolute error (MAE) between predicted and true final displacements with 95% confidence interval except for the eNav (w/o GPS) algorithm, which is a map-aided algorithm without tracking real trajectories. The results about mean absolute errors are illustrated in the second column of Table 1.1.

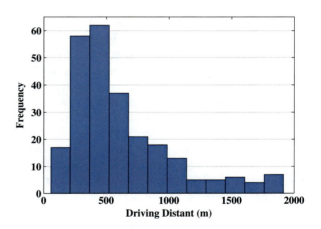

Fig. 1.2 Histogram of driving distance

Table 1.1 CarTrack task accuracy

	MAE (meter)	Map-aided accuracy
DeepSense	**40.43 ± 5.24**	**93.8%**
DS-SingleGRU	44.97 ± 5.80	90.2%
DS-noIndvConv	52.15 ± 6.24	88.3%
DS-noMergeConv	53.06 ± 6.59	87.5%
Sensor-fusion	606.59 ± 56.57	
eNav (w/o GPS)		6.7%

Bold values reflect the evaluation performance of techniques that are presented in the current section

Compared with senior-fusion algorithm, DeepSense reduces the tracking error by an order of magnitude, which is mainly attributed to its capability to learn the composition of noise model and physical laws. Then, we compare our DeepSense model with three variants as mentioned before. The results show the effectiveness of each designing component of our DeepSense model. The individual and merge convolutional subnets learn the interaction within and among sensor measurements respectively. The stacked recurrent structure increases the capacity of model more efficiently. Removing any component will cause performance degradation.

DeepSense model achieves 40.43 ± 5.24 m mean absolute error. This is almost equivalent to half of traditional city blocks (80 m $\times$ 80 m), which means that, with the aid of map and the assumption that car is driving on roads, DeepSense model has a high probability to provide accurate trajectory tracking. Therefore, we propose a naive map-aided track method here. For each segment of original tracking trajectory, we assign them to the most probable road segment on map (i.e., the nearest road segment on map). We then compare the resulted trajectory with ground truth. If all the trajectory segments are the same as the ground truth, we regard it as a successful tracking trajectory. Finally, we compute the percentage of successful tracking trajectories as accuracy. eNav (w/o GPS) is a map-aided algorithm, so we directly compare the trajectory segments. Sensor-fusion algorithm generates tracking errors that are comparable to driving distances, so we exclude it from the comparison. We show the accuracy of map-aided versions of algorithms in the third column of Table 1.1. DeepSense outperforms eNav (w/o GPS) with a large margin, because eNav (w/o GPS) intrinsically depends on occasional on-demand GPS samples to correct tracking error.

We next examine how tracking performance is affected by driving distances. We first sort all evaluation samples according to driving distance. Then we separate them into 10 groups with 200 m step size. Finally, we compute mean absolute error and accuracy of map-aided track for DeepSense algorithm separately for each group. We illustrate the results in Fig. 1.3. For the mean absolute error metric, driving longer distance generally results in large error, *but the error does not accumulate linearly over distance*. There are mainly two reasons for this phenomenon. On one hand, we observe that the error of our predicted trajectory usually occurs during the beginning of the driving, where uncertainty in predicting driving direction is the major cause. This is also the motivation that we add the penalty term for cost function in Sect. 3.2. On the other hand, longer-driving cases in our testing samples are more stable, because we extract the trajectory from zero-speed to zero-speed. For the map-aided track, longer driving distances even yields slightly better accuracy. This is because long-distance trajectory usually contains long trajectory segments, which can help to find the ground truth on the map.

Finally, some our DeepSense tracking results (without the help of map and with downsampling) are illustrated in Fig. 1.4.

1 Neural Network Models for Time Series Data

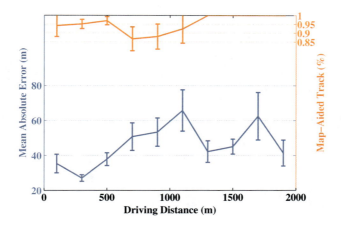

Fig. 1.3 Performance over driving distance

4.4.2 HHAR

For HHAR task, we perform leave-one-user-out evaluation (i.e., leaving the whole data from one user as testing data) on datasets consisting of 9 users, which are labelled from a to i. We illustrate the result of evaluations according to three metrics: accuracy, macro F_1 score, and micro F_1 score with 95% confidence interval in Fig. 1.5.

The DeepSense based algorithms (including DeepSense and three variants) outperform other baseline algorithms with a large margin (i.e., at least 10%). Compared with two hand-crafted feature based algorithms HAR-RF and HAR-SVM, DeepSense model can automatically extract more robust features, which generalize better to the user who does not appear in the training set. Compared with a deep model, such as HAR-RBM and HAR-MultiRBM, DeepSense model exploit local structures within sensor measurements, dependency along time, and relationships among multiple sensors to generate better and more robust features from data. Compared with three variants, DeepSense still achieves the best performance (accuracy: 0.942 ± 0.032, macro F_1: 0.931 ± 0.041, and micro F_1: 0.942 ± 0.032). This reinforces the effectiveness of our design components in DeepSense model.

Then we illustrate the confusion matrix of best-performing DeepSense model in Fig. 1.6. Predicting Sit as $Stand$ is the largest error. It is hard to classify these two, because two activities should have similar motion sensor measurements by nature, especially when we have no prior information about testing users. In addition, the algorithm has a minor error about misclassification between $ClimbStair$-up and $ClimbStair$-$down$.

Fig. 1.4 Examples of tracking trajectory without the help of map: Blue trajectory (DeepSense) and Red trajectory (GPS)

4.4.3 UserID

This task focuses on user identification with biometric motion analysis. We evaluate all algorithms with 10-fold cross validation. We illustrate the result of evaluations according to three metrics: accuracy, macro F_1 score, and micro F_1 score with 95% confidence interval in Fig. 1.7. Specifically, The figure on the left shows the results when algorithms observe 1.25 seconds of evaluation data, the figure on the right shows the results when algorithms observe 5 seconds of evaluation data.

DeepSense and three variants outperform other baseline algorithms with a large margin again (i.e. at least 20%). Compared with the template extraction and matching method, GaitID, DeepSense model can automatically extract distinct

1 Neural Network Models for Time Series Data

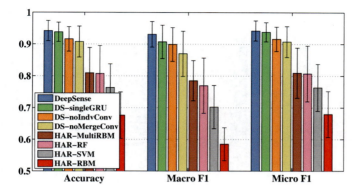

Fig. 1.5 Performance metrics of HHAR task

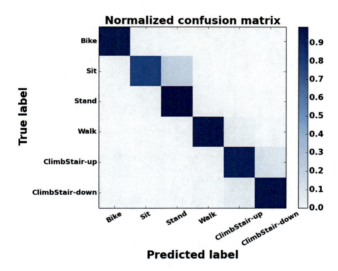

Fig. 1.6 Confusion matrix of HHAR task

features from data, which fit well to not only walking but also all other kinds of activities. Compared with method that first extracts templates and then apply neural network to learn features, IDNet, DeepSense solves the whole task in the end-to-end fashion. We eliminate the manually processing part and exploit local, global, and temporal relationships through our architecture, which results better performance. In this task, although the performance of different variants is similar when observing data with 5 seconds, DeepSense still achieves the best performance (accuracy: 0.997 ± 0.001, macro F_1: 0.997 ± 0.001, and micro F_1: 0.997 ± 0.001).

We further compare DeepSense with three variants by changing the number of evaluation time intervals from 5 to 20, which corresponds to around 1 to 5 seconds. We compute the accuracy for each case. The results illustrated in Fig. 1.8 suggest

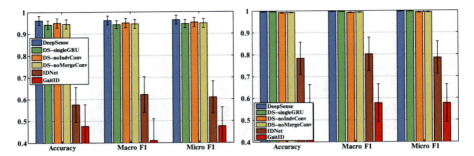

Fig. 1.7 Performance metrics of UserID task for different time intervals: 1.25 s (left) and 5 s (right)

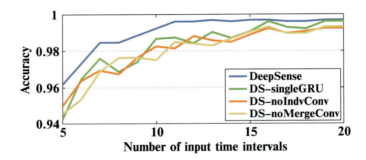

Fig. 1.8 Accuracy over input measurement length of UserID task

that DeepSense performs better than all the other variants with a relatively large margin when algorithms observe sensing data with shorter time. This indicates the effectiveness of design components in DeepSense.

Then we illustrate the confusion matrix of best-performing DeepSense model when observing sensing data with 5 seconds in Fig. 1.9. It shows that the algorithm gives a pretty good result. On average, only about two misclassifications appear during each testing.

4.5 Latency and Energy

Final, we examine the computation latency and energy consumption of DeepSens stereotypical deep learning models are traditionally power hungry and time consuming. We illustrate, through our careful measurements in all three example application scenarios, the feasibility of directly implementing and deploying DeepSense on mobile devices without any additional optimization.

Experiments measure the whole process on smart devices including reading the raw sensor inputs and are conducted on two kinds of devices: Nexus 5 and Intel

1 Neural Network Models for Time Series Data

Fig. 1.9 Confusion matrix of UserID task

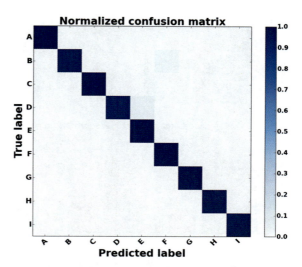

Fig. 1.10 Test platforms: Nexus5 and Intel Edison

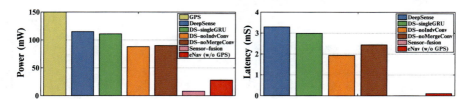

Fig. 1.11 Power and Latency of carTrack solutions on Nexus 5

Edison, as shown in Fig. 1.10. The energy consumption of applications on Nexus 5 is measured by PowerTutor [46], while the energy consumption of Intel Edison is measured by an external power monitor. The evaluations of energy and latency on Nexus 5 are shown in Figs. 1.11, 1.12, and 1.13, and Intel Edison Figs. 1.14, 1.15, and 1.16. Since algorithms for carTrack are designed to report position every second,

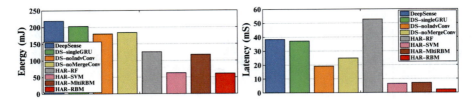

Fig. 1.12 Energy and Latency of HHAR solutions on Nexus 5

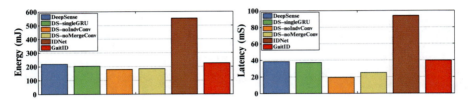

Fig. 1.13 Energy and Latency of UserID solutions on Nexus 5

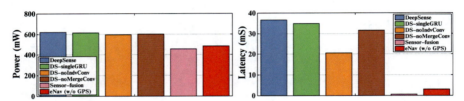

Fig. 1.14 Power and Latency of carTrack solutions on Edison

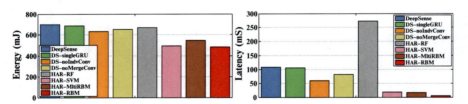

Fig. 1.15 Energy and Latency of HHAR solutions on Edison

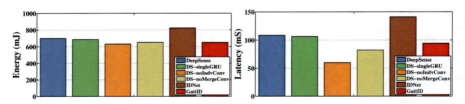

Fig. 1.16 Energy and Latency of UserID solutions on Edison

we show the power consumption in Fig. 1.14. Other two tasks are not periodical tasks by nature. Therefore, we show the per-inference energy consumption in Figs. 1.15 and 1.16. For experiments on Intel Edison, notice that we measured total energy consumption, containing 419 mW idle-mode power consumption.

For the carTrack task, all DeepSense based models consume a bit less energy compared with 1-Hz GPS samplings on Nexus 5. The running times are measured in the order of microsecond on both platforms, which meets the requirement of per-second measurement.

For the HHAR task, all DeepSense based models take moderate energy and low latency to obtain one classification prediction on two platforms. An interesting observation is that HHAR-RF, a random forest model, has a relatively longer latency. This is due to the fact that random forest is an ensemble method, which involves combining a bag of individual decision tree classifiers.

For the UserID task, except for the IDNet baseline, all other algorithms show similar running time and energy consumption on two platforms. IDNet contains both a multi-stage pre-processing process and a relative large CNN, which takes longer time and more energy to compute in total.

References

1. N.D. Lane, E. Miluzzo, H. Lu, D. Peebles, T. Choudhury, A.T. Campbell, A survey of mobile phone sensing. IEEE Commun. Mag. (2010)
2. Y. Ren, Y. Chen, M. C. Chuah, J. Yang, Smartphone based user verification leveraging gait recognition for mobile healthcare systems, in *SECON* (2013)
3. A. Stisen, H. Blunck, S. Bhattacharya, T.S. Prentow, M.B. Kjærgaard, A. Dey, T. Sonne, M.M. Jensen, Smart devices are different: Assessing and mitigatingmobile sensing heterogeneities for activity recognition, in *Sensys* (2015)
4. S. Nath, Ace: exploiting correlation for energy-efficient and continuous context sensing, in *MobiSys* (2012)
5. C. Xu, S. Li, G. Liu, Y. Zhang, E. Miluzzo, Y.-F. Chen, J. Li, B. Firner, Crowd++: unsupervised speaker count with smartphones, in *UbiComp* (2013)
6. J. Ko, C. Lu, M.B. Srivastava, J.A. Stankovic, A. Terzis, M. Welsh, Wireless sensor networks for healthcare. Proc. IEEE (2010)
7. M. Rabbi, M.H. Aung, M. Zhang, T. Choudhury, Personal sensing: understanding mental health using ubiquitous sensors and machine learning, in *UbiComp* (2015)
8. C.-Y. Li, C.-H. Yen, K.-C. Wang, C.-W. You, S.-Y. Lau, C. C.-H. Chen, P. Huang, H.-H. Chu, Bioscope: an extensible bandage system for facilitating data collection in nursing assessments, in *UbiComp* (2014)
9. T. Li, C. An, Z. Tian, A. T. Campbell, X. Zhou, Human sensing using visible light communication, in *MobiCom* (2015)
10. Y. Zhu, Y. Zhu, B.Y. Zhao, H. Zheng, Reusing 60ghz radios for mobile radar imaging, in *MobiCom* (2015)
11. E. Miluzzo, A. Varshavsky, S. Balakrishnan, R.R. Choudhury, Tapprints: your finger taps have fingerprints, in *MobiSys* (2012)
12. C. Wang, X. Guo, Y. Wang, Y. Chen, B. Liu, Friend or foe? Your wearable devices reveal your personal pin, in *AsiaCCS* (2016)

13. S. Hu, L. Su, S. Li, S. Wang, C. Pan, S. Gu, M.T. Al Amin, H. Liu, S. Nath, et al., Experiences with enav: a low-power vehicular navigation system, in *UbiComp* (2015)
14. L. Kang, B. Qi, D. Janecek, S. Banerjee, Ecodrive: a mobile sensing and control system for fuel efficient driving, in *MobiCom* (2015)
15. Y. Zhao, S. Li, S. Hu, L. Su, S. Yao, H. Shao, T. Abdelzaher, Greendrive: a smartphone-based intelligent speed adaptation system with real-time traffic signal prediction, in *ICCPS* (2017)
16. W.T. Ang, P.K. Khosla, C.N. Riviere, Nonlinear regression model of a low-g mems accelerometer. IEEE Sensors J. (2007)
17. M. Park, *Error Analysis and Stochastic Modeling of MEMS-Based Inertial Sensors for Land Vehicle Navigation Applications*. (Library and Archives Canada, 2005). Bibliothèque et Archives Canada
18. G. Chandrasekaran, T. Vu, A. Varshavsky, M. Gruteser, R.P. Martin, J. Yang, Y. Chen, Tracking vehicular speed variations by warping mobile phone signal strengths, in *PerCom* (2011)
19. K. Lin, A. Kansal, D. Lymberopoulos, F. Zhao, Energy-accuracy aware localization for mobile devices, in *MobiSys* (2010)
20. M. Gadaleta, M. Rossi, Idnet: Smartphone-based gait recognition with convolutional neural networks (2016). arXiv:1606.03238
21. I.G.Y. Bengio, A. Courville, *Deep Learning* (MIT Press, 2016). Book in preparation
22. K. He, X. Zhang, S. Ren, J. Sun, Deep residual learning for image recognition (2015). *arXiv:1512.03385*
23. G.E. Dahl, D. Yu, L. Deng, A. Acero, Context-dependent pre-trained deep neural networks for large-vocabulary speech recognition. IEEE TASLP (2012)
24. D. Bahdanau, K. Cho, Y. Bengio, Neural machine translation by jointly learning to align and translate (2014). *arXiv:1409.0473*
25. J. Donahue, L. Anne Hendricks, S. Guadarrama, M. Rohrbach, S. Venugopalan, K. Saenko, T. Darrell, Long-term recurrent convolutional networks for visual recognition and description, in *CVPR* (2015)
26. K. Greff, R.K. Srivastava, J. Koutník, B.R. Steunebrink, J. Schmidhuber, Lstm: a search space odyssey (2015). *arXiv:1503.04069*
27. N. Srivastava, R.R. Salakhutdinov, Multimodal learning with deep boltzmann machines, in *NIPS* (2012)
28. N.D. Lane, P. Georgiev, L. Qendro, Deepear: robust smartphone audio sensing in unconstrained acoustic environments using deep learning, in *UbiComp* (2015)
29. S. Bhattacharya, N.D. Lane, From smart to deep: robust activity recognition on smartwatches using deep learning, in *PerCom Workshops* (2016)
30. V. Radu, N.D. Lane, S. Bhattacharya, C. Mascolo, M.K. Marina, F. Kawsar, Towards multimodal deep learning for activity recognition on mobile devices, in *UbiComp: Adjunct* (2016)
31. N.D. Lane, S. Bhattacharya, P. Georgiev, C. Forlivesi, L. Jiao, L. Qendro, F. Kawsar, Deepx: a software accelerator for low-power deep learning inference on mobile devices, in *IPSN* (2016)
32. R. LiKamWa, Y. Hou, J. Gao, M. Polansky, L. Zhong, Redeye: analog convnet image sensor architecture for continuous mobile vision, in *ISCA* (2016), pp. 255–266
33. F.J.O. Morales, D. Roggen, Deep convolutional feature transfer across mobile activity recognition domains, sensor modalities and locations, in *ISWC* (2016)
34. O. Rippel, J. Snoek, R.P. Adams, Spectral representations for convolutional neural networks, in *NIPS* (2015)
35. S. Ioffe, C. Szegedy, Batch normalization: accelerating deep network training by reducing internal covariate shift (2015). *arXiv:1502.03167*
36. J. Chung, C. Gulcehre, K. Cho, Y. Bengio, Empirical evaluation of gated recurrent neural networks on sequence modeling (2014). *arXiv:1412.3555*
37. M. Schuster, K.K. Paliwal, Bidirectional recurrent neural networks. IEEE Trans Sig. Process. (1997)
38. W. Zaremba, I. Sutskever, O. Vinyals, Recurrent neural network regularization (2014). *arXiv:1409.2329*

1 Neural Network Models for Time Series Data

39. T. Cooijmans, N. Ballas, C. Laurent, A. Courville, Recurrent batch normalization (2016). *arXiv:1603.09025*
40. Qualcomm Snapdragon 800 Processor, https://www.qualcomm.com/products/snapdragon/processors/800
41. Intel Edison Compute Module, http://www.intel.com/content/dam/support/us/en/documents/edison/sb/edison-module_HG_331189.pdf
42. G. Milette, A. Stroud, *Professional Android Sensor Programming* (John Wiley & Sons, Hoboken, 2012)
43. D. Figo, P.C. Diniz, D.R. Ferreira, J.M. Cardoso, Preprocessing techniques for context recognition from accelerometer data. Pers. Ubiquit. Comput. (2010)
44. N.Y. Hammerla, R. Kirkham, P. Andras, T. Ploetz, On preserving statistical characteristics of accelerometry data using their empirical cumulative distribution, in *ISWC* (2013)
45. H.M. Thang, V.Q. Viet, N.D. Thuc, D. Choi, Gait identification using accelerometer on mobile phone, in *ICCAIS* (2012)
46. L. Zhang, B. Tiwana, Z. Qian, Z. Wang, R.P. Dick, Z.M. Mao, L. Yang, Accurate online power estimation and automatic battery behavior based power model generation for smartphones, in *CODES+ISSS* (2010)

Chapter 2
Self-Supervised Learning from Unlabeled IoT Data

Dongxin Liu and Tarek Abdelzaher

Abstract With a network architecture selected and before elaborating other inference challenges in a book on Edge AI, it behooves us to consider the question of training neural networks for edge AI applications. Training is a pre-requisite of all that comes next. The key training bottleneck in AI is traditionally the cost of data labeling. Unlabeled data are widely available in the IoT space, but labeling is expensive. This leads to the central question covered in this chapter: how best to exploit the widely-available unlabeled data to reduce labeling cost while improving inference quality? An inspiration for the answer comes from a key recent development in artificial intelligence—the introduction of foundation models for different categories of applications (such as ChatGPT for conversational interfaces, and CLIP for images). The hallmark of foundation models is their ability to learn in an unsupervised fashion (i.e., without an explicit need to label the data). While many approaches were proposed for training foundation models, two of the most important ones are (1) contrastive learning, and (2) masking. Briefly, contrastive learning teaches the model a notion of semantic similarity by presenting it with similar data samples (e.g., an image and its rotation) and contrasting those with randomly chosen samples. Masking forces the model to extract high-level semantic structure from data by masking parts of the input and asking the model to reproduce it from the remaining (unmasked) data. Pretraining a model using either approach allows it to leverage large amounts of unlabeled data. How do these two techniques transfer to the IoT domain and how effective are they at reducing the need for data labeling? The chapter answers these questions from recent work on contrastive learning and masked auto-encoding for the IoT domain that can be thought of as a precursor of the emergence of foundation models for IoT applications.

D. Liu (✉) · T. Abdelzaher
University of Illinois at Urbana Champaign, Urbana, IL, USA
e-mail: dongxin3@illinois.edu; zaher@illinois.edu

© The Author(s), under exclusive license to Springer Nature Switzerland AG 2023
M. Srivatsa et al. (eds.), *Artificial Intelligence for Edge Computing*,
https://doi.org/10.1007/978-3-031-40787-1_2

1 Introduction

The Internet of Things (IoT) describes physical objects (or groups of such objects) that are embedded with sensors, processing ability, software, and other technologies that connect and exchange data with other devices and systems over the Internet or other communications networks.[1] Due to the development of ubiquitous computing, embedded systems, wireless sensor networks, control systems, and machine learning, the field of IoT has got great evolvement. Most of the current IoT applications measure physical phenomenas (e.g., acceleration, vibration, or wireless signal propagation) with corresponding sensors, and the measurements can be used for different tasks including health and wellness [1–3], behavior and activity recognition [4–7], context sensing [8–11], and object tracking and detection [12–17].

Motivated by the increasing popularity of deep neural networks [18] that have demonstrated outstanding performance on various machine learning tasks, such as image classification [19, 20], object detection [21, 22], and natural language processing [23]. Deep learning techniques have also been widely studied in the field of IoT and got great success in IoT applications. Compared with the traditional approaches which designed machine learning models (such as SVM [24]) based on the manually designed features, the deep neural network models directly take the sensing signals as input and demonstrate much higher performance. Specific neural network models have been designed to fuse multiple sensory modalities and extract temporal relationships for sensing applications. These models have shown significant improvements on multiple IoT applications such as audio sensing[25, 26], tracking and localization [27, 28], and human activity recognition [29, 30].

Deep neural networks are traditionally trained using *supervised* learning algorithms that need a large amount of labeled training data. As deep learning techniques mature, the underlying neural network models often become deeper, thereby exacerbating the need for large-scale labeled datasets. Even when collecting the training data is not difficult, labeling or annotating the datasets usually requires extensive human labor work. As a result, self-supervised learning algorithms, which is a kind of unsupervised learning technique and build models only based on *unlabeled data*, have drawn more attention in recent years [31–35]. These techniques obviate extensive labeling burden, while attaining comparable performance with their supervised learning counterparts. In self-supervised learning scenarios, an encoder (that maps original inputs into lower-dimensional latent features or representations) is trained using *unlabeled data*. The learned latent feature can then be further used to accomplish a variety of downstream tasks, such as classification or regression by training a much simpler model (*e.g.*, a linear model) that takes the latent feature (instead of the original data) as input. Only a small amount of labeled data is needed to train such (much simpler) models. As alluded to earlier, the big win

---

[1] https://www.techtarget.com/iotagenda/definition/Internet-of-Things-IoT

2 Self-Supervised Learning from Unlabeled IoT Data

lies in the fact that the bulk of the training (to generate the underlying latent representation) uses *unlabeled data*. Self-supervised learning has gotten outstanding performance in computer vision and NLP, but still not got a lot of attentions in IoT applications. Considering the unique frequency characteristics of sensing data for IoT applications, directly applying self-supervised learning framework from computer vision or NLP to IoT context would not get the best performance.

In this chapter, we introduce the work we have done on customizing the self-supervised learning frameworks to IoT applications by carefully taking the frequency domain characteristics into consideration. The central design philosophy is inspired by the unique frequency domain characteristics that exist in the sensing signals of the IoT applications. IoT applications often measure physical phenomena, where the underlying processes (such as acceleration, vibration, or wireless signal propagation) are fundamentally a function of signal frequencies and thus have sparser and more compact representations in the frequency domain. We believe that designing the self-supervised learning frameworks for IoT applications from the perspective of frequency domain would largely improve their performance. A general workflow for self-supervised learning can be divided into two stages:

- **Pre-training:** Train the parameters of an encoder with large amount of unlabeled training data. Then freeze the encoder.
- **Downstream tasks:** The encoder trained in the pre-training stage would map the original inputs to their corresponding representations. A simple model can be built on the representations and then a small labeled dataset can be used to train the model.

Our customization would mainly focus on the pre-training stage. However, before the customization, we need to verify the feasibility of self-supervised learning frameworks on IoT applications. This means that we need to answer the question: "Does those frameworks designed for computer vision and NLP still work on dealing with sensing signals of the IoT applications?" Hence, we begin with studying the performance of self-supervised learning framework, which is a popular and widely-utilized self-supervised learning framework, on sensing signals, and then carried out the customizations from different aspects:

- **Frequency domain:** We designed the data augmentation (an important part of the self-supervised contrastive pre-training) and encoder from the frequency domain.
- **Labeled data:** We observed that the labeled data, even through with a small amount, can help to train a better encoder. And thus we proposed a semi-supervised (instead of self-supervised) contrastive pre-training framework.
- **Data Augmentation:** The performance of self-supervised contrastive learning largely rely on the choice of the data augmentation algorithm. A bad choice of data augmentation algorithm would reduce the quality of the trained encoder a lot. We thus designed a self-supervised learning framework for IoT application that does not rely on data augmentation.

In the following parts, we propose a brief introduction of how we customize the self-supervised learning frameworks to IoT applications.

1.1 Time-Domain Self-Supervised Contrastive Learning

In order to learn whether the self-supervised contrastive learning framework, which demonstrated outstanding performance in computer vision tasks, would perform well when dealing with sensing data, we directly applied the self-supervised contrastive learning framework to the sensing applications by designing time domain encoder and data augmentations, and then evaluated its performance. In this work, we built a *Semi*-supervised *A*utomatic *M*odulation *C*lassification framework, namely, *SemiAMC*, to efficiently utilize unlabeled radio signals. SemiAMC consists of two parts: (1) self-supervised contrastive pre-training and (2) a downstream classifier. In self-supervised contrastive pre-training, we apply the design of SimCLR [34], which is an effective self-supervised contrastive learning framework, to train an encoder using a large amount of unlabeled training data. Then, we freeze the parameters of the encoder and train a classifier, which takes the representations (output of the encoder) as input, based on a small amount of labeled training data.

We evaluated the performance of SemiAMC on a widely-used modulation classification dataset RadioML2016.10a [36]. The evaluation results demonstrated that SemiAMC efficiently utilizes unlabeled training data to improve classification accuracy. Compared with previous supervised neural network models, our approach achieves better accuracies given the same number of labeled training samples. The evaluation results verified that the time domain approach do work on sensing data. In this work, we get a basic version of self-supervised contrastive learning framework for IoT applications, and in the next steps, we would show how our customization strategies help to improve the performance of it.

1.2 Frequency-Domain Self-Supervised Contrastive Learning

Existing self-supervised contrastive learning frameworks build their encoders with convolutional neural network (CNN) or recurrent neural network (RNN) which take the time domain data as input. However, our recent work has shown that for many IoT applications, the essential features of interest live in the frequency domain [37]. This insight calls for adapting the existing contrastive learning frameworks to frequency domain. In this part, we built the encoder from a time-frequency perspective using Short-Time Fourier Neural Networks (STFNet) [37] as basic building block. STFNet is a new neural network model specially designed for IoT applications. It operates directly in the frequency domain by mapping sensor measurements to the frequency domain with the Short-Time Fourier Transform (STFT), and was shown to perform much better than CNNs in many physical applications. We also design data augmentation operations from both time domain and frequency domain that enrich the contrastive prediction tasks for (a category of) IoT applications.

We evaluated the performance of our STFNet-based contrastive self-supervised learning framework on several human activity recognition (HAR) datasets, where the measurements of multiple sensors collected by different devices are used to determine the activities of a diverse group of participants. The experiments demonstrated obvious performance gains when we build the contrastive self-supervised learning framework from a time-frequency perspective instead of purely from the time-domain.

1.3 Semi-Supervised Contrastive Learning

From the label perspective, self-supervised learning is a special kind of unsupervised learning algorithm, which means that no label is required while training a model under self-supervised learning framework. A lot of NLP tasks are trained under this way. Not like NLP tasks, most IoT sensing tasks are classification or recognition tasks which do need labels. The labels here could provide extra information for the self-supervised contrastive learning frameworks. If we can find a way to make use of the labels during the self-supervised training, we would learn better latent features and hence improve the performance of the whole system.

To leverage this insight, we proposed *SemiC-HAR*, a semi-supervised contrastive learning framework for HAR. SemiC-HAR takes both labeled and unlabeled data into consideration. The key challenge of this work lies in how to efficiently utilize both labeled and unlabeled data in building the contrastive pre-training framework. To extract features from original inputs, previous research has studied self-supervised contrastive learning [34] which uses only unlabeled data and supervised contrastive learning [38] which uses only labeled data. In SemiC-HAR, we proposed a semi-supervised contrastive learning framework to efficiently utilize both labeled and unlabeled data in the pre-training process by carefully re-designing the contrastive loss function.

1.4 Spectrogram Masked Autoencoder for IoT Applications

The performance of self-supervised contrastive learning highly relies on the choice of the data augmentation strategy. Different data augmentation strategies would lead to very different qualities of the trained encoder. To make things worse, different from computer vision (where the input is image or video) and NLP (where the input the natural language), the inputs for IoT applications are very different because different applications use different type sensors and thus yield different types of sensing signals, such as audio signals, seismic signals, radio signals, and measurements of motion sensors. Considering the difference of the physical phenomena behind the sensing tasks, the strategies to augment different sensing signals are very different. In this way, designing self-supervised contrastive learning frameworks requires a lot of human expertise on data augmentation. Hence, it would be very helpful if we could design a self-supervised learning framework for IoT applications that does not rely on data augmentation.

In this work, we proposed a new self-supervised learning framework based on the masked autoencoding approach [39], and re-designed the loss function to make it working in a better way to deal with spectrograms instead of images. The evaluation on both of the human activity recognition task and target detection task demonstrated the effectiveness of our customization.

1.5 A Case Study: Self-Supervised Learning on IoBT-OS

Finally, we studied the performance of our self-supervised learning framework on a real system instead of just running evaluations on different datasets. In this work, we built an operating system for the Internet of Battlefield Things (IoBT-OS) where our self-supervised learning framework plays an important role on improving decision accuracy and thus optimizing the latency of the decision loop. Then, we use a case study of seismic and acoustic based target detection to evaluate the performance of our proposed framework, and the evaluation result verified the effectiveness of our self-supervised learning framework.

1.6 Chapter Organization

The rest of this chapter is organized as follows: In Sect. 2, we introduce the self-supervised contrastive learning framework, and show its performance on dealing with wireless radio signals from the time domain. We begin our customization by re-designing the data augmentation and structure of encoder from the perspective of frequency domain. This part of work is put in Sect. 3. After that, we propose our semi-supervised contrastive learning framework in Sect. 4. In Sect. 5, we show the work we have done to reduce the dependency on data augmentation. We then put the IoBT-OS case study in Sect. 6. Finally, we summarize this chapter in Sect. 7.

2 Time-Domain Self-Supervised Contrastive Learning Framework for IoT

2.1 Overview

In this section, we use the modulation classification problem as an example to study the performance of the time-domain self-supervised contrastive learning framework. Automatically recognizing or classifying the radio modulation is a key step for many commercial and military applications, such as dynamic spectrum access, radio fault detection, and unauthorized signal detection in battlefield scenarios. The modulation classification problem has gotten widely studied in the past few years. Two general types of algorithms, likelihood-based (LB) and feature-based (FB), have been applied to solve the modulation classification problem. Likelihood-based methods [40, 41] make decisions based on the likelihood of the radio signal and achieve optimal performance in the Bayesian sense. However, they suffer from high computational complexity. Feature-based methods [42–44], on the other hand, make decision based on the manually-extracted features from the radio signal. Given the input radio signal, various features related to phase, amplitude or frequency are manually extracted and then used as inputs of the classification algorithms. Machine learning algorithms, such as support vector machine (SVM) and decision trees, are popular candidates for the classification algorithms.

Recently, with advances in deep learning techniques, deep neural networks achieves a great success in multiple fields, such as image classification [20] and natural language processing [23]. Deep neural network models, such as convolutional neural networks (CNNs) [45, 46] and recurrent neural networks (RNNs) [47, 48], have also been applied to the modulation classification problem. Such neural network models directly feed the raw signal data or its transforms as input and generally achieve much better performance than previous approaches.

However, training a deep neural network model requires a large amount of training data. In practice, it's usually difficult to collect a large volume of high quality and reliable radio signals as well as their modulations (labels) as training data. One common way to deal with the lack of training data is data augmentation. Previous studies proposed several data augmentation strategies [49–51] to avoid overfitting caused by the lack of training data. Another way is to utilize *unlabeled data*. The difficulty in collecting a large training dataset for radio modulation classification mainly comes from the labeling burden for radio signals. It is much easier to collect the radio signals without labeling them. Hence, it would help a lot if we could efficiently utilize *unlabeled* radio signals. As far as we know, the benefits of using unlabeled data have not yet been studied for the modulation classification problem.

Self-supervised learning, as a kind of unsupervised learning, obviates much of the labeling burden by extracting information from unlabeled data. Self-supervised learning algorithms have had great success in areas of computer vision [34], natural language processing [23], and IoT applications [30]. The goal of self-supervised learning is to train an encoder to extract useful intrinsic information (or features) from unlabeled data and use that information as the inputs to a classifier or predictor in a downstream task. Since these features already store a lot of intrinsic information about the original input, a simple classifier (*e.g.*, linear classifier) is enough for the downstream task. Using this approach, most training uses unlabeled data to learn the intrinsic features. Only a small amount of labeled data is then needed to train the downstream classifier.

In this section, we build a *Semi*-supervised *A*utomatic *M*odulation *C*lassification framework, namely, *SemiAMC*[52], to efficiently utilize unlabeled radio signals.

SemiAMC consists of two parts: (1) self-supervised contrastive pre-training and (2) a downstream classifier. In self-supervised contrastive pre-training, we apply the design of SimCLR [34], which is an effective self-supervised contrastive learning framework, to train an encoder using a large amount of unlabeled training data as well as data augmentation. Then, we freeze the parameters of the encoder and train a classifier, which takes the representations (output of the encoder) as input, based on a small amount of labeled training data.

We evaluated the performance of SemiAMC on a widely-used modulation classification dataset RadioML2016.10a[36]. The evaluation results demonstrated that SemiAMC efficiently utilizes unlabeled training data to improve classification accuracy. Compared with previous supervised neural network models, our approach achieved a better accuracy given the same number of labeled training samples.

2.2 Signal Model

We consider a single-input single-output communication system where we sample the in-phase and quadrature components of a radio signal through an analog to digital converter. The received radio signal $r(t)$ can be represented as

$$r(t) = c * s(t) + n(t), \tag{2.1}$$

where $s(t)$ refers to the modulated signal from the transmitter, c refers to the path loss or gain term on the signal, and $n(t)$ is the Gaussian noise. The received radio signal $r(t)$ is sampled N times at some sampling rate to obtain a length N vector of complex values. In this work, we treat the complex valued input as a 2-dimensional real-valued input. We denote it as X, where X is a $2 \times N$ matrix containing the in-phase (I) and quadrature (Q) components of the N received signal samples. The goal of modulation classification is to determine the modulation type for any given radio signal X.

2.3 Architecture of SemiAMC

SemiAMC aims at training a classifier to accurately recognize the modulation type for any given radio signal. As a semi-supervised framework, SemiAMC is trained with both labeled and unlabeled data. We show the architecture of SemiAMC in Fig. 2.1. The illustrated workflow is as follows.

The first step is called self-supervised contrastive pre-training, where we train an encoder to map the original radio measurements into low-dimensional representations. This is done in a self-supervised manner, with unlabeled data only. The supervision here comes from optimizing the *contrastive loss* function, that

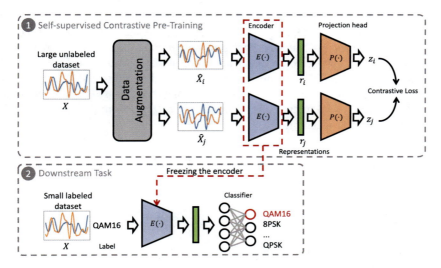

Fig. 2.1 Overview of SemiAMC

maximizes the agreement between the representations of differently augmented views for the same data sample.

In step two, we freeze the encoder learned during the self-supervised contrastive pre-training step, and map the labeled input radio signals to their corresponding representations in the low-dimensional space. The classifier can be trained based on these representations and their corresponding labels. Here a relatively simple classifier (*e.g.*, linear model) usually work well, because the latent representation has already extracted the intrinsic information from the signal input. In this way, a small number of labeled data samples is enough to train the classifier. When we have enough labeled data, we can also fine-tune the last one or more layers of the encoder to further improve the performance of SemiAMC.

2.4 Self-Supervised Contrastive Pre-Training

We applied SimCLR [34] as our self-supervised contrastive pre-training framework. As shown in Fig. 2.1, our self-supervised contrastive pre-training mainly consists of four components.

2.4.1 Data Augmentation

The first component is a stochastic data augmentation module. Given any signal input X, two views of X, that are denoted as $\hat{X}_i$ and $\hat{X}_j$, are generated from the same family of data augmentation algorithms. The data augmentation algorithms are highly application dependent. For example, in image related applications, random color distortion, cropping, and Gaussian blur are commonly used data augmentation algorithms. In our approach, we augment the I/Q signal with the rotation operation [49] which could keep the features for classification. For a modulated signal $X = [I, Q]^T$, where I and Q refer to $1 \times N$ vectors storing the in-phase (I) and quadrature (Q) signals, we rotate it with an angle θ randomly selected from $\{0, \pi/2, \pi, 3\pi/2\}$. The augmented signal sample is

$$\hat{X} = \begin{bmatrix} \hat{I} \\ \hat{Q} \end{bmatrix} = \begin{bmatrix} \cos\theta & -\sin\theta \\ \sin\theta & \cos\theta \end{bmatrix} \begin{bmatrix} I \\ Q \end{bmatrix}. \quad (2.2)$$

2.4.2 Encoder

The second part is a neural network based encoder E that extracts intrinsic information from the augmented examples $\hat{X}_i$ and $\hat{X}_j$, and stores them in the latent representations r_i and r_j:

$$r_i = E(\hat{X}_i), \ r_j = E(\hat{X}_j). \quad (2.3)$$

Various choices of the network architectures can be used to design the encoder. For example, CNN-based encoders are widely utilized in learning representations of visual data and RNN-based encoders are usually utilized to deal with time-series. Figure 2.2 shows the architecture of the encoder we use in our approach. The I/Q signal input is first passed through a 1D convolutional layer (32 kernels with size 24) to extract the spatial characteristics. The following two LSTM layers (with 128 units) are used to extract the temporal characteristics. In the end, the 1D convolutional layer (128 kernels with size 8) and the max pooling layer are used to generate the representations.

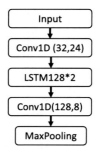

Fig. 2.2 The architecture of the encoder

2.4.3 Projection Head

The third part is a small neural network projection head P that maps representations to the space where contrastive loss is applied. In our approach, we use a multilayer perceptron (MLP) with one hidden layer as the projection head, which means

$$z_i = P(r_i) = W^{(2)}\sigma(W^{(1)}r_i), \tag{2.4}$$

where σ is the ReLU activation function. It has been proved that calculating the contrastive loss on z_i's (instead of on the representations r_i's directly) improves the performance of self-supervised contrastive learning [34].

2.4.4 Contrastive Loss

The last part is the contrastive loss function that is defined for a contrastive prediction task aiming at maximizing the agreement between examples augmented from the same signal input. We use the normalized temperature-scaled cross entropy loss (NT-Xent) [53] as the loss function. For a given mini-batch of M examples in the training process, since for each example X, we will generate a pair of augmented examples, there will be $2M$ data points. The two augmented versions $\hat{X}_i$ and $\hat{X}_j$ of the same input X are called a positive pair. All remaining $2(M-1)$ in this batch are negative examples to them. Cosine similarity is utilized to measure the similarity between two augmented data samples. The similarity is calculated on z_i and z_j:

$$\text{sim}(z_i, z_j) = \frac{z_i^T z_j}{\|z_i\| \|z_j\|}. \tag{2.5}$$

Here, $\|z_i\|$ refers to the ℓ_2 norm of z_i. The loss function for a positive pair of examples (i, j) is defined as

$$L_{i,j} = -\log \frac{\exp(\text{sim}(z_i, z_j)/\tau)}{\sum_{k=1}^{2M} \mathbb{1}_{[k \neq i]} \exp(\text{sim}(z_i, z_k)/\tau)}, \tag{2.6}$$

where τ refers to the temperature parameter of softmax and $\mathbb{1}_{[k \neq i]} \in \{0, 1\}$ is an indicator function equaling to 1 iff $k \neq i$. The loss for all positive pairs (i, j) is calculated and the average is used as the final contrastive loss.

2.5 Evaluation

We evaluated the performance of SemiAMC based on RadioML2016.10a [36], which is a public modulation classification dataset. We first introduce the dataset we use and then show the experimental setup, including data preprocessing and detailed implementation. Finally, we analyze the performance of SemiAMC.

2.5.1 Dataset

We use RadioML2016.10a to study the performance of SemiAMC. RadioML2016.10a is a synthetic dataset including radio signals of different modulations at varying signal-to-noise ratios (SNRs). It consists of 11 commonly used modulations (8 digital and 3 analog): WBFM, AM-DSB, AM-SSB, BPSK, CPFSK, GFSK, 4-PAM, 16-QAM, 64-QAM, QPSK, and 8PSK. For each modulation, there are 20 different SNRs from -20dB to $+18$dB and there are 1000 signals under each SNR. Hence, the RadioML2016.10a has $1000 \times 20 \times 11 = 220,000$ signal examples in total. Each signal in RadioML2016.10a has 128 I/Q samples. In our approach, we put each signal in a 2×128 matrix X.

2.5.2 Experimental Setup

We split the dataset into three parts: training, validation, and testing by a ratio of 2:1:1. Specifically, for each modulation type and SNR, we randomly divide the 1000 signals into 500 signals for training, 250 signals for validation, and 250 signals for testing. Normalized signals are used as the input.

In the self-supervised contrastive pre-training part, we use a two-layer projection head. The first layer is a fully connected layer with 128 hidden units and ReLU activation. The second is a fully connected layer with 64 hidden units without an activation function. The encoder, the output of which is 128, is trained under a batch size of 512, with a total of 100 batches, and initial learning rate $1e^{-4}$ with cosine decay. In the downstream task part, we freeze the encoder and build a two-layer classifier on the representations. The first layer of the classifier is a fully connected layer with 128 neurons and ReLU activation. The second is a Softmax layer with 11 neurons, one for each modulation scheme. We apply droupout and L2 regularization to mitigate over-fitting.

We first study the performance when all the training and validation data have labels. We then study the performance when only part of the training and validation data have labels. We also study the performance of our approach when different amounts of unlabeled data are used to train the encoder. We further compute the classification accuracy separately for each SNR and modulation scheme. We run five times with different random seeds and take the average as the final performance.

2.5.3 Comparison with Supervised Frameworks

In this part, we compare the performance of our semi-supervised learning framework to that of previous supervised frameworks when we have enough labeled training data. Specifically, here we assume that we have labels for all the training and validation data. We first train the encoder with all signal samples in the training dataset. Then, we freeze the encoder and train the classifier based on both of the signals and labels in the training set. During this process, we also fine-tune the parameters of the encoder to get better result. We stop the training process when the validation loss does not decrease for 30 epochs and use the model with minimum validation loss to predict the classification accuracy on test set.

We compared the performance of SemiAMC against three supervised algorithms named CNN2 [45], ResNet [46], and LSTM2 [47]. As shown in Fig. 2.3, SemiAMC performs the best even through there is no extra unlabeled data. The performance gain here mainly comes from the architecture of our encoder design and the data augmentation while training the encoder in self-supervised contrastive pre-training. SemiAMC clearly outperforms the supervised baselines above −2dB SNRs and achieves a maximum accuracy of 93.45% under 12dB SNR.

2.5.4 Performance under Different Amount of Labeled Data

In this part, we studied the performance of SemiAMC given a large amount of unlabeled data and a small amount of labeled data. For each modulation type under each SNR, we have 500 signal samples for training, 250 signal samples for validation, and 250 signal samples for testing. We assume that for each modulation type under each SNR, only $n = 1, 2, 5(1\%), 10, 20, 30, 40, 50(10\%)$ signal samples have labels in the training set, and $\lceil n/2 \rceil$ signal samples have labels in the validation set. We train the encoder with all signal samples (without labels) in the training set, then we train the classifier and fine-tune the encoder with the labeled signal samples in the training set. The model with minimum validation loss is used to predict the classification accuracy on the test set.

To study the effectiveness of the encoder learned in the self-supervised contrastive pre-training, we directly train the encoder and classifier under a supervised

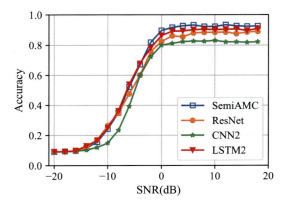

Fig. 2.3 Comparison with other frameworks

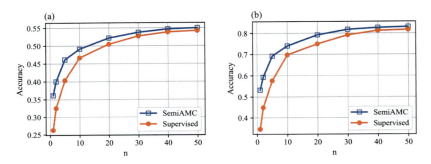

Fig. 2.4 Accuracy under different amount of labeled data. (**a**) Accuracy for all SNRs. (**b**) Accuracy for SNRs > 0

way with the labeled data in the training set, and then use the model with minimum validation loss to recognize the signals in the test set. We denote this approach as "Supervised". The classification accuracy for test signals under all SNRs and SNRs larger than zero are shown in Fig. 2.4. We can see that SemiAMC outperforms its corresponding supervised version. This confirms the effectiveness of the encoder trained with our self-supervised contrastive learning framework. We also observed that the gap between SemiAMC and the supervised version becomes larger with the decrease in labeled data. It suggests that our framework can effectively utilize unlabeled data. Naturally, the impact of unlabeled data is larger when less data are labeled. The performance gain of SemiAMC for SNRs larger than zero (Fig. 2.4b) are larger than those computed across all SNRs (Fig. 2.4a). It illustrated improved effectiveness of SemiAMC in dealing with higher SNR signals.

To further understand the classification accuracy under different SNR signals when different amounts of labeled data are given, we calculate the accuracy distribution under all SNR signals when $n = 1, 5, 10$. The results are shown in Fig. 2.5. We observed that the performance gain of SemiAMC compared with the supervised approach mainly comes from the signals with high SNR while SemiAMC performs similarly with the supervised approach when the SNRs are small. We also studied the performance of SemiAMC of each modulation type when $n = 1, 10$ and the SNR is 10dB. We draw the corresponding confusion matrices in Fig. 2.6. We observed similar accuracy patterns for different modulation types between SemiAMC and its corresponding supervised approach. For example, both of them would incorrectly recognize WBFM modulated signals as AM-DSB.

2.5.5 Performance under Different Amount of Unlabeled Data

Finally, we studied the recognition accuracy when different amounts of *unlabeled* data are given. Unlabeled data is used to train the encoder in the self-supervised pre-training part. In this experiment, we have 500 samples per SNR per modulation type for training. We assume that $n = 10$ of the samples have labels, and used $u =$

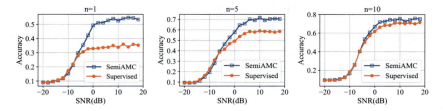

Fig. 2.5 Accuracy for different SNRs when $n = 1, 5, 10$

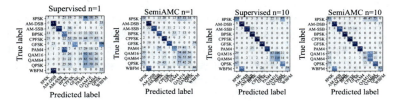

Fig. 2.6 Confusion matrix of SemiAMC and supervised when $n = 1, 10$

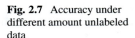

Fig. 2.7 Accuracy under different amount unlabeled data

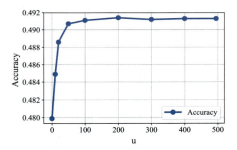

0, 10, 20, 50, 100, 200, 300, 400, 490 extra unlabeled samples besides the $n = 10$ labeled samples to train SemiAMC. Figure 2.7 plots the overall accuracy versus the change in the amount of unlabeled data. It is observed that the classification accuracy increases with increasing amounts of unlabeled data (up to a certain level).

3 Frequency-Domain Self-Supervised Contrastive Learning Framework for IoT

3.1 Overview

In this section, we introduce the use of contrastive representation learning techniques to signals whose essential features are best represented in the *frequency domain*. Contrastive representation learning allows automatic extraction of essential features from complex signals by projecting them *in an unsupervised fashion* onto appropriate lower-dimensional spaces. Such projections can then be used for such purposes as signal classification, clustering, and entity detection. Prior work focused on time-domain signals. Our recent work has shown that, for many applications, the essential features of interest live in the frequency domain [37]. This insight calls for adapting the existing contrastive learning framework to frequency domain signals, which is the main contribution of this work.

This work is motivated by the increasing popularity of deep neural networks [18] that have demonstrated outstanding performance on various machine learning tasks, such as image classification [19, 20], natural language processing [23], and IoT applications [54, 55]. Deep neural networks are traditionally trained using *supervised* learning algorithms that need a large amount of annotated training data. As deep learning techniques mature, the underlying neural network models often become deeper, thereby exacerbating the need for large-scale annotated datasets. Even when collecting the training data is not difficult, annotating the datasets usually requires extensive human labor. As a result, self-supervised learning algorithms have drawn more attention in recent years [31–35]. These techniques obviate extensive labeling, while attaining comparable performance with their supervised learning counterparts. In self-supervised learning scenarios, an encoder (that maps original inputs into lower-dimensional latent representations) is trained using *unlabeled* data. The learned latent representation can then be further used to accomplish a variety of downstream tasks, such as classification or regression by training a much simpler model (e.g., a linear model) that takes the latent representation (instead of the original data) as input. Only a small amount of labeled data is needed to train such (much simpler) models. As alluded to earlier, the big win lies in the fact that the bulk of the training (to generate the underlying latent representation) uses *unlabeled data.*

The key to successful latent representation learning is to ensure that the learned latent representation captures the essential properties of the data, such as "similarity" in some appropriate semantic space. Teaching the encoder the right notion of similarity is often achieved by *data augmentation*, where input data samples are locally perturbed in a manner that does not substantially change their semantic interpretation. A cost function (used in training) ensures that such local perturbations are projected to *nearby locations* in the latent space. Designing the appropriate data perturbation for an application domain thus becomes an interesting research problem.

Recently, self-supervised learning (using data augmentation) has seen great success in computer vision [31, 33, 34, 56] and NLP [23, 35]. In IoT applications, however, the study of self-supervised learning has not gotten enough attention. Compared with computer vision or NLP tasks, the notion of "semantic similarity" in IoT applications is less clear. For example, in vision, an upside-down or mirror-image picture of a dog is still a dog, but what is a similarity notion in an accelerometer time-series or a backscatter signal received by a WiFi radio or RFID reader?

In this work, we apply the design of SimCLR [34], an effective framework for contrastive self-supervised representation learning (shown to perform well on the ImageNet dataset), but carefully redesign the structure of the encoder and the data augmentation operations. In lieu of using a convolutional neural network, we build the encoder using STFNet [37]. STFNet (or Short-Time Fourier Neural Networks) is a new neural network model specially designed for IoT applications. It operates directly in the frequency domain by mapping sensor measurements to the frequency domain with the Short-Time Fourier Transform (STFT), and was shown to perform much better than CNNs in many physical applications [37]. We also design data augmentation operations from both time domain and frequency domain that enrich the contrastive prediction tasks for (a category of) IoT applications.

We evaluate the performance of our STFNet-based contrastive self-supervised learning framework, we call STF-CSL, on several human activity recognition datasets, where the measurements of multiple sensors collected by different devices are used to determine the activities of a diverse group of participants. The experiments demonstrate obvious performance gains when we build the contrastive self-supervised learning framework from a time-frequency perspective (STF-CSL) instead of purely from time-domain (CNN-base approach).

3.2 Background and Related Work

We first briefly overview related background on learning paradigms, IoT applications, self-supervision, and representation learning. These pieces, as we show later, serve as the building blocks of our contrastive self-supervised representation learning framework.

3.2.1 Deep Neural Network for IoT Applications

The impressive success in using deep neural networks for image classification [20] precipitated a reemergence of interest in deep learning. In recent years, deep neural networks have achieved significant performance improvements in multiple areas, including computer vision [19, 20], natural language processing [23, 57], and IoT applications [54, 55, 58].

In the context of IoT, researchers employed deep learning models to improve the predictive accuracy when time-series measurements (from sensors, such as accelerometers, gyroscopes, and magnetometers) generate inputs for various estimation and classification applications. Examples include applications in health and wellness [59, 60] and human activity recognition [61–63]. General frameworks were also proposed for integrating convolutional and recurrent neural networks for sensing applications [54]. To improve system efficiency at executing neural networks on low-end IoT devices, researchers have investigated means to compress structures and/or parameters of the neural network model, while keeping the accuracy almost the same [64, 65].

Inspired by the physics of measured processes, Yao et al. [37] addressed the customization of learning machinery to the frequency domain, and proposed a new foundational neural network building block, namely, the ShortTime Fourier Neural Network (STFNet). STFNet integrates neural networks with traditional time-frequency analysis, and performs the inner-layer operations in the frequency domain. STFNet outperformed previous CNN and/or RNN based approaches on a category of IoT applications. In this section, we build our self-supervised learning model based on STFNet, and evaluate its performance in self-supervised scenarios.

3.2.2 Self-Supervised Learning

Self-supervised learning, where the model is trained on unlabeled data, has become an increasingly popular framework for training deep neural networks. In a self-supervised learning framework, self-supervised tasks, also known as pretext tasks, are designed to generate locally perturbed data samples (with labels that capture the notion of local similarity in the perturbed unannotated data). A loss function can thus be generated that prefers mapping such similar inputs to nearby locations in the latent space.

Self-supervised learning has been widely studied in computer vision and NLP, and many ideas have been proposed to design the pretext tasks. For example, in computer vision, tasks such as predicting image rotations [31], solving jigsaw puzzles game [66], and colorful image colorization [32] are used to generate the self-supervision. The pretext tasks can also be implemented by a generative model. The generative modeling based self-supervised model can be trained to reconstruct the original input to learn the meaningful representations. For example, autoencoder [67] is trained to reconstruct the input images. The context encoder [33] is designed to recover the missing part of the input image when a mask is applied. Another commonly used framework in training self-supervised models is contrastive learning, where the task is designed to discriminate positive and negative samples. Chen et al. [34] designed a simple framework for contrastive learning of visual representations, namely, SimCLR and got impressive performance results on ImageNet. SimCLR learns representations for the input images by maximizing the agreement between the latent expression of two augmentations generated from the same input. Due to the simplicity and effectiveness of SimCLR, we apply it as our fundamental framework when customizing self-supervised learning framework to IoT applications.

In NLP research, self-supervised methods have also gotten widely studied. The original Word2Vec paper [68] proposed two architectures to compute continuous representations of words, where the pretext tasks are predicting the center word given the surrounding words, or, conversely, predicting the surrounding words given the center one. Devlin et al. proposed BERT [23] to pretrain deep bidirectional representations from unlabeled text by predicting randomly masked words in the text. It achieves excellent performance in language representation learning.

Customizing self-supervised learning models to the needs of IoT applications requires researchers to carefully design the encoder and data augmentation algorithms that work well for IoT applications. Existing work [4–6] implemented self-supervised learning frameworks for human activity recognition tasks. They designed convolutional or recurrent neural-network-based encoders that extract spatial and temporal properties of inputs, and applied data augmentation on the time-domain time-series inputs. Saeed et al. [4] proposed a multi-task self-supervised learning framework for sensory data and applied transformation discrimination as the pretext tasks. The model was trained to discriminate the transformed signals from the original inputs. Haresamudram et al. [6] introduced a masked reconstruction based self-supervised learning model for human activity recognition, where the self-supervision was generated by predicting the randomly masked timestep of the input sensor signal based on the remaining values. Tang et al. [5] explored the adoption and adaptation of SimCLR to human activity recognition, and designed a contrastive learning framework to learn the representations for sensor signal inputs. We advance this work by exploiting the conjecture that the underlying physics of measured phenomena in IoT applications are best expressed (e.g., are sparser and more compactly expressed) in the *frequency domain* [37], thus redesigning SimCLR to work with both time domain and frequency domain data. To make the scope of this work manageable, we focus on a subset of IoT applications exemplified by human activity recognition tasks.

3.2.3 Representation Learning

Representation learning, also called feature learning, is a way to automatically learn intrinsic structure and extract useful information from raw data that can effectively support impending machine learning tasks, such as regression and classification. The traditional manual feature engineering, which is labor-intensive, extracts or organizes useful information from data by relying on human ingenuity and prior knowledge. Representation learning [69, 70], however, allows the features to be learned by machines in an automatic way.

Principal component analysis (PCA) [71] and linear discriminant analysis (LDA) [72] are two early data representation learning algorithms. They learn low-dimensional representations of data with linear transformation techniques. In 2000, Roweis and Saul proposed locally linear embedding (LLE) [73], which is a nonlinear learning approach for generating low-dimensional neighbor-preserving representations from the high-dimensional input. With the development of deep learning, self-supervised neural networks have become widely utilized to extract the intrinsic representations in many domains, including images [32–34, 66, 74], video [75–77], audio [78, 79], and natural language processing [23, 80]. In this work, we show how self-supervised learning is customized to extract representations of sensor data by taking a frequency domain perspective.

3.3 Design of STFNet

In this part, we introduce the technical details of STFNets. We separate the technical descriptions into six parts. In the first two subsections, we provide some background followed by a high-level overview of STFNet components, including (1) hologram interleaving, (2) STFNet-filtering, (3) STFNet-convolution, and (4) STFNetpooling. In the remaining four subsections, we describe the technical details of each of these components, respectively.

3.3.1 STFNet Overview

IoT devices sample the physical environment generating time-series data. Discrete Fourier Transform (DFT) is a mathematical tool that converts n samples over time (with a sampling rate of f_s) into a n components in frequency (with a frequency step of f_s/n). The more samples are selected, the finer the component resolution is in frequency. We can always transform the whole sequence of data with DFT, achieving a high frequency resolution. However, we then lose information on signal evolution over time, or the time resolution. In order to solve this problem, Short-Time Fourier Transform (STFT) divides a longer time signal into shorter segments of equal length and computes DTF separately on each shorter segment. By losing a certain degree of frequency resolution, STFT helps us regain the time resolution to some extent. In choosing n, there arises a fundamental trade-off between the attainable time and frequency resolution, which is called the *uncertainty principle* [81]. For the purposes of learning to predict a given output, the optimal trade-off point depends on the time and frequency granularity of the features that best determine the outputs we want to reproduce. The goal of STFNets is thus to learn frequency domain features that predict the output, while at the same time learn the best resolution trade-off point in which the relevant features exist.

The building component of an STFNet is an *STFNet block*, shown in Fig. 2.8. An STFNet block is the layer-equivalent in our neural network. The larger network would normally be composed by stacking such layers. Within each block, STFNet circumvents the uncertainty principle by computing multiple STFT representations with different time-frequency resolutions. Collectively, these representations constitute what we call the *time-frequency hologram*. And we call an individual time-frequency signal representation, a hologram representation. They are then used to mutually enhance each other by filling-in missing frequency components in each.

Candidate frequency-domain features are then extracted from these enhanced representations via general spectral manipulations that come in two flavors; filtering and convolution. They represent global and local feature extraction operations, respectively. The filtering and convolution kernels are learnable, making each STFNet layer a building block for spectral manipulation and learnable frequency domain feature extraction. In addition, there's a new mechanism, called pooling, for frequency domain dimensionality reduction in STFNets. Combinations of features

Fig. 2.8 Overview design of STFNet block

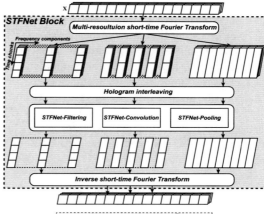

Fig. 2.9 Data flow within a block of STFNet

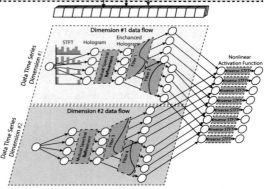

extracted using the above manipulations then pass through activation functions and an inverse STFT transform to produce (filtered) outputs in the time domain. Stacking STFNet blocks has the effect of producing progressively sharper (i.e., higher order) filters to shape the frequency domain signal representation into more relevant and more fine-tuned features.

Figure 2.9 gives an example of an SFTNet block that accepts as input a two-dimensional time-series signal (e.g., 2D accelerometers data). Each dimension is then transformed to the frequency domain at four different resolutions using STFT, generating four different internal nodes, each of which representing the signal in the frequency domain at a different time-frequency resolution. Collectively, the four representations constitute the hologram. In the next step, mutual enhancements are done improving all representations. Each representation then undergoes a variety of alternative spectral manipulations (called "filters" in the figure). Two filters are shown in the figure for each dimension. The parameters of these filters are the weights multiplied by the frequency components of the filter input; a different weight per component. These parameters are what the network learns. Note that, a filter does not change the time-frequency resolution of the corresponding input. Filter outputs of the same time-frequency resolution are then combined additively across all dimensions and passed through a non-linear activation function (as in a conventional convolutional neural network). An inverse STFT brings each such combined output back to the time domain, where it becomes an input to the next STFNet block. (Alternatively, the inverse STFT can be applied after dimension combination and before the activation function.) Hence, each output time-series is produced by applying spectral manipulation and fusion to one particular time frequency resolution of all input time-series. Once converted to the time domain, however, the output time-series can be resampled in the next block at different time-frequency resolutions again. The goal of STFNet is to learn the weighting of different frequency components within each filter in each block such that features are produced that best predict final network outputs.

3.3.2 STFNet Block Fundamentals

In this subsection, we introduce the formulation of the design elements within each STFNet block. In the rest of this section, all vectors are denoted by lower-case letters (*e.g.*, $\mathbf{x}$ and $\mathbf{y}$), while matrices and tensors are represented by bold upper-case letters (*e.g.*, $\mathbf{X}$ and $\mathbf{Y}$). For a vector $\mathbf{x}$, the jth element is denoted by $\mathbf{x}_{[j]}$. For a tensor $\mathbf{X}$, the tth matrix along the third axis is denoted by $\mathbf{X}_{[\cdot,\cdot,t]}$, and other slicing denotations are defined similarly. We use calligraphic letters to denote sets (e.g., $\mathcal{X}$ and $\mathcal{Y}$). For set $\mathcal{X}$, $|\mathcal{X}|$ denotes the cardinality.

We denote the input to the STFNet layer as $\mathbf{X} \in \mathbb{R}^{T \times D}$, where the input D-dimension time-series are divided into windows of size T samples. We call T the signal length and D the signal dimension. Since STFNet concentrates on sensing signals, we assume that all the raw and internal-manipulated sensing signals are real-valued in time domain.

As shown in Fig. 2.8, the input signal $\mathbf{X}$ first goes through a multi-resolution short-time Fourier transform (Multi_STFT), which is a compound traditional short-time Fourier transform (STFT), to provide a time-frequency hologram of the signal. STFT breaks the original signal up into chunks with a sliding window, where sliding window $\mathbf{W}(t)$ with width τ only has non-zero values for $1 \leq t \leq \tau$. Then each chunk is Discrete-Fourier transformed,

$$\mathbf{STFT}^{(\tau,s)}(\mathbf{X})_{[m,k,d]} = \sum_{t=1}^{T} \mathbf{X}_{[t,d]} \cdot \mathbf{W}(t - s \cdot m) \cdot e^{-j\frac{2\pi k}{\tau}(t - s \cdot m)}, \tag{2.7}$$

where $\mathbf{STFT}^{(\tau,s)}(\mathbf{X}) \in \mathbb{C}^{M \times K \times D}$ denotes the short-time Fourier transform with width τ and sliding step s. M denotes the number of time chunks. K denotes the number of frequency components. Since input signal $\mathbf{X}$ is real-valued, its discrete Fourier transform is conjugate symmetric. Therefore, we only need the $\lfloor \tau/2 \rfloor + 1$ frequency components to represent the signal, *i.e.*, $K = \lfloor \tau/2 \rfloor + 1$. STFNet uses sliding chunks with a rectangular window and no overlaps to simplify the formulation, *i.e.*, $s = \tau$ and $M = T/\tau$. Therefore the short-time Fourier transform can be denoted as $\mathbf{STFT}^{(\tau)}(\mathbf{X})$.

2 Self-Supervised Learning from Unlabeled IoT Data

The Multi_STFT operation is composed of multiple short-time Fourier transform with different window widths $\mathcal{T} = \{\tau_i\}$. The window width, τ_i, determines the time-frequency resolution of STFT. Larger τ_i provides better frequency resolution, while smaller τ_i provides better time resolution. In STFNet, the window widths are set to be powers of 2, *i.e.*, $\tau_i = 2^{p_i} \ \forall p_i \in \mathbb{Z}_0^+$. We can thus formulate Multi_STFT as:

$$\mathbf{Multi\_STFT}^{(\mathcal{T})}\{\mathbf{X}\} = \{\mathbf{STFT}^{(\tau_i)}(\mathbf{X})\} \text{ for } 2^{p_i} \in \mathcal{T}. \tag{2.8}$$

Next, according to Fig. 2.8, the multi-resolution representations go into the hologram interleaving component, which enables the representations to compensate and balance their time-frequency resolutions with each other. The technical details of the hologram interleaving component are introduced in Sect. 3.3.3.

The STFNet layer then manipulates multiple hologram representations with the same set of spectral-compatible operation(s), including STFNet-filtering, STFNet-convolution, and STFNet-pooling. We will formulate these operations in Sects. 3.3.4, 3.3.5, and 3.3.6, respectively.

Finally, the STFNet layer converts the manipulated frequency representations back into the time domain with the inverse short-time Fourier transform. The resulting representations from different views of the hologram are weighted and merged as the input "signal" for the next layer. Since we merge the output representations from different views of the hologram, we reduce the output feature dimension of STFNet-filtering and convolution operations by the factor of $1/|\mathcal{T}|$ to prevent the dimension explosion.

3.3.3 STFNet Hologram Interleaving

In this subsection, we introduce the formulation of an innovative time-frequency domain operation proposed in STFNets: hologram interleaving. Due to the Fourier uncertainty principle, the representations in time-frequency hologram either have high time resolution or high frequency resolution. The hologram interleaving tries to use representations with high time resolution to instruct the representations with low time resolution to highlight the important components over time. This is done by two steps:

1. Revealing the mathematical relationship of aligned time-frequency components among different representations in the time-frequency hologram.
2. Updating the original relationship in a data-driven manner through neural-network attention components.

We start from the definition of time-frequency hologram, generated by Multi_STFT defined in (2.8). Note that, the window width set $\mathcal{T}$ is defined as $\{2^{p_i}\}$, $\forall p_i \in \mathbf{Z}_0^+$. Without loss of generality, an illustration of multi-resolution short-time Fourier transformed representations with input signal having length 16 and signal dimension 3 as well as $\mathcal{T} = \{4, 8, 16\}$ are illustrated in Fig. 2.10.

Fig. 2.10 The design of hologram interleaving

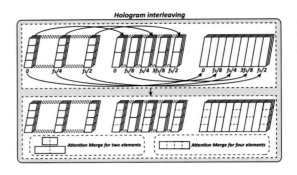

In order to find out the relationship of aligned time-frequency components, we start with the frequency-component dimension. Since different representations only change the window width τ_i of STFT but not the sampling frequency f_s of input signal, these frequency components represent frequencies from 0 to $f_s/2$ (Nyquist frequency) with step f_s/τ_i. Then we can first obtain the relationship of frequency ranging steps among different representations,

$$\forall p_i > p_j, \frac{f_s/\tau_j}{f_s/\tau_i} = 2^{p_i - p_j} \in \mathbf{Z}_0^+. \tag{2.9}$$

Therefore, a low frequency-resolution representation (with window width 2^{p_j}) can find their frequency-equivalent counterparts for every $2^{p_i - p_j}$ frequency components in a high frequency-resolution representation (with window width 2^{p_i}). The upper part of Fig. 2.10 provides a simple illustration of such relationship. In the following analysis, we will use the original index k and corresponding frequency $k \cdot f_s/\tau_i$ interchangeably to recall the frequency component from the time-frequency hologram $\mathbf{STFT}^{(\tau)}(\mathbf{X})_{[m,k,d]}$.

Next, we analyze the relationship over the time-chunk dimension, when two representations have frequency-equivalent components. Note that time chunks in $\mathbf{STFT}^{(\tau)}(\mathbf{X})$ are generated by sliding rectangular window without overlap. Based on (2.7), for representations having window widths $\tau_i = 2^{p_i}$ and $\tau_j = 2^{p_j}$ ($p_i > p_j$),

$$\mathbf{STFT}^{(\tau_i)}(\mathbf{X})_{[m, 2^{p_i - p_j} k, d]} = \sum_{t=2^{p_i} m+1}^{2^{p_i}(m+1)} \mathbf{X}_{[t,d]} \cdot e^{-j \frac{2\pi 2^{p_i - p_j} k}{2^{p_i}}(t - m \cdot 2^{p_i})},$$

$$= \sum_{m_j = 2^{p_i - p_j} m}^{2^{p_i - p_j}(m+1)-1} \sum_{t=m_j+1}^{2^{p_j}(m_j+1)} \mathbf{X}_{[t,d]} \cdot e^{-j \frac{2\pi k}{2^{p_j}}(t - m \cdot 2^{p_j})},$$

$$= \sum_{m_j = 2^{p_i - p_j} m}^{2^{p_i - p_j}(m+1)-1} \mathbf{STFT}^{(\tau_j)}(\mathbf{X})_{[m_j, k, d]}.$$

$$\tag{2.10}$$

2 Self-Supervised Learning from Unlabeled IoT Data

Therefore, given the equivalent frequency component, a time component in low time-resolution representation (with window width 2^{p_i}) is the sum of $2^{p_i - p_j}$ aligned time components of the high time-resolution representation (with window width 2^{p_j}). As a toy example in Fig. 2.10, the first row of the middle tensor is equal to the sum of first two rows of the left tensor for frequencies 0, $f_s/4$, and $f_s/2$. The row of the right tensor is equal to the sum of four rows of the left tensor for frequencies 0, $f_s/4$, and $f_s/2$. The row of the right tensor is equal to the sum of two rows of the middle tensor for frequencies $f_s/8$ and $3f_s/8$, etc.

According to the analysis above, the high frequency-resolution representations lose their fine-grained time resolutions at certain frequencies by summing the corresponding frequency components up over a range of time. However, the high time-resolution representations preserve these information.

The idea of hologram interleaving is to replace the sum operation in high frequency-resolution representation with a weighted merge operation to highlight the important information over time. For a certain frequency component, the weight of merging is learnt through the most fine-grained information preserved in the time-frequency hologram. In this work, we implement the weighted merge operation as a simple attention module. For a merging input $\mathbf{z} \in \mathbb{C}^{S \times 1}$, where S is the number of elements to be merged, the merge operation is formulated as:

$$
\begin{aligned}
\mathbf{a} &= \mathrm{softmax}(|\mathbf{W}_m \mathbf{z}|), \\
y &= S \times \mathbf{a}^\mathsf{T} \mathbf{z},
\end{aligned}
\tag{2.11}
$$

where $|\cdot|$ is the piece-wise magnitude operation for complex-number vector; and $\mathbf{W}_m \in \mathbb{C}^{S \times S}$ is the learnable weight matrix. Notice that the final merged result is rescaled by the factor S to imitate the "sum" property of Fourier transform.

3.3.4 STFNet-Filtering Operation

Spectral filtering is a widely-used operation in time-frequency analysis. The STFNet-filtering operation replaces the traditional manually designed spectral filter with a learnable weight that can update during the training process. Although the spectral filtering is equivalent to the time-domain convolution according to convolution theorem[2], the filtering operation helps to handle the multi-resolution time-frequency analysis, and facilitates the parameterization and modelling. We denote the input tensor as $\mathbf{X} \in \mathbb{C}^{M \times K \times D}$, where M is the number of time chunk, K is frequency component number, and D is input feature dimension. The STFNet-filtering operation is formulated as:

$$
\mathbf{Y}_{[m,k,\cdot]} = \mathbf{X}_{[m,k,\cdot]} \mathbf{W}_{f[k,\cdot,\cdot]},
\tag{2.12}
$$

---

[2] https://en.wikipedia.org/wiki/Convolution_theorem

where $\mathbf{W}_f \in \mathbb{C}^{K \times D \times O}$ is the learnable weight matrix, O the output feature dimension, and $\mathbf{Y} \in \mathbb{C}^{M \times K \times O}$ the output representation. The function of STFNet-filtering operation is providing a set of learnable global frequency template matchings over the time. However, it is not straightforward to extend the matching operation to the representations with different time-frequency resolutions. Although we can create multiple $\mathbf{W}_f$ with different frequency resolutions K, it can introduce unnecessary complexity and redundancy. STFNet-filtering solves this problem by interpolating the frequency components in weight matrix. As we mentioned in Sect. 3.3.3, data in hologram with different frequency resolutions have the same frequency range (from 0 to $f_s/2$) but different frequency steps (f_s/τ). Therefore, STFNet-filtering operation has only one weight matrix $\mathbf{W}_f$ with $K = \lfloor \tau/2 \rfloor + 1$ frequency components. When the operation input has $K' = \lfloor \tau'/2 \rfloor + 1$ frequency components with $K' < K$, we can subsample the frequency components in $\mathbf{W}_f$. When $K' > K$, we interpolate the frequency components of $\mathbf{W}_f$. STFNet provides two kind of interpolation methods: (1) linear interpolation and (2) spectral interpolation.

The linear interpolation generates the missing frequency components in extended weight matrix $\mathbf{W}'_f \in \mathbb{C}^{K' \times D \times O}$ from the two neighbouring frequency components in $\mathbf{W}_f$:

$$
\begin{aligned}
k_l &= \left\lfloor k' \frac{\tau}{\tau'} \right\rfloor \quad k_r = k_l + 1, \\
\mathbf{W}'_{f[k',\cdot,\cdot]} &= \mathbf{W}_{f[k_l,\cdot,\cdot]} \left(k_r - k' \frac{\tau}{\tau'} \right) + \mathbf{W}_{f[k_r,\cdot,\cdot]} \left(k' \frac{\tau}{\tau'} - k_l \right).
\end{aligned}
\tag{2.13}
$$

The spectral interpolation utilizes the relationship between discrete-time Fourier transform (DTFT) and discrete Fourier transform (DFT). For a time-limited signal (with length τ), DTFT regards it as a infinite-length data with zeros outside the time-limited range, while DFT regards it as a τ-periodic data. As a result, DTFT generates a continuous function over the frequency domain, while DFT generates a discrete function. Therefore, DFT can be regarded as a sampling of DTFT with step f_s/τ. In order to increase the frequency resolution of $\mathbf{W}_f$, we can increase the sampling step from f_s/τ to f_s/τ', which is called spectral interpolation. Spectral interpolation can be done through zero padding in the time domain [81],

$$
\mathbf{W}'_{f[\cdot,d,o]} = \mathbf{DFT}\left(\mathbf{ZeroPad}_{\tau'-\tau} \mathbf{IDFT}\left(\mathbf{W}_{f[\cdot,d,o]} \right) \right),
\tag{2.14}
$$

where $\mathbf{ZeroPad}_t$ denotes padding t zeros at the end of sequence, and $\mathbf{IDFT}(\cdot)$ denotes the inverse discrete Fourier transform. Please note that, if we pad infinite zeros to the IDFT result, then DFT turns into DTFT. An simple illustration of STFNet-filtering operation is shown in Fig. 2.11.

Fig. 2.11 The STFNet-filtering operation

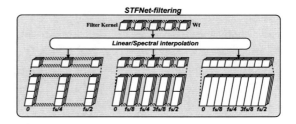

3.3.5 STFNet-Convolution Operation

Other than the filtering operation that handles global pattern matching, we still need the convolution operation to deal with local motifs in the frequency domain. We denote the input tensor as $\mathbf{X} \in \mathbb{C}^{M \times K \times D}$, where M is the number of time chunk, K number of frequency component, and D input feature dimension. The convolution operation involves two steps: (1) padding the input data, and (2) convolving with kernel weight matrix $\mathbf{W}_c \in \mathbb{C}^{1 \times S \times D \times O}$, where S is the kernel size along the frequency axis and O is still the output feature dimension.

Without the padding step, the output of convolution operation will shrink the number of frequency components, which may break the underlying structure and information in the frequency domain. Therefore, we need to pad extra "frequency component" to keep the shape of output tensor unchanged compared to that of the input data. In the deep learning research, padding zeros is a common practice. Zero padding is reasonable for inputs such as images and signal in the time domain, meaning no additional information in the padding range. However, padding zero-valued frequency component introduces additional information in the frequency domain.

Therefore, STFNet-convolution operation proposes the spectral padding for time-frequency analysis. According to the definition of DFT, transformed data is periodic within the frequency domain. In addition, if the original signal is real-valued, then the transformed data is conjugate symmetric within each period.

Previously, we cut the number of frequency components of a τ-length signal to $K = \lfloor \tau/2 \rfloor + 1$ for reducing the redundancy. In the spectral padding, we add these frequency components back according to the rule

$$\mathbf{X}_{[\cdot, \tau-k, \cdot]} = \mathbf{X}_{[\cdot, -k, \cdot]} = \mathbf{X}^*_{[\cdot, k, \cdot]}, \tag{2.15}$$

where $\mathbf{X}^*$ denotes complex conjugation. In addition, the number of padding before and after the input tensor is same as the previous padding techniques. Then we can define the basic convolution operation in STFNet

$$\mathbf{Y} = \mathbf{SpectralPad}(\mathbf{X}) \circledast \mathbf{W}_c, \tag{2.16}$$

where **SpectralPad**$(\cdot)$ denotes our spectral padding operation, and $\circledast$ denotes the convolution operation.

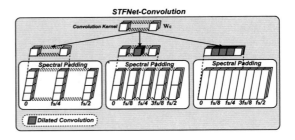

Fig. 2.12 The STFNet-convolution operation with dilated configuration

Next, we discuss the way to share the kernel weight matrix $\mathbf{W}_c$ with multi-resolution data. Other than interpolating the kernel weight matrix as shown in (2.13) and (2.14), we propose another solution for the STFNet-convolution operation. The convolution operation concerns more about the pattern of relative positions on the frequency domain. Therefore, instead of providing additional kernel details on fine-grained frequency resolution, we can just ensure that the convolution kernel is applied with the same frequency spacing on representations with different frequency resolutions. Such idea can be implemented with the dilated convolution [82]. If $\mathbf{W}_c$ is applied to a input tensor with $K = \lfloor \tau/2 \rfloor + 1$ frequency components, for a input tensor with $K' = \lfloor \tau'/2 \rfloor + 1$ frequency components ($\tau' > \tau$), the dilated rate r is set to $\tau'/\tau - 1$.

$$r = \tau'/\tau - 1. \qquad (2.17)$$

A simple illustration of STFNet-convolution with dilated configuration is shown in Fig. 2.12.

3.3.6 STFNet-Pooling Operation

In order to provide a dimension reduction method for sensing series within STNet, we introduce the STFNet-pooling operation. STFNet-pooling truncates the spectral information over time with a pre-defined frequency pattern. As a widely-used processing technique, filtering zeroes unwanted frequency components in the signal. Various filtering techniques have been designed, including low-pass filtering, high-pass filtering, and band-pass filtering, which serve as templates for our STFNet-pooling. Instead of zeroing unwanted frequency components, STFNet-pooling removes unwanted components and then concatenates the left pieces. For applications with domain knowledge about signal-to-noise ratio over the frequency domain, specific pooling strategy can be designed. In this work, we focus on low-pass STFNet-pooling as an illustrative example.

To extend the STFNet-pooling operation to multiple resolutions and preserving spectral information, we make sure that all representations have the same cut-off frequency according to their own frequency resolutions. A simple example of low-pass STFNet-pooling operation is shown in Fig. 2.13. We can see that our three tensors are truncated according to the same cut-off frequency, $f_s/4$.

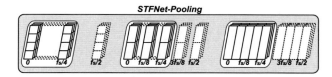

Fig. 2.13 The low-pass STFNet-pooling operation

3.4 Design of STF-CLS

Below, we first give an overview of our STFNet-based contrastive self-supervised representation learning framework, namely STF-CLS [30], in Sect. 3.4.1, and then provide the details of our design. We introduce the contrastive self-supervised learning framework (SimCLR) in Sect. 3.4.2. Section 3.4.3 covers both of the time-domain and frequency-domain data augmentation operations. Finally, in Sect. 3.4.4, we describe the design of STFNet and the corresponding encoders.

3.4.1 Overview

Our approach, STF-CSL, aims at learning good representations that store the intrinsic information of sensor measurement inputs in the latent space. The learned representations can be further utilized as the input of the target tasks such as classification or regression. Figure 2.14 presents an overview of STF-CSL. The illustrated workflow is as follows.

As a first step, called contrastive self-supervised pre-training, we train an encoder to map the original sensor measurements into a low-dimensional representation. This is done in a self-supervised manner, with unlabeled data only. Here, the supervision comes from optimizing a cost function, called *contrastive loss*, that aims at maximizing the agreement between the representations learned from locally perturbed data examples that are randomly generated by the same data augmentation operation.

We then freeze the parameters of the encoder learned during the contrastive self-supervised pre-training. Given any sensor measurement input, the encoder maps it to a corresponding low-dimensional representation. This representation facilitates downstream learning. A relatively simple model (*e.g.*, linear model) usually work wells, relating the latent representation to various desired outputs. Thus, we only need a small labeled training dataset to train the linear model. The last one or more layers of the encoder can also be fine-tuned during this supervised training process.

Next we show how to customizes this contrastive self-supervised representation learning framework to include frequency domain data.

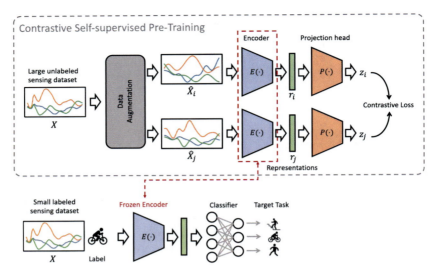

Fig. 2.14 Overview of our designed STFNet-based contrastive self-supervised representation learning approach (STF-CSL). We train an STFNet-based encoder with a contrastive self-supervised learning framework using only unlabeled sensing data. And then, the learned encoder and the corresponding representations can be utilized to train the target model (*e.g.*, classification) with a small amount of labeled data

3.4.2 Contrastive Self-Supervised Learning Framework

As shown in the contrastive self-supervised pre-training part of Fig. 2.14, there are four major components.

The first part is a stochastic data augmentation module. Let X refers to the original input, which is the time-series measured from one or multiple sensors. $\hat{X}_i$ and $\hat{X}_j$ are two samples randomly generated from the same data augmentation family. Here, we define $\hat{X}_i$ and $\hat{X}_j$ as a positive pair of examples since they are generated from the same input X. In this work, we apply data augmentation operations from both of the time-domain and frequency-domain. The details are shown in Sect. 3.4.3.

The second part is a neural network based encoder $E(\cdot)$ that maps the augmented example to the latent representations. Various choices of the network architectures can be used to design the encoder. For example, CNN-based encoders like ResNet [83] are widely utilized in learning representations of visual data. The encoder in our approach is built with STFNet and maps the data samples generated by data augmentation $\hat{X}_i$ and $\hat{X}_j$ to the corresponding representations $\mathbf{r}_i = E(\hat{X}_i)$ and $\mathbf{r}_j = E(\hat{X}_j)$.

The third part is a small neural network projection head $P(\cdot)$ that maps representations to the space where contrastive loss is applied. We use a multilayer perceptron (MLP) with one hidden layer to obtain $\mathbf{z}_i = P(\mathbf{r}_i) = W^{(2)}\sigma(W^{(1)}\mathbf{r}_i)$, where σ is ReLU activation function. The reason we apply the projection head

2 Self-Supervised Learning from Unlabeled IoT Data

before calculating the contrastive loss is that it has been proven that it is beneficial to define the contrastive loss on z_i's rather than r_i's [34].

The last part is a contrastive loss function defined for a contrastive prediction task. We use the normalized temperature-scaled cross entropy loss (NT-Xent) [53] as the loss function. For a given mini-batch of N examples in the training process. Since for each example X, we will generate a pair of augmented examples, and there will be $2N$ data points. The two augmented versions $\hat{X}_i$ and $\hat{X}_j$ of the same input X are called positive pair. All remaining $2(N - 1)$ in this batch are negative examples. Cosine similarity is utilized to measure the similarity between two augmented examples $\hat{X}_i$ and $\hat{X}_j$. The similarity is calculated on z_i and z_j:

$$\text{sim}(z_i, z_j) = \frac{z_i^T z_j}{\|z_i\| \|z_j\|}. \tag{2.18}$$

Here, $\|z_i\|$ refers to the ℓ_2 norm of z_i. The loss function for a positive pair of examples (i, j) is defined as

$$L_{i,j} = -\log \frac{\exp(\text{sim}(z_i, z_j)/\tau)}{\sum_{k=1}^{2N} \mathbb{1}_{[k \neq i]} \exp(\text{sim}(z_i, z_k)/\tau)}, \tag{2.19}$$

where τ refers to the temperature parameter of softmax and $\mathbb{1}_{[k \neq i]} \in \{0, 1\}$ is an indicator function

$$\mathbb{1}_{[k \neq i]} = \begin{cases} 1 \text{ for } k \neq i, \\ 0 \text{ for } k = i. \end{cases} \tag{2.20}$$

The loss for all positive pair (i, j) would be calculated and the average of them is used as the final loss functions.

3.4.3 Data Augmentation

Data augmentation plays an important role in our contrastive self-supervised learning framework. Different data augmentation operations would lead to very different performances in learning the representations.

Existing work applies different time-domain data augmentation operations in their self-supervised learning frameworks for sensor measurements [4–6]. Frequency domain data augmentation, however, has not been studied in self-supervised learning although it was shown to be beneficial in a lot of time-series related applications such as time-series classification [84] and anomaly detection [85]. We apply both time-domain and frequency-domain data augmentation operations in our approach. Some data augmentation operations (*e.g.*, adding noise) can be used in many tasks, while others are more task-specific. In this work, we use several wearable device based human activity recognition datasets to demonstrate the effectiveness of our approach. Data augmentation operations are described below.

For the time-domain data augmentation, we use similar time-domain data augmentation operations with previous work [4–6].

- **Jitter**: Add random Gaussian noise to the input signals. This simulates the heterogeneity of the sensor hardware.
- **Rotation**: Apply a rotation matrix to the multi-dimensional input. Rotation of the input is related to differences in sensor placement with respect to measured objects, which is label-invariant.
- **Permutation**: This transformation first slices the input data into multiple segments along the time axis, and then permutes the segments to generate a new data example.
- **Time-warping**: This transformation generates a new data example with the same label by stretching and warping the time intervals between the values of the input time-series. Small changes on the temporal locations of the input would not change its label.
- **Scaling**: This transformation multiplies the data with a random scalar.
- **Inversion**: This transformation multiplies the input data with -1.
- **Horizontal flipping**: This transformation reverses the input data along the time-direction. Horizontally flipping generates a mirror version of the input data in the opposite time direction.
- **Channel-shuffling**: This transformation randomly permutes the channels of the multi-channel sensor input data.

Besides the time-domain data augmentation operations used by previous approaches, we also study the performance of the following three frequency-domain data augmentation operations in our contrastive self-supervised representation learning framework.

- **Low-pass filter**: This transformation is application specific, but for most physical systems with inertia, the key information in the sensor measurement time-series is stored in the low frequency part. This is because system inertia attenuates high-frequency signal components more so than low-frequency components. Thus, the presence of significant high-frequency components is often attributed to noise. The label of the input should therefore not change if we pass the input through a low-pass filter (which often increases signal-to-noise ratio). We randomly select the cutoff frequency in a reasonable range that is related to (and slightly higher than) the rate at which significant dynamics occur in the measured process.
- **Amplitude and phase perturbation**: For a given time series $\mathbf{x} = [x_1, x_2, \ldots, x_n]^T$, we first get its frequency domain expression by applying discrete Fourier transform on it

$$F(\omega_k) = \frac{1}{n} \sum_{t=1}^{n} x_t e^{-j\omega_k t} = \Re[F(\omega_k)] + j\Im[F(\omega_k)] = A(\omega_k) \exp[j\theta(\omega_k)]$$

(2.21)

where $A(\omega_k)$ is the amplitude spectrum and $\theta(\omega_k)$ refers to the phase spectrum. In this transformation, the amplitude and/or the phase values of randomly selected segments of the frequency domain data are perturbed by Gaussian noise.

- **Phase shift**: Similar to phase perturbation, where the phase spectrum values are perturbed by Gaussian noise, the phase shift augmentation adds a random value between $-\pi$ and π to the phase values.

3.4.4 Design of the STFNet-Based Encoder

Our encoder is based on the STFNets architecture for learning in the frequency domain [37]. An STFNet block is equivalent to one layer in CNNs, and the encoder in our approach is built by stacking multiple STFNet blocks. Each layer takes multi-dimensional time-series data as input, and computes multiple Short Term Fourier Transform (STFT) representations with different time-frequency resolutions that differ in the size of the input sliding window in the time domain and correspondingly the frequency resolution produced in the frequency domain. We set the multiresolution sliding window to $\{16, 32, 64, 128\}$. Frequency-domain features are then extracted through spectral manipulations (STFNet-Filtering, STFNet-Convolution, and STFNet-Pooling). The kernels of STFNet-Filtering and STFNet-Convolution are learnable. Our encoder uses three STFNet layers. Each STFNet layer is followed by a dropout layer with a keep-probability of 0.8. An 1D global average pooling layer is put to follow the last STFNet layer.

3.5 Evaluation

We evaluate the STF-CSL system using several public datasets. We focus on device-based and device-free human activity recognition with motion sensors (accelerometer and gyroscope) and WiFi. We first introduce the datasets we use and then show the experimental setup, including data preprocessing and detailed implementation. Finally, we analyze the quality of representations studied by STF-CSL.

3.5.1 Datasets

We use human activity recognition data for the evaluation. We chose human activity recognition not because of its application value (it's an old application with existing off-the-shelf systems that already do very well), but rather because it is a widely-researched application with lots of available empirical data and many prior results that allow a comprehensive comparison of the proposed scheme to prior work. It allows us to demonstrate (using apples-to-apples comparisons) that extending contrastive learning to the frequency domain is a good idea. We use six human activity recognition datasets that cover a wide range of sensing signals (accelerometer, gyroscope, WiFi), devices (smartphone, smartwatch, WiFi transmitter and receiver), data collection protocols, and activity recognition tasks. Below, we introduce these datasets:

HHAR

HHAR [86] is a human activity recognition dataset devised to benchmark human activity recognition algorithms in real-world context. The dataset is gathered with a variety of different device models including different smartphones and smart-watches. In this dataset, a total of six different activities are studied: biking, sitting, standing, walking, upstairs and downstairs. Four smartwatches (2 LG watches, 2 Samsung Galaxy Gears) and eight smartphones (2 Samsung Galaxy S3 mini, 2 Samsung Galaxy S3, 2 LG Nexus 4, 2 Samsung Galaxy S+) are used by 9 users to record accelerometer and gyroscope readings under different activities. The sampling rate of signals varied across phones and watches with values between 50–200 Hz.

MobiAct

MobiAct [87] is a benchmark dataset for smartphone-based human activity recognition. It collects signals from montion sensors (3D accelerometer, gyroscope, orientation) in a Samsung Galaxy S3 smartphone to perform fall detection and recognize activities of daily living. It records 4 types of falls and 12 types of activities of daily living (ADL) from a total of 66 participants of different age, height and weight. More than 3200 trials were collected, all captured by the same Samsung Galaxy S3 smartphone. The smartphone was freely placed in the participants' pockets with random orientation. In our evaluation, we use the data related to 10 activities to evaluate the performance of our approach. The considered activities are: standing (STD), walking (WAL), jogging (JOG), jumping (JUM), stairs up (STU), stairs down (STN), standing to sitting on a chair (SCH), sitting on a chair (SIT), car-step in (CSI) and car-step out (CSO).

MotionSense

MotionSense dataset [88] includes time-series data generated by accelerometer and gyroscope sensors. It is collected with an iphone 6, kept in the participants's front pocket. A total of 24 participants in a range of gender, age, weight, and height perform six activities, including downstairs, upstairs, walking, jogging, sitting, and standing in the same environment and condition.

UCI HAR

UCI HAR [89] is a public dataset which studies human activity recognition using smartphones. A total of 30 participants with ages from 19 to 48 years old were selected in this dataset. Each participant wears a Samsung Galaxy S II smartphone on their waist. Six activities were performed by the participant: standing, sitting, laying down, walking, walking downstairs and walking upstairs. The readings of the accelerometer and gyroscope were collected at a sampling rate 50 Hz.

WISDM

WISDM [90] dataset is released by the Wireless Sensor Data Mining Lab. It's an early dataset which studies human activity prediction with mobile devices. It contains data collected from 29 participants. The participants take a smartphone including Nexus One, HTC Hero, and Motorola Backflip and perform six different activities: walking, upstairs, jogging, downstairs, sitting, and standing. The outputs of the accelerometers are collected in a 20 Hz sampling rate.

EI

EI [91] is a deep learning based device-free human activity recognition framework. It made use of Channel State Information (CSI) to analyze human activity. The CSI data of 11 participants with different ages, height, and weight was collected from 6 different rooms in two buildings. The CSI values of 30 OFDM subcarriers were recorded to analyze six different types of human activities: wiping the whiteboard, walking, moving a suitcase, rotating the chair, sitting, as well as standing up and sitting down.

We summarize the information of the datasets in Table 2.1.

3.5.2 Experiment Setup

In this part, we first introduce how we preprocess the datasets. Next, we show the detailed implementation of our approach.

Table 2.1 A summary of the datasets

Dataset	No. of users	No. of activities
HHAR	9	6
MobiAct	66	10
MotionSense	24	6
UCI HAR	30	6
WISDM	29	6
EI	11	6

Data Preprocessing

For HHAR, MobiAct, MotionSense, UCI HAR, and WISDM, we use the accelerometer data. We use all the measured CSI when considering the EI dataset. We linearly interpolate the reading to 50 Hz and then segment the datasets into sliding windows of size 512 with 50% overlap. For each dataset, we use the data from approximately 75% of users as the training data and the remaining as the test set.

Implementation Details

We follow the general architecture in Fig. 2.14. The encoder consists of three STFNet layers with {64, 64, 256} kernels. We set the sliding window for multiresolution short-time Fourier transform to be {16, 32, 64, 128} for all the three layers. Each STFNet layer is followed by a dropout layer with keep probability 0.8. A global average pooling layer is put in the end of the encoder. We utilize a two-layer projection head. The first layer is a fully connected layer with 256 hidden units and ReLU activation function. The second is a fully connected layer with 64 hidden units without activation function. We apply a combination of rotation and amplitude and phase perturbation as our default data augmentation operation on HHAR, MobiAct, MotionSense, UCI HAR, and WISDM dataset. For EI, we use the amplitude and phase perturbation as augmentation operation.

Our model was trained using an Adam optimizer [92] with learning rate 10^{-4} and decay rate 0.9 and 0.99 for the first and second moment, respectively. We add L2 regularization to the loss function in order to mitigate over-fitting. We use a batch size of 64, and train the model for 50 epochs. In the evaluation, we run 5 times for each experiment and take the average as the final performance.

Linear Classification

To evaluate the quality of the representations learned by our approach, we build a linear layer on the representations to do the classifications. After the contrastive self-supervised pre-training, we freeze the parameters of the learned encoder, and train the linear classifier to implement human activity recognition.

Models in Comparison

In this evaluation, we are trying to show the superiority of STFNet, compared with CNN, on contrastive self-supervised learning framework. Hence, we build a three-layer CNN followed by a global average pooling layer. The CNN based encoder has the same structure with our encoder. We summarize the models we used in evaluation as follows:

- **STF-CSL:** In this model, we use our STFNet-based encoder in the contrastive self-supervised pre-training and then freeze the parameters of the encoder. A linear layer built on the output of the encoder, which means the representations we learned, is used to do the classification. The labeled dataset is used to train the linear layer.
- **CNN-CSL:** This model uses CNN instead of STFNet to build the encoder. The remaining parts are the same with STF-CSL. Tang et al. also applied this approach in their work [5].
- **Supervised:** We compare with performance of our approach with supervised model. In this way, we train the STFNet-based encoder and the linear layer directly in a supervised way.

3.5.3 Results

We show the evaluation results in this part. We begin with the comparison between CNN and STFNet, and then study the performance under different data augmentation operations. At last, we show the visualization of the learned representations.

Comparison Between CNN and STFNet

We first compare the performance of CNN and STFNet in the contrastive self-supervised learning framework. We train the CNN-based encoder (CNN-CSL) and the STFNet-based encoder (STF-CSL) with the whole training set (without labels) in a self-supervised way. And then we train the linear classifier when 1%, 2%, 5%, 10%, 20%, 50% and 100% labels are given, and calculate the corresponding F1 scores on the test sets. The results are shown in Fig. 2.15. We can

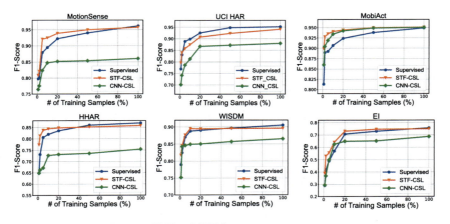

Fig. 2.15 Comparison between CNN and STFNet

observe that STF-CSL (orange curves) obviously outperforms CNN-CSL (green curves) on different datasets. This means that STFNets could extract the intrinsic information from the inputs in a better way that CNNs. To illustrate the effectiveness of the contrastive self-supervised learning framework, we also compare STF-CSL with a supervised model. In the supervised model, the inputs first pass through the STFNet-bases encoder, and then the linear classifier. We train the whole model using the given labeled training data. The F1 scores of the supervised model are shown as blue curves in Fig. 2.15. We observed that STF-CSL performs better than the supervised model when only a small number of labeled data items is given. This is because self-supervised pre-training has already extracted the key information of the input data. When there're enough labeled training data items, STF-CSL performs on par with the supervised model. STF-CSL usually performs not as well as the supervised model when 100% of the labels are given. Under this scenario, the performance of STF-CSL can be further improved by fine-tuning the encoder with the labeled data. We do not cover this part in our evaluation since we mainly focus on illustrating the advantages of STFNet over CNN.

Performance Under Different Data Augmentation Operations

Data augmentation is an important part of the contrastive self-supervised learning framework. Different data augmentation operations would lead to very different performances. In this part, we study the performance of STF-CSL under different data augmentation operations, including both time-domain and frequency-domain data augmentations, on MontionSense, MobiAct, and UCI HAR dataset. We perform the contrastive self-supervised pre-training with different data augmentation operations mentioned in Sect. 3.4.3 and then train the linear classifier with 100% of labels. The results are shown in Table 2.2. We found that the performance of STF-CSL varies a lot under different data augmentation operations. For example, in MobiAct

Table 2.2 Performance under different data augmentation

Data augmentation	MotionSense	MobiAct	UCIHAR
Jitter	0.8448	0.8455	0.7084
Rotation	0.9485	0.9410	0.9367
Permutation	0.8885	0.8848	0.7490
Time-warping	0.8685	0.8661	0.7770
Scaling	0.8726	0.8612	0.8717
Inversion	0.9159	0.9015	0.8062
Horizontally flipping	0.8946	0.8281	0.8493
Channel-shuffling	0.8421	0.9020	0.7684
Low-pass filter	0.8854	0.8441	0.8773
Amp. and phase perturbation	0.9397	0.9441	0.9396
Phase shift	0.8360	0.8578	0.8898

dataset, rotation performs the best among time-domain data augmentations, leading to a 0.9410 F1-score. For frequency-domain data augmentations, amplitude and phase perturbation performs the best, leading to a 0.9441 F1-score. An interesting observation is that perturbation on amplitude and phase values on the frequency-domain performs much better than directly adding Gaussian noise on time-domain values. This is because perturbing the amplitude and phase values could generate more diverse data samples with the same label, which would improve the scalability of the encoder.

Visualization of the Learned Representation

We visualize the representations learned by STF-CSL through t-Distributed Stochastic Neighbor Embedding (t-SNE) [93]. t-SNE is statistical technique for visualizing high-dimensional data. It embeds high-dimensional data to a low-dimensional space of two or three dimensions by minimizing the KL divergence between them. We run the self-supervised pre-training of STF-CSL on the six HAR datasets using only unlabeled data and embed the 256-dimension output of the encoder (the activation from the global average pooling layer) to a 2D space. Then we dye each node with different colors according to their real label. The results are shown in Fig. 2.16. We observe that samples with the same label could group together. This means that the contrastive self-supervised pre-training could extract the intrinsic information from the signal input without the help of labels.

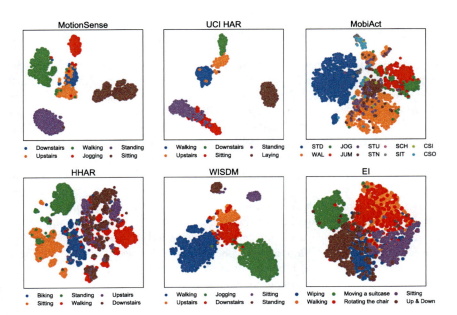

Fig. 2.16 Visualization of the learned representations

3.6 Discussion and Limitations

The work presented in this section shows promise in using frequency domain knowledge for contrastive learning. Improvements of up to 10% in classification accuracy are shown when we build the contrastive learning framework from frequency domain compared to contrastive learning solutions with encoders and data augmentation in the time-domain only. However, the work is far from being a comprehensive frequency domain contrastive learning framework. As such, it invites future work in multiple avenues as presented below.

Application Domain Limitations

The results presented in this section are computed for activity recognition datasets. While many datasets are used, a significant similarity exists in the time-scales of explored activities. This similarity makes it easier to tune data augmentation to the needs of this application category. It is likely that the augmentation solutions described in this work will not work out-of-the-box for other categories of IoT applications. For example, a significantly broader application domain might be one that uses RF-sensing to measure activities at different timescales from breathing and heartbeats (*e.g.*, with mm-wave radars) to various instances of vibrometry (using reflected WiFi, LoRa, ultrasonic, LiDAR, 5G, or other signals). This remains a rich avenue for future exploration.

Frequency-Domain Perturbation Design

The frequency-domain perturbations described in this work are far from generalizable across different IoT application categories. For example, the phase shift perturbation operator essentially means that downstream AI tasks cannot use phase as input since this the data augmentation will teach the encoder to put signals of different phase shifts into the same latent cluster. For applications where phase shift measurements are important (which indeed is the case in a large fraction of RF-sensing applications), the phase shift operator should not be used for data perturbation. Alternative perturbations must be found. Similarly, the assumption that low-pass filters preserve the gist of the input signal are only true when detecting activities that evolve much slower compared to the sampling frequency used. For several categories of IoT applications, this assumption is not true. For example, in compressive sensing, measurements occur at a much lower frequency compared to the Nyquist frequency of the measured phenomenon. A low-pass filter will likely substantially degrade signal representation. A different form of filter needs to be designed for data augmentation purposes.

2 Self-Supervised Learning from Unlabeled IoT Data

Time-Domain Perturbation Design

The work inherits time-domain perturbations from prior work. There is room for significant customization of time-domain perturbations. For example, rotation and permutation operators do not really correspond to valid physical transformations on the data. Changing sensor locations does not actually result in geometric signal rotation. Also, strong auto-correlations in sensor data make arbitrary permutations invalid. For another example, reversing time by horizontal flipping often does not have a meaningful physical interpretation. Humans will not walk backwards and shattered objects will not self-reassemble. Understanding proper time-domain perturbations for IoT applications remains an open problem.

Limitations of Contrastive Learning

Not all local data perturbations are equally "similar" to the unperturbed signal. Understanding proper notions of similarity is very important to the design of better cost functions to use for contrastive loss in the contrastive learning framework. This work did not investigate different options for design of contrastive loss functions.

STFNet Limitations

The contrastive learning framework used in this work leverages STFNet for encoder design. Different from the original STFNet approach, which works is supervised, without data augmentation, our approach combines it with the self-supervised contrastive learning framework and data augmentation. STFnet uses the Short Time Fourier Transform. Signal processing literature offers many other transforms, such as Laplace Transform and Wavelet Transform. It is interesting to investigate which transform offers better results.

Transfer Learning

Self-supervised learning can be interpreted as a form of transfer learning. We can run the contrastive learning framework on a large domain-agnostic measurement dataset to train the encoder, and then reuse the parameters of the encoder as the starting point when training the downstream tasks with domain-specific labels. We leave this part as future work.

The above discussion suggests that this work is more of a conversation starter. Initial evidence suggests (using a limited study on human activity recognition applications) that frequency domain data augmentation is helpful. Perhaps the impressive part is that, despite all of the above limitations, we show accuracy gains of up to 10%. How much can be gained further if the above items are addressed?

The work suggests that frequency domain data augmentation is a fertile area for future work, but such future work is beyond the scope of this section.

4 Frequency-Domain Semi-Supervised Contrastive Learning Framework for IoT

4.1 Overview

In this section, we introduce another way we apply the customization to the self-supervised contrastive learning framework for IoT applications. Specifically, we train the encoder during the self-supervised contrastive pre-training using both labeled and unlabeled data instead of only using unlabeled data as previous work did. We designed a semi-supervised (instead of self-supervised) contrastive learning framework and evaluated its performance on Human Activity Recognition (HAR) application. The evaluation shows that our approach, namely SemiC-HAR [29], demonstrates better performance compared with the original self-supervised contrastive learning framework.

The increasing number of smart phones and wearable devices yield large volumes of sensory data which can be utilized for multiple applications, including behavior and activity recognition [4–6], health and well-being [94–96], and localization and tracking [97, 98]. Specially, *Human Activity Recognition (HAR)* has been a popular application since the introduction of motion sensors (such as accelerometers and gyroscopes) on smart phones and wearable devices, offering an opportunity for understanding physical activities. Human activity recognition aims at recognizing human physical activities, such as walking, running, sitting, going downstairs, and going upstairs, based on the time-series sensory data collected from the accelerometers and gyroscopes. Traditional approaches classify human activities based on manually extracted features from the time-series data. Recent developments in deep learning demonstrated outstanding performance on various machine learning tasks [19, 20, 23], thereby inspiring the application of deep neural networks for recognizing human activities [99–101]. Novel learning paradigm emerged, such as STFNet [37] that learns sensing signals in the frequency domain, proving to be more effective compared with the traditional time-domain approaches.

Training a deep neural network to accurately recognize human activities requires a large amount of labeled data. However, it is difficult to collect labeled data for human activity recognition because the collection of the time-series sensory data and the corresponding manual annotation with activity labels must happen at the same time. This makes the HAR datasets usually laboratory-based. Although it needs a lot of human labor to label a HAR dataset, collecting raw time-series measurements from motion sensors is much easier. Hence, in order to reduce the labeling burden and effectively take advantage of the massive unlabelled sensory data, semi-supervised deep learning frameworks [4–6, 102] for HAR have been widely studied in recent years.

The semi-supervised framework for HAR is a straightforward application of general self-supervised learning techniques [103]. Self-supervised learning is a kind of unsupervised learning technique that requires mostly unlabeled data. It aims at learning representations or features that extract intrinsic structure (*e.g.*, clusters) from the original inputs with no need for labels. A downstream stage can then learn to map elements of that intrinsic structure to output categories with minimal supervision. Self-supervised contrastive learning [34] is a representative self-supervised learning framework that has gotten wide use in multiple areas due to its outstanding performance and simplicity of implementation. Self-supervised contrastive learning could help HAR applications to effectively utilize unlabeled sensory data [5]. This framework first performs self-supervised contrastive pre-training to extract intrinsic structure from the original sensory inputs, and then trains a simple (usually linear) classifier on the learned features. The contrastive pre-training runs in a self-supervised manner, where only unlabeled sensory data are used. A small amount of labeled data are used to train the classifier. Previous research has shown that labeled data can help improve the quality of features learned by contrastive learning [38]. Hence, we can get better performance if we carefully utilize both labeled and unlabeled data during the contrastive pre-training process.

To leverage this insight, we propose *SemiC-HAR*, a semi-supervised contrastive learning framework for HAR. SemiC-HAR takes both labeled and unlabeled data into consideration. The key challenge lies in how to efficiently utilize both labeled and unlabeled data in building the contrastive pre-training framework. To extract features from original inputs, previous research has studied self-supervised contrastive learning [34] which uses only unlabeled data and supervised contrastive learning [38] which uses only labeled data. In SemiC-HAR, we propose a semi-supervised contrastive learning framework to efficiently utilize both labeled and unlabeled data in the pre-training process by carefully re-designing the contrastive loss function. As far as we know, we are the first to design the semi-supervised contrastive learning framework and apply it to HAR.

In SemiC-HAR, we first train the parameters of an encoder that extracts intrinsic structure from sensory inputs to produce low-dimensional features, using the proposed semi-supervised contrastive learning framework. In this process, the exploitation of both labeled and unlabeled data and our design of the contrastive loss lead to a clear performance gain, compared with purely supervised or self-supervised approaches. After the semi-supervised contrastive pre-training, we build a linear classifier on the learned features to implement HAR. The labeled data are used to train the classifier.

The contribution of this work can be thus summarized as follows:

- We propose SemiC-HAR, a semi-supervised contrastive learning framework for HAR. SemiC-HAR first extracts features from original sensory inputs in a semi-supervised way, and then trains a classifier to recognize human activities with

the labeled data. As far as we know, SemiC-HAR is the first application of a semi-supervised contrastive learning framework for HAR.
- We evaluate the performance of SemiC-HAR on six different HAR datasets, featuring different numbers of participants, devices (including different smartphones and wearable devices), and activities. We demonstrate that our semi-supervised contrastive learning framework outperforms supervised and self-supervised contrastive learning frameworks in learning features for sensory data when limited amounts of labeled data are available.
- We compared the performance of SemiC-HAR with previous supervised and semi-supervised frameworks for HAR when different amounts of labeled data are given. The evaluation results show that SemiC-HAR offers the best trade-off, compared with the previous approaches, between the amount of labeled data and the quality of activity recognition.

4.2 Preliminary and Motivation

We first give a preliminary overview of self-supervised contrastive learning and supervised contrastive learning, and then describe the motivation why we propose our semi-supervised contrastive learning framework for HAR.

4.2.1 Self-Supervised Contrastive Learning

Self-supervised contrastive learning is an unsupervised learning framework that learns representations or features from the inputs by maximizing agreement between differently augmented views of the same data sample via a contrastive loss in the latent space. SimCLR [34] is the representative and most commonly used self-supervised contrastive learning framework that has gotten outstanding performance in different areas like computer vision, automatic speech recognition, as well as IoT applications [5, 34, 104, 105]. The framework of SimCLR is shown in Fig. 2.17. SimCLR mainly consists of four components.

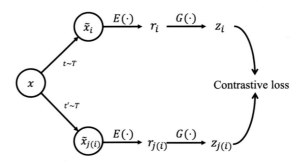

Fig. 2.17 The framework of self-supervised contrastive learning (SimCLR)

2 Self-Supervised Learning from Unlabeled IoT Data

The first part is a stochastic data augmentation module, where different views of the same data sample are generated from the same family of data augmentation operations. As shown in Fig. 2.17, x refers to the original input. Two views of x, here are $\tilde{x}_i$ and $\tilde{x}_{j(i)}$, are generated through two separate operations (t and t') from the same data augmentation family. The data augmentation operations are highly application dependent. For example, in image classification applications, random cropping, random color distortion, and Gaussian blur are commonly used data augmentation operations; in IoT applications, instead, time-warping, jitter, and channel-shuffling are popular augmentation operations when dealing with sensory data.

The second part is an encoder, E, that extracts intrinsic structure from the augmented views, leading to appropriate features or representations. Here r_i and $r_{j(i)}$ refer to the representations of $\tilde{x}_i$ and $\tilde{x}_{j(i)}$:

$$r_i = E(\tilde{x}_i), \ r_{j(i)} = E(\tilde{x}_{j(i)}). \tag{2.22}$$

The third part is a projection head, G, that maps the representations to a latent space, where the contrastive loss is calculated. A common choice for the projection head is multilayer perceptron (MLP) with one or two hidden layers. For example, SimCLR applies a MLP with one hidden layer as the projection head, which means

$$z_i = G(r_i) = W^{(2)}\sigma(W^{(1)}r_i), \tag{2.23}$$

where σ refers to the ReLU activation function. It has been proved in [34] that calculating the contrastive loss on z_i's instead of on the representations r_i's directly could improve the performance of contrastive learning. Hence most of the contrastive learning framework would need the projection head, G.

The last part is the contrastive loss function. The contrastive loss is defined for the contrastive prediction task which aims at maximizing the agreement between positive pairs of examples ($\tilde{x}_i$ and $\tilde{x}_{j(i)}$). Specifically, given a mini-batch of N data examples, the data augmentation would generate $2N$ data points. The data points generated from the same example are called a pair of positive examples. Let $i \in I \equiv \{1, 2, \ldots, 2N\}$ be the index of an arbitrary augmented data sample, and let $j(i)$ be the index of the corresponding positive examples of i. The remaining $2(N-1)$ augmented data points in the same mini-batch are called negative examples. The aim of the contrastive task is to discriminate positive pairs of examples from negative ones. The loss function for each augmented data sample i is defined as

$$\ell_i^{self} = -\log \frac{\exp(\text{sim}(z_i, z_{j(i)})/\tau)}{\sum_{k \in K(i)} \exp(\text{sim}(z_i, z_k)/\tau)}, \tag{2.24}$$

where τ refers to the temperature parameter that controls the strength of penalties on hard negative samples and $K(i) \equiv I \setminus \{i\}$. $\text{sim}(z_i, z_{j(i)})$ defines the similarity between z_i and $z_{j(i)}$. A common choice is the cosine similarity:

$$\text{sim}(z_i, z_{j(i)}) = \frac{z_i^T z_{j(i)}}{\|z_i\| \|z_{j(i)}\|}, \tag{2.25}$$

where $\|z_i\|$ refers to the ℓ_2 norm of z_i. The final loss function is defined as the sum of the contrastive loss across all positive pairs in a mini-batch:

$$\mathcal{L}^{self} = \sum_{i \in I} \ell_i^{self}. \tag{2.26}$$

The parameters of the encoder can be trained by minimizing the contrastive loss function in a self-supervised training process using unlabeled data. In this way, the encoder could extract the intrinsic structure from the original inputs and store it in the corresponding representations. After the self-supervised training, the parameters of the encoder are frozen. In the downstream tasks, the original inputs would pass through the encoder and a classification or regression model would build on the computed structured representations instead of the original inputs.

4.2.2 Supervised Contrastive Learning

The idea of self-supervised contrastive learning is pulling together an anchor and a single positive example (an augmented version of the same input sample) in the latent space and pushing apart the anchor from a set of negative examples consisting of the entire remainder of the mini-batch. However, among such negative examples, there may exist data samples which are in the same class as the anchor. Those samples from the same class should be considered as positive examples instead of negative ones. The self-supervised contrastive learning frameworks takes all of them as negatives because there's no labeling information and we do not know whether a data sample is from the same class with the anchor or not. If labeling information is given, the contrastive loss can be re-designed in a better way by carefully selecting the positive and negative examples for the anchor. This inspires a supervised contrastive learning framework.

Prannay et al. propose a supervised learning framework that builds on the self-supervised contrastive learning literature by leveraging label information [38]. The contrastive loss in their supervised learning framework considers many positive examples per anchor instead of only a single positive in the self-supervised contrastive learning. Specifically, the positive examples are those in the same class with the anchor, rather than the single one which is augmented from the same input with the anchor, as done in the self-supervised contrastive learning.

Supervised contrastive learning applies the same framework with self-supervised contrastive learning, which is shown in Fig. 2.17, but re-designs the contrastive loss. The contrastive loss proposed by Khosla et al. [38], called SupCon loss, considers the scenario that more than one sample is known to belong to the same class due to the existence of labels. SupCon loss follows the following forms

2 Self-Supervised Learning from Unlabeled IoT Data

$$\mathcal{L}^{sup} = \sum_{i \in I} \ell_i^{sup}, \tag{2.27}$$

where

$$\ell_i^{sup} = \frac{-1}{|P(i)|} \sum_{p \in P(i)} \log \frac{\exp(\mathrm{sim}(z_i, z_p)/\tau)}{\sum_{k \in K(i)} \exp(\mathrm{sim}(z_i, z_k)/\tau)}. \tag{2.28}$$

Here, $P(i) \equiv \{p \in K(i) : \tilde{\mathbf{y}}_p = \tilde{\mathbf{y}}_i\}$ refers to the set of indices of all positive examples in the mini-batch which have the same label $\tilde{\mathbf{y}}_p$ $(= \tilde{\mathbf{y}}_i)$ with i. $|P(i)|$ is the cardinality of $P(i)$.

4.2.3 Motivation

In the HAR applications, we have a large amount of unlabeled sensory data and a limited amount of labeled sensory data. Early solutions used only labeled data to build supervised models for HAR. To efficiently apply unlabeled data, semi-supervised and self-supervised techniques were introduced, including contrastive learning [5]. Supervised contrastive learning has proven that labeled data could help to improve the quality of the representations learned by the contrastive learning framework. This motivates us to combine both labeled and unlabeled data in the contrastive pre-training process.

We cannot directly use the framework of supervised contrastive learning, since only a part of the data (instead all of them) is labeled. The challenge lies in defining the positive examples in a mini-batch for an anchor under a semi-supervised (instead of fully supervised) scenario. In this work, we propose a semi-supervised contrastive learning framework with a carefully designed contrastive loss to make it work in a semi-supervised scenario.

4.3 Design of SemiC-HAR

In this section, we describe the design of our semi-supervised contrastive learning framework for HAR, namely SemiC-HAR. We first give an overview of SemiC-HAR, and then introduce the details.

4.3.1 Overview

SemiC-HAR focuses on HAR applications. Human activities can be accurately recognized through the time-series measurements of the motion sensors (*e.g.*, accelerometer and gyroscope) deployed on smartphones or wearable devices. The output of the motion sensors can be divided into segments with fixed time windows. To take advantage of unlabeled data, we consider two parts of the dataset: (1) a small labeled HAR dataset $D = \{(x_1, y_1), \ldots, (x_m, y_m)\}$ and (2) a large unlabeled HAR dataset $U = \{x_{m+1}, \ldots, x_{m+n}\}$, where x_i refers to the time-series motion sensor measurements with fixed window and y_i refers to the corresponding label. The goal of our approach is to build a function $f(x_i)$, trained on both of the labeled and unlabeled data (i.e., $D \cup U$), to accurately predict the activity label of each window of the time-series data x_i.

Figure 2.18 illustrates the framework of SemiC-HAR. There are mainly four components:

- **Supervised Training**: We first train a "temporal classifier" C based on the labeled dataset D.
- **Self-Labeling**: The "temporal classifier" can be used to predict labels for the unlabeled dataset U. The predicted labels and the raw time-series data in U can then form a new labeled dataset U'. We then merge the two datasets D and U' and put them together to form the self-labeled dataset S.

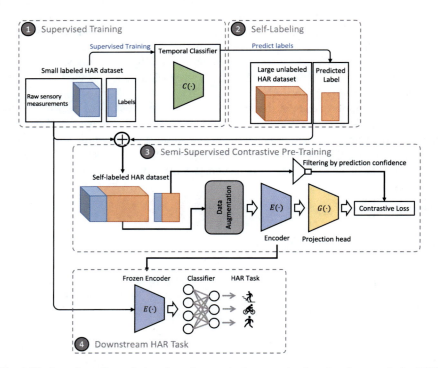

Fig. 2.18 Overview of our designed semi-supervised contrastive learning framework for HAR (SemiC-HAR). SemiC-HAR consists of four components: (1) Supervised training, (2) Self-labeling, (3) Semi-supervised contrastive learning, and (4) Downstream HAR task

- **Semi-Supervised Contrastive Pre-Training**: Based on S, we train the encoder E, that maps the raw sensory data to low-dimensional representations, through our semi-supervised contrastive learning algorithm. The encoder is then frozen and passed to the downstream HAR task.
- **Downstream HAR Task**: Given the encoder trained by the semi-supervised contrastive pre-training, we pass the time-series data from the labeled dataset D to the encoder and get the corresponding representations. Then, we train a simple classifier on the representations under a supervised way.

We will describe the details of the four components in the following subsections.

4.3.2 Supervised Training

We first train the temporal classifier C on the small labeled dataset, D. We apply STFNet [37] as the basic building block of the temporal classifier. STFNet is the Short-Time Fourier Neural Network, specially designed to deal with time-series inputs. An STFNet layer is equivalent to one layer in CNNs but it processes the input in the frequency-domain through Short-Time Fourier Transform instead of in the time-domain like CNNs. Since many human activities have cyclical motion patterns, they have clear and distinct representations in the frequency domain. STFNet has therefore shown obvious performance gains in HAR applications, compared to the traditional convolutional and recurrent neural network approach.

The temporal classifier comprises three STFNet layers, each of which followed by a dropout layer. A 1D global average pooling layer is used after the last STFNet layer. Finally, a softmax output layer is put to follow the global average pooling layer.

4.3.3 Self-Labeling

Given the temporal classifier, we label the large unlabeled dataset $U = \{x_{m+1}, \ldots, x_{m+n}\}$. Let y_i denote the label of x_i predicted by the temporal classifier, and $w_i \in [0, 1]$ be the corresponding prediction confidence. After the self-labeling, we can get a labeled dataset $U' = \{(x_{m+1}, y_{m+1}, w_{m+1}), \ldots, (x_{m+n}, y_{m+n}, w_{m+n})\}$. For each data sample x_i in the the labeled dataset D, we also give it a prediction confidence, $w_i = 1$. We extend the dataset D in a similar way to form $D' = \{(x_1, y_1, w_1), \ldots, (x_m, y_m, w_m)\}$, where $w_i|_{i=1 \ldots m} = 1$.

We merge the two datasets to generate a new one, which we call the "Self-labeled HAR DataSet":

$$S = D' \cup U'. \tag{2.29}$$

There are raw sensor measurements, labels, and the corresponding prediction confidences in dataset S.

4.3.4 Semi-Supervised Contrastive Pre-Training

Given the self-labeled HAR dataset S, an intuitive way is to directly use the supervised contrastive learning framework we introduce in Sect. 4.2.2 to train the encoder E. However, the labels in S are not 100% correct because some of them are predicted by our temporal classifier. The wrong labels, that are not correctly predicted by the temporal classifier, would have side effects when designing the contrastive loss. Hence, in order to efficiently utilize the correct labels to improve the performance of contrastive learning while eliminating the side effect caused by wrong labels, we propose a modified semi-supervised contrastive learning framework.

Our approach applies the same architecture as the supervised contrastive learning (Sect. 4.2.2) but a different design of the contrastive loss function. Recall that in the SupCon loss function (Eqs. (2.27) and (2.28)) in supervised contrastive learning, all the data samples in the mini-batch that have the same label with the anchor are considered positive examples of the anchor, and the remaining data are considered negative. To eliminate the side effect caused by wrong labels, we use a filter to separate the labels into two groups: labels with high prediction confidence and labels with low prediction confidence. For those data samples with high-confidence labels, we use both of the raw sensory measurements and the labels; for those with low-confidence labels, we only use their raw sensory measurements.

Specifically, For each x_i in a mini-batch of N data examples, two views of augmented data examples, denoted by $\tilde{x}_i$ and $\tilde{x}_{j(i)}$, are generated from the same family of data augmentation operations. The labels and prediction confidences of $\tilde{x}_i$ and $\tilde{x}_{j(i)}$ are the same with x_i:

$$\tilde{y}_i = \tilde{y}_{j(i)} = y_i, \tag{2.30}$$

$$\tilde{w}_i = \tilde{w}_{j(i)} = w_i. \tag{2.31}$$

Let $I \equiv \{1, 2, \ldots, 2N\}$ be the index of the $2N$ augmented data examples, and $\tilde{x}_i$ ($i \in I$) be an arbitrary augmented data example. After $\tilde{x}_i$ passing through the encoder E, we get the representation $r_i = E(\tilde{x}_i)$. The contrastive loss is calculated on the latent space of $z_i = G(r_i)$ after applying the projection head G. We consider two scenarios when designing our contrastive loss, for each anchor $\tilde{x}_i$ ($i \in I$),

- If the prediction confidence $\tilde{w}_i$ of its label $\tilde{y}_i$ is less than or equal to a threshold TH, then $\tilde{x}_i$ will be processed as an unlabeled data example. The contrastive loss for anchor $\tilde{x}_i$ would be the same as the sefl-supervised contrastive learning where there is only one positive example $\tilde{x}_{j(i)}$ which is augmented from the same data sample as $\tilde{x}_i$.
- If the prediction confidence is larger than the threshold TH, $\tilde{x}_i$ would be considered as a labeled data example. In this scenario, we apply a contrastive loss function which is similar but not the same with the supervised contrastive learning. We choose such data example $p \in I$ that satisfies the following two

2 Self-Supervised Learning from Unlabeled IoT Data

conditions as the positive examples: (1) $\tilde{x}_p$ in the same class with $\tilde{x}_i$ ($\tilde{y}_p = \tilde{y}_i$), and (2) the prediction confidence $\tilde{w}_p$ is larger than the threshold TH. We also use $\tilde{w}_p$ as the weight for each component corresponding to the positive example $\tilde{x}_p$.

Our contrastive loss can then be defined as

$$\mathcal{L}^{semi} = \sum_{i \in I} \ell_i^{semi}. \tag{2.32}$$

If $\tilde{w}_i \leq TH$,

$$\ell_i^{semi} = -\log \frac{\exp(\mathrm{sim}(z_i, z_{j(i)})/\tau)}{\sum\limits_{k \in K(i)} \exp(\mathrm{sim}(z_i, z_k)/\tau)}. \tag{2.33}$$

If $\tilde{w}_i > TH$,

$$\ell_i^{semi} = \frac{-1}{|P'(i)|} \sum_{p \in P'(i)} \log \frac{\tilde{w}_p \exp(\mathrm{sim}(z_i, z_p)/\tau)}{\sum\limits_{k \in K(i)} \exp(\mathrm{sim}(z_i, z_k)/\tau)}. \tag{2.34}$$

Here, $K(i) \equiv I \setminus \{i\}$; $P'(i) \equiv \{p \in K(i) : \tilde{y}_p = \tilde{y}_i \text{ and } \tilde{w}_p > TH\}$ refers to the set of indices of all positive examples for anchor i in the mini-batch, $|P'(i)|$ is the cardinality of $P'(i)$; and $\mathrm{sim}(z_i, z_k)$ defines the cosine similarity between z_i and z_k.

By optimizing our contrastive loss $\mathcal{L}^{semi}$, we can get the parameters of the encoder, which can be then used in the downstream HAR task. The details of our semi-supervised contrastive learning framework, including the data augmentation we used, the architectures of the encoder and projection head, are described in more detail in the evaluation Sect. 4.4.1.

4.3.5 Downstream HAR Task

After the optimization of the semi-supervised contrastive learning, we freeze the parameters of the encoder. In the downstream HAR task, we first map the raw measurements in the labeled dataset D to their representations, and then build a linear classifier on the representations. Since the encoder has already extracted the intrinsic information of the raw measurements, the linear classifier trained with the small labeled dataset could accurately recognize human activities. In this way, a relatively simple model trained on a small labeled dataset could accurately recognize human's activities.

4.4 Evaluation

In this section, we introduce the experiments we have done to evaluate the performance of SemiC-HAR. We use the same datasets as mentioned in Sect. 3.5.1. We first describe the experiment setup including the data pre-processing and details of our approach, SemiC-HAR, and then we show the evaluation results.

4.4.1 Experiment Setup

Data Pre-Processing

For EI dataset, we use all the CSI data of 30 OFDM subcarriers as the input. For the remaining 5 datasets, we apply the measurements from the accelerometers as the input. For all sensing data, we linearly interpolate them to 50 Hz and segment them into sliding windows of size 512 with 50% overlap. Data from about 75% of the users in each dataset is used to training our model and the remaining is used as the test set.

Model Details

The temporal classifier consists of three STFNet layers with the same kernel size 64. We set 4 resolutions {16, 32, 64, 128} for the short-time Fourier transform of each layer. A dropout layer with a keep probability 0.8 is put after each STFNet layer. After the last STFNet layer, we use a 1D global average pooling layer followed by a softmax output layer. The encoder E has a similar structure with the temporal classifier except that E has no softmax layer and the kernel sizes of the three STFNet in E are {64, 64, 256}. This means that the representation has a length of 256. The projection head G comprises two fully connected layers with {256, 64} hidden units. The first layer uses ReLU as its activation function and the second layer has no activation function. We set the default prediction confidence threshold to $TH = 0.9$. In the downstream HAR task, we build a linear classifier on the learned representations. We also fine-tune the last layer of the encoder when training the linear classifier. We apply a combination of rotation [4] and frequency-domain perturbation [85] as the data augmentation operation for the 5 motion sensor based datasets, and use frequency-domain perturbation when dealing with the WiFi based dataset EI.

We train our semi-supervised contrastive learning model using an Adam optimizer [92] with learning rate 10^{-4} and default decay rate. L2 regularization is added to the contrastive loss function in order to mitigate over-fitting. We train our model with a batch size 64 and 50 epochs. We have 5 runs for each experiment and calculate the average as the final performance.

4.4.2 Results

Comparison with Other HAR Frameworks

We first compared SemiC-HAR with existing HAR frameworks. We consider two semi-supervised frameworks (SelfHAR[102], Transformation Discrimination[4]) and a supervised framework (STFNet[37]). Transformation Discrimination is a semi-supervised frameworks for HAR which first trains an encoder to learn the representations under a self-supervised learning model and then trains a simple classifier on the representations. Transformation Discrimination implements the self-supervision by discriminating the data examples before and after the data augmentation. SeflHAR is a semi-suerpvised HAR framework with a teacher-student model. STFNet learns sensing signals from the Time-Frequency perspective and has shown in prior work to do better for labeled time-series data. We keep 10% of the data labeled in the training set and the remaining 90% unlabeled. We use both of the labeled and unlabeled data when training semi-supervised models and use only the labeled data when training the supervised model, STFNet. Then we compare the weight-average F1-score for each of them. The results are shown in Table 2.3. We observe that SemiC-HAR performs the best compared with the previous HAR models when only 10% of the training data are labeled.

Comparison with Self-Supervised and Supervised Contrastive Learning

We then compare the performance of our semi-supervised contrastive learning framework with the existing supervised and self-supervised contrastive learning frameworks when different amounts of the labeled data is given. For the self-supervised contrastive learning model, we train the encoder E based on all the raw sensory measurements in $D \cup U$. For the supervised contrastive learning model, we train the encoder based on the small labeled dataset D. The remaining parts, including the encoder, projection head, and the linear classifier, except the contrastive loss function of those three models (self-supervised, supervised, and SemiC-HAR) are the same.

Table 2.3 Comparison with existing models

Dataset	STFNet	SelfHAR	Transformation discrimination	SemiC-HAR
HHAR	0.8188	0.7739	0.7753	0.8510
MobiAct	0.9067	0.9138	0.8877	0.9479
MotionSense	0.8947	0.9312	0.9062	0.9393
UCI HAR	0.8985	0.8927	0.8915	0.9264
WISDM	0.8885	0.8809	0.8780	0.9006
EI	0.5611	0.6293	0.6049	0.6758

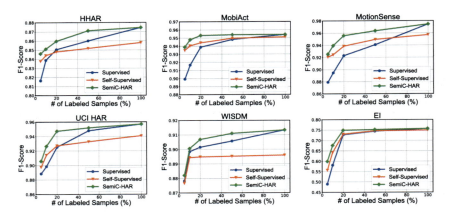

Fig. 2.19 Comparison with supervised and self-supervised contrastive learning

The results are shown in Fig. 2.19. We train the models when 5%, 10%, 20%, 50% and 100% of the data have labels, which means $\frac{|D|}{|D \cup U|} = 5\%, 10\%, 20\%, 50\%, 100\%$. We observe that the self-supervised contrastive learning model performs better than the supervised one when the size of D is small. This is because supervised contrastive learning can only use the labeled data in D. When the size of D is larger, the supervised contrastive learning model would outperform the self-supervised model. Our semi-supervised contrastive learning framework perform better than both of them, because SemiC-HAR takes both of the labeled (D) and unlabeled (U) data when training the encoder.

Performance of Self-Labeling

The performance of SemiC-HAR is highly dependent on the accuracy of the self-labeling, that is, whether the temporal classifier C trained on the small labeled dataset D could accurately annotate the raw measurements in the unlabeled dataset U. After annotating U, we can get a new labeled dataset U'. U' can be divided into two parts $U' = U'_+ \cup U'_-$ according to the prediction confidence. Here U'_+ refers to a subset of U' where the elements in U'_+ have prediction confidence larger than the threshold. U'_- refers to the subset of U' with prediction confidence less than the threshold. Since U'_+ (as well as D) provides the supervision while optimizing the contrastive loss, the size of U'_+ and the accuracy of the predicted labels in U'_+ would have impact on the performance of our approach. If most of the elements in U can be annotated with high prediction confidence (i.e., $|U'_+| \approx |U'|$) and most of the highly confident elements are correctly annotated (i.e., labels of most elements in U'_+ are correct), SemiC-HAR would perform well. We study the size of U'_+ and the accuracy of the predicted labels in U'_+ under different values of threshold (from 0.5 to 0.9) when 5%, 10%, 20%, and 50% of the training data have labels, which means $\frac{|D|}{|D \cup U|} = 5\%, 10\%, 20\%, 50\%$.

2 Self-Supervised Learning from Unlabeled IoT Data 81

Table 2.4 MotionSense: Size of the high confident prediction set

	0.5	0.6	0.7	0.8	0.9
5%	0.9998	0.9904	0.9819	0.9735	0.9645
10%	0.9995	0.9971	0.9917	0.9871	0.9804
20%	0.9988	0.9962	0.9921	0.9866	0.9782
50%	0.9995	0.9981	0.9967	0.9926	0.9879

Table 2.5 MotionSense: accuracy of the high confident predictions

	0.5	0.6	0.7	0.8	0.9
5%	0.9456	0.9513	0.9558	0.9592	0.9652
10%	0.9757	0.9764	0.9789	0.9806	0.9834
20%	0.9913	0.9918	0.9932	0.9944	0.9951
50%	0.9935	0.9935	0.9935	0.9948	0.9953

The results of "MotionSense" dataset are shown in Tables 2.4 and 2.5. Table 2.4 shows the size of U'_+ under different thresholds when 5%, 10%, 20%, and 50% of the training data have labels. Here we use the ratio between U'_+ and U' ($|U'_+|/|U'|$) to describe the size of U'_+. We can observe that most of the elements in U can be annotated with high prediction confidence even though we use a high threshold like 0.9. Table 2.5 shows the accuracy of the predicted labels in U'_+. We observe that the temporal classifier accurately annotate the unlabeled dataset even though it is trained based on a small number of labeled dataset. The performance of the temporal classifier is good because (1) STFNet performs well when limited labeled training data is given, (2) data in U and D belong to the same set of users.

A higher threshold would lead to higher accuracy but smaller size of U'_+. So there's a trade-off here. In this work, we use a fix threshold 0.9. It would be an interesting future work to study the selection of the threshold.

5 Spectrogram Masked Autoencoder for IoT

5.1 Overview

In this section, we introduce our augmentation-free self-supervised learning framework for IoT. The increasing number of smart sensing devices such as smart phones and smart wearable devices yields a large amount of sensing data which can be used to build multiple intelligent applications. Great processes have been made on studying such data, including target detection [106, 107], health and well-being [94–96], activity recognition [4–6], and localization[97, 98]. Inspired by the success gotten by the deep learning techniques on various areas like image processing [19, 20] and natural language processing (NLP) [23], a lot of deep neural network models have been designed to deal with sensing data.

As the development of deep learning techniques, deeper and deeper neural network models were designed and applied to sensing tasks. However, training such deep models requires a huge amount of labeled training data. Collecting a large dataset is not an easy job and would require a lot of human labor work. But it's observed that the workload of collecting a training dataset mainly comes from the data annotation while collecting unlabeled data is a relatively easier job. Facing this phenomenon, self-supervised learning, which is a kind of unsupervised learning technique and builds models only based on unlabeled data, has became more and more popular in dealing with sensing data.

Self-supervised learning technique was first studied in NLP and computer vision areas and has gotten great successes [23, 34, 38, 39, 108]. The goal of self-supervised learning is to learn an encoder which maps the inputs to the features or representations in some latent space. This process is only based on the unlabeled data. The encoder is then freezed and can be used in the downstream tasks like classification and regression. This means that the downstream tasks are built on the learned features instead of the original inputs. The self-supervised learning algorithms would let the features store the intrinsic information of the original input, and thus training models based on the features would need much less number of labeled training data.

Besides computer vision and NLP tasks, self-supervised learning techniques have also gotten studied in sensing tasks [29, 30, 52, 102]. Most of the previous works customized the self-supervised contrastive learning models (*e.g.*, SimCLR [34]) to sensing tasks by carefully the unique characteristics (*e.g.*, frequency domain characteristics) of sensing time-series data. The key idea of self-supervised contrastive learning is to pull together the similar samples in the latent space while pushing apart the diverse samples. Data augmentation plays an important role on the self-supervised contrastive learning technique. And it has been proved that the qualities of the features learned by the self-supervised contrastive learning largely depends on the data augmentation algorithms [5, 29]. For the same task, different data augmentation strategies would lead to very different performance. Hence, we have to carefully design the data augmentation strategy when applying self-supervised contrastive learning for a specific sensing task. What makes things worse is that different sensing tasks with different sensors usually need different data augmentation strategies. Different from computer vision and NLP tasks where the input data is image (video) or natural languages, the sensing data can be very different for different tasks. They can be measurements from accelerator, gyroscope, geophone, and microphone. Data augmentation for those data are quite different considering the different background physical phenomenon. And those data strategies that work for multiple sensing tasks, such as adding Gaussian noise, do not perform well on self-supervised contrastive learning [5]. This makes the design of self-supervised learning framework stuck in the chosen of data augmentation strategies. Hence, it would be very helpful if we could design a self-supervised learning framework for sensing tasks that does not depend on data augmentation. And that motivates this work.

The success of BERT [23] on NLP brought significant interest in the autoencoding method. And recently, [39] presented a simple, effective and scalable form of a masked autoencoder (MAE) for visual feature learning by carefully taking the difference of vision and language into consideration. Unlike the previous contrastive learning models, the MAE model does not rely on data augmentations. Hence, in this work, we built the MAE model for physical sensing data. Since the underlying processes (such as vibration or acceleration) of physical sensing are fundamentally a function of signal frequencies, the sensing data usually has a sparser and more compact representation in the frequency domain. And hence, its a good choice to use spectrograms as the input of the MAE model. The key difference between spectrogram and vision data (such as image) is that we have some prior knowledge about the spectrogram. We know that the high frequency components in the spectrogram are usually noise in the physical sensing tasks while the information mainly stores in the low frequency components. However, we do not have such information in images. In this way, the MAE should be trained by carefully reconstructing the low frequency components of the spectrograms while reconstructing the high frequency components would not help a lot. We thus re-designed the loss function of MAE by assigning different weights on different frequency components. That is, we use the weighted mean squared error (WMSE) between the original and reconstructed spectrograms as the loss function of our version of MAE. This is different from the loss function of vision based MAE which use the same weight for each pixel while calculating the mean squared error. We believe that the encoder trained with our weighted mean squared error would extract the intrinsic information from the spectrograms of sensing data in a better way, and our evaluation results confirm it.

The main contribution of this work can be summarized as follows:

- We proposed SMAE, which a Spectrogram Masked AutoEncoder framework that specially designed for physical sensing tasks. Different from the previous vision MAE, we carefully re-designed the loss function by calculating the weighted mean squared error between the original and reconstructed spectrograms. As far as we know, SMAE is the first masked autoencoder framework that designed for physical sensing applications.
- We evaluated the performance of SMAE on different datasets, where measurements of different physical sensors (such as accelerometer, gyroscope, gephpone, and microphone) are used to implement different sensing tasks like vehicle detection and human activity recognition. The experiments demonstrated the performance gain of SMAE compared with previous vision MAE. We also compared the performance of SMAE and self-supervised contrastive learning, which illustrated that SMAE does not rely on the choice of data augmentations.

5.2 Self-Supervised Learning for Sensing Data

Although self-supervised learning techniques have been widely studied in computer vision and NLP areas, the process of self-supervised leaning methods in sensing lags behind. In 2019, Saeed et al. proposed a multi-task self-supervised learning framework for human activity detection [4], which is, as far as we know, the first attempt of self-supervision for sensing representation learning. Multiple pretext tasks were designed to learn the representations. Self-supervised contrastive learning has also been applied to deal with sensing data such as measurements from accelerometer and gyroscope [5, 29, 102] and WiFi signals [30].

However, the previous models highly rely on the design of data augmentations, and the commonly used data augmentation (such as adding Gaussian noise) does not perform well. This motivates our work based on the masked autoencoder model which does not rely on the data augmentations. To the best of our knowledge, the work presented in this section (SMAE) is the first study of mask autoencoder on physical sensing data.

5.3 Design of SMAE

In this part, we first give an overview of our spectrogram masked autoencoder (SMAE), and then we present the detailed design including the masking, encoder, decoder, and the loss function.

5.3.1 Overview of SMAE

In SMAE, we use a same architecture as MAE [39], which is shown in Fig. 2.20. Different from MAE, our model SMAE takes the spectrograms of sensing data as input. The goal of SMAE is to reconstruct the original spectrogram given its partial observation. The first step is masking, where we randomly mask part of the input spectrogram and get a partial observation. Then, the encoder would take the partial observation as input and generate the corresponding representations of the partial observation. After that, the decoder would try to take the representations as input and

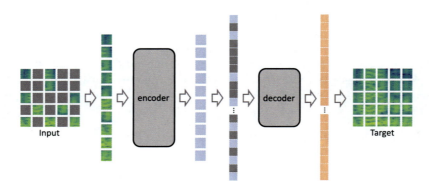

Fig. 2.20 The architecture of SMAE

then reconstruct the original input spectrogam. The SMAE is trained by minimizing the difference between the reconstructed spectrogram and the original one.

SMAE adopts a similar asymmetric design with MAE, where SMAE uses a deep and complicated encoder and a lightweight decoder. After the training process, the decoder would be dropped and only the encoder would be used in the downstream tasks such as classification or regression. In the downstream tasks, the input spectrograms are first mapped to the latent representations and then neural network models are built on the representations.

5.3.2 Masking

SMAE adopts a similar masking strategy with MAE. We first divide a spectrogram into several non-overlapping patches. Then, we randomly mask a subset of the patches, shown as gray blocks in Fig. 2.20. The remaining patches are kept as the input of the encoder. In order to largely eliminate the redundancy in the input spectrogram, a large masking ratio should be used (in this work 75%). In this way, the reconstruction task cannot be easily solved by simply extrapolation from the nearby unmasked patches. This makes the reconstruction task hard and thus would help the encoder to distill the intrinsic information of the input.

Different masking strategies can be used to mask the input spectrogram. In our approach, we adopt a uniform distribution to randomly mask the patches without replacement. Other mask strategies can also be adopted. For a spectrogram, different frequency bands store different information. For example, a female voice frequency range covers fairly from 350 Hz upto 17 kHz, and male voice covers a frequency range of 100 Hz–8 kHz.[3] The frequencies of human activity are between 0 and 20 Hz [109]. This means that the information density at different frequency bands are different and the redundancies of different frequency bands are different. In this way, we may design a masking strategy that samples patches of different frequency bands with different probabilities. This kind of masking strategy could eliminate the redundancy of the spectrogram in a more fine-grained way and thus help to train an efficient encoder. In this work, we focus on the design of the loss function of SMAE and leave the design of mask strategy in the future work.

5.3.3 SMAE Encoder

In SMAE, we use vision transformer [110] as the encoder. The encoder takes the embeddings of the unmasked patches with added positional embeddings as input. The input would be passed through several layers of transformer blocks and then generates the representations. In SMAE, only a small ratio of the patches is

---

[3] https://seaindia.in/sea-online-journal/human-voice-frequency-range/

unmasked. Even through the encoder is a very large network, it would not consume a lot of memory and computation resources.

For each unmasked patch of the input, the encoder would generate a representation vector. After training SMAE, the parameters of the encoder would be frozen and be used in the downstream tasks. In the downstream task, the encoder would take the full set of patches as input and generate representation vector for the whole input spectrogram (without masking). The target model can be built on those representation vectors.

5.3.4 SMAE Decoder

The encoder in SMAE generate a representation vector for each of the unmasked spectrogram patch. For each of the masked patch, we define a mask token [23], which is shared, learned vector that shows the presence of missing spectrogram patch. We concatenate the mask tokens and the representation vectors for unmasked patches together, and then add the positional embeddings of the full set to all of them. The decoder would take the sum of tokens (as well as representations) and positional embeddings as input and generate the reconstructed spectrogram.

SMAE adopts an asymmetric architecture where the designs of the encoder and decoder can be independent. In SMAE, we use vision transformer block to build the decoder, which is similar with the design of the encoder. However, the decoder is shallower and narrower than the encoder. This is because that the input size of the decoder is larger than that of the encoder. The decoder takes both of the masked tokens and unmasked representation vectors as input, but the encoder only takes the unmasked patches. The design of lightweight decoder would significantly reduce the pre-training time and computational resource consumption.

The decoder is only used in the pre-training process. After the pre-training, the decoder would be dropped, and only the encoder would be used in the downstream tasks.

5.3.5 SMAE Loss Function

The output of the SMAE decoder is a set of vectors. Each vector represents a patch in the input spectrogram. The vectors are reshape to the same size with the input spectrogram. In this way, the goal of SMAE is to predict the values of the masked (missing) patches. In our implementation, we designed weighted mean squared error (WMSE), and used it as the loss function. Here, assumes that X and $\hat{X}$ refer to the original and reconstructed spectrograms with the same size $m \times n \times c$ (c refers to the number of channels), the MAE between X and $\hat{X}$ can be computed as

$$\text{MSE} = \frac{1}{m \times n \times k} \sum_{i=1}^{m} \sum_{j=1}^{n} \sum_{k=1}^{c} (X_{ijk} - \hat{X}_{ijk})^2. \qquad (2.35)$$

In the MAE model, only the loss on the masked part of the input is computed. Let's define the mask as a tensor M which has the same size with the spectrogram. If the elements of M are 0s, this means that the corresponding elements of the spectrogram are sampled and 1s in M means the corresponding elements in the spectrogram are masked. In this way, the MAE becomes

$$\text{MSE} = \frac{1}{m \times n \times k} \sum_{i=1}^{m} \sum_{j=1}^{n} \sum_{k=1}^{c} M_{ijk}(X_{ijk} - \hat{X}_{ijk})^2. \qquad (2.36)$$

In SMAE, different from the original MAE implementation, we designed the weighted MSE (WMSE) and use it as the loss functions. This design is inspired the observation that different frequency components in a spectrogram store different information and have different importance. And the key difference between spectrogram and image is that we do have the preliminary knowledge about which part is important and which part is not. This is because of the characteristics of the physical sensing data. Minimizing the difference on the "important" components would help to train the encoder while minimizing the difference on the components that are not so important would not improve the quality of the encoder. For example, in most physical sensing tasks, such as human activity recognition and target recognition, most of the useful information locates at the low frequency parts of the spectrogram, and very high frequency parts are usually noise. Reconstructing those noise would not help a lot while training the encoder in the pre-training process. Hence, we can multiply smaller weights with those parts while calculating MSE. And for those low frequency components, we multiply them with larger weights to emphasize their importance. The WMSE thus can be defined as

$$\text{WMSE} = \frac{1}{m \times n \times k} \sum_{i=1}^{m} W_i \sum_{j=1}^{n} \sum_{k=1}^{c} M_{ijk}(X_{ijk} - \hat{X}_{ijk}))^2, \qquad (2.37)$$

where $i = 1, 2 \ldots, m$ refer to different frequencies in the spectrogram and W_i refers to the corresponding weights. As shown in Fig. 2.21, the weight for the highest frequency is the minimum and gradually increase as the decreasing of the frequency. In detail, we set

Fig. 2.21 Weighted mean squared error

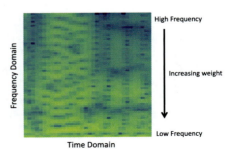

$$W_m = W_{min}, \tag{2.38}$$

$$W_1 = W_{max}, \tag{2.39}$$

and the weights increase in a linear way, which means

$$W_i = W_{max} - \frac{(i-1)(W_{max} - W_{min})}{m-1}. \tag{2.40}$$

In the evaluation sections, we illustrate that WMSE would help to train a better encoder in the pre-training process.

5.4 Evaluation

We study the performance of SMAE using two datasets with different sensing data. We first introduce the dataset and then describe the experiment setup. Finally, we show the evaluation results.

5.4.1 Datasets

We studied the performance of SMAE on different modalities of sensing data for two commonly studied tasks: human activities recognition and vehicle recognition. For the human activities recognition, the accelerometers and gyroscope measurements of smartphone or wearable devices can be collected and used to recognize the activities (*e.g.*, walking, sitting, running) of human beings. For the vehicle recognition task, the acoustic data (from microphone) and seismic data (from geophone) are collected and used to detect and recognize the different types of vehicles. Studying the performance of SMAE on multiple types of sensing data could demonstrate the scalability of SMAE. Now we introduce the two datasets in the evaluation.

MotionSense

MotionSense [88] is a human activity recognition dataset which consists of the measurements of accelerometer and gyroscope sensors of an iphone 6s. Different activities of 24 participants, with an iphone 6s kept in their front pocket, were recorded, and at the same time, the measurements of the two mentioned sensors were collected. To make the dataset general, the participants vary from ages, genders, heights and weights. For example, 14 males and 10 females were included in the dataset. A total of six activities were considered, including walking, jogging, sitting,

2 Self-Supervised Learning from Unlabeled IoT Data

standing, upstairs, and downstairs. All the accelerometer and gyroscope data was collected in 50 Hz.

ACVR

The $\underline{A}$coustic and $\underline{S}$eismic based $\underline{V}$echicle $\underline{R}$ecognition dataset (ACVR) is a dataset collected by ourselves. It's a commonly studied topic where the acoustic and seismic data can be used to do vehicle detection and classification. To collect our detaset, we build the sensor device using a Raspberry Pi connected with a microphone array and a geophone. We deployed our sensor on the grounds and collected both of the acoustic and seismic data while vehicles were driven around. We consider three types of vehicles with different engines, shapes, weights, and tires: Warthog UGV, Chevrolet Silverado, and Polaris all-terrain vehicle. The the acoustic data was collected by the microphone under 16 kHz and the seismic data was collected by the geophone under a sampling rate of 100 Hz. The data collection lasted for 115 minutes rough average split by the three types of targets.

5.4.2 Experiment Setup

In this part, we introduce the experiment setup including the data pre-processing and the implementation details of SMAE.

Data Pre-Processing

For MotionSense dataset, we use both of the accelerometer and gyroscope data. First, we segment the data using a sliding window of size 512 and step size of 256. 75% of the data is used as training set and the remaining 25% is used as testing.

For the acoustic and seismic dataset, considering the difference in sampling rates of the microphone and geophone, we first downsample the acoustic data to 100 Hz, which is the same as the seismic data. We performance the downsampling in the follow way: we first passed the acoustic data through a low-pass filter of 400 Hz to reduce the high frequency noise and then passed it through a high pass filter of 25 Hz in order to reduce the influence of wind. After that, we sample one point from every 160 points to get the 100 Hz acoustic data. Similar with MotionSense, we use a segment length of 512 and the ratios of training and testing sets are 75%, 25%.

We generate the spectrogram for each segment of sensing data through Short-Time Fourier Transform (STFT) with a window size 64 and step 14, and then resize the spectrogram to size 32×32. There are three dimensions for the accelerometer and gyroscope data, one dimension for the seismic data, and two channels for the acoustic data. For each dimension (or channel) of one segment sensing data, a spectrogram of size $32 \times 32 \times 2$ is generated. Here the channel size 2 refers to the real and imaginary parts of the spectrogram. Then the spectrograms of different dimensions and modalities are concatenated in the channel dimension. We then normalize the content of the spectrograms to values between 0 and 1.

Implementation Details

In our SMAE model, following the architecture in Fig. 2.20, we build the encoder with six layers of vision transformer and decoder with two layers. We set the length of the representation vector as 128. In each layer of the encoder and decoder, we use a same number of attention heads, which is four. In the pre-training, we use a batch size of 128. We apply AdamW [111] with a basic learning rate 10^{-3} and weight decay 10^{-4} as the optimizer. We run the pre-training a total of 200 epochs and 15% of them are used as warmup epochs.

5.4.3 Comparison with Previous Self-Supervised Approaches

We first study the performance of SMAE with previous self-supervised learning approaches on the about mentioned two datasets. To study the advantage of our weighted mean squared error, we compared the performance of SMAE with the original MAE approach which led great performance on computer vision [39]. And we also compare the performance of SMAE with the previous commonly used self-supervised contrastive learning approach SimCLR [34]. For completeness, we directly train the encoder with a linear classifier under a supervised way and compare its performance with SMAE. For the self-supervised approaches SMAE, MAE, and SimCLR, we first perform pre-training and then fine-tune the encoder as well as the linear classifier. We use the classification accuracy as the metric to study their performance.

Data augmentation is an essential part of the self-supervised contrastive learning like SimCLR. MAE and our SMAE can work without data augmentation. We can also add the data augmentation part in MAE and SMAE and the performance would get slightly improved. In this work, we use "rotation" [30] as the data augmentation algorithm for MotionSense, and "timewarp" as the data augmentation algorithm for ASVR.

The results are shown in Table 2.6. We can observe that SMAE performs the best for both of the two dataset, it shows a 95.55% accuracy for MotionSense and 95.61% accuracy for ASVR. Specially, our SMAE performs better that the supervised approach with the same structure of classifier, this means that the masked autoencoder approach can work well on dealing with sensing data. And SMAE also perform better than directly applying the original MAE approach, which designed for computer vision tasks, on sensing data (spectrogram), this means that using our weighted mean squared error would help to train a better encoder in the pre-training process.

2 Self-Supervised Learning from Unlabeled IoT Data

Table 2.6 Comparison with previous approaches

	MotionSense	ASVR
Supervised	94.38%	93.15%
SimCLR	95.25%	93.22%
MAE	95.10%	93.95%
SMAE	95.55%	94.61%

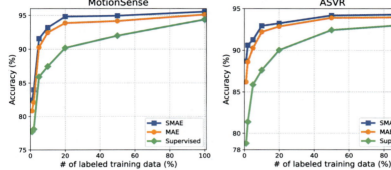

Fig. 2.22 Performance under different number of labeled data

5.4.4 Performance Under Different Number of Training Data

The goal of self-supervised learning is to efficiently utilize the unlabeled data. In this part, we study the performance of SMAE when there are large amount of unlabeled data and only a small part of them has labels. We assume that only 1%, 2%, 5%, 10%, 20%, 50%, 100% of the training data has labels. The encoders of MAE, and SMAE were trained with all the training data (without label), and the models as well as the encoder in the downstream task were fine-tuned using the labeled training data. To study the influence of unlabeled data, we also train the encoder as well as the linear classifier only based on the labeled training data and compared its performance with MAE and SMAE.

Figure 2.22 illustrates the results. We observed that both MAE and SMAE perform better than the pure supervised approach under different number of labeled training data. And the gaps between the self-supervised approaches and the supervised approach is larger when we have smaller number of labeled training data. If we compare the performance between MAE and SMAE, we found that SMAE always perform better than MAE under different number of labeled training data. This verifies the effectiveness of our designed loss function, which is weighted mean squared error between the original and reconstructed spectrograms..

5.4.5 Performance Under Different Augmentation Strategies

One reason that we study the masked autoencdoer based self-supervised learning frameworks is that the performance of the previous commonly studied self-supervised contrastive learning approaches (*e.g.*, SimCLR) are highly dependent on the choice of data augmentation strategies. As an essential part of the contrastive learning, different data augmentations would lead to very different performances. In this part, we compare the performance between SimCLR and SMAE under different choice of data augmentation strategies. We use the MotionSense dataset in this part.

We study eight data augmentation strategies which have been widely utilized for accelerometer and gyroscope based human activity recognition [4, 5, 30].

- **Jitter**: Add random Gaussian noise to the sensing data. Jitter simulates the noise generated by the sensors.
- **Time-warp**: Time-warp operation could stretch or warp the time intervals between the sensing time-series and generate new sensing data time-series with same label as the original one.
- **Scaling:** Scaling operation generate new time-series by multiplying a random scalar to the original time-series.
- **Rotation:** The accelerometer and gyroscope are three-dimensional sensors. Rotate them should not change the labels of human activities. This data augmentation imitates the rotation fo the sensors.
- **Horizontal flipping:** Reverse the sensing data along time direction.
- **Channel-shuffling:** Randomly shuffle the channels (dimenstions) of the multi-channel sensor measurements.
- **Inversion**: Mulptiply the sensing data with -1.
- **Permutation**: First cut the sensing time-series into multiple segments along the time direction, and then permute them.

We train SimCLR and our SMAE under the above eight data augmentation strategies and compare the corresponding performance. Since data augmentation is not an essential part of SMAE, we also study the performance of SMAE without any data augmentation. We show the results in Table 2.7. We observe that the accuracies of SimCLR differ a lot under different data augmentation strategies. However, the accuracies of SMAE keep almost the same under different data augmentations. This illustrates the advantage of SMAE that it needs minimal or no data augmentation.

Table 2.7 Performance under different data augmentation

Data augmentation	SimCLR	SMAE
No augmentation	–	94.48%
Jitter	84.81%	95.10%
Time-warp	87.22%	94.98%
Scaling	87.71%	95.30%
Rotation	95.25%	95.55%
Horizontally flipping	89.79%	95.48%
Channel-shuffling	84.70%	94.82%
Inversion	92.08%	95.40%
Permutation	87.30%	95.01%

6 A Case Study: Self-Supervised Learning on IoBT-OS

6.1 Overview

In this part, we introduce IoBT-OS, an operating system for the Internet of Battlefield Things where our self-supervised learning framework plays an important role on improving decision accuracy and thus optimizing decision latency. Then, we use a case study of seismic and acoustic based target detection to evaluate the performance of our proposed framework, and the evaluation result verified the effectiveness of our self-supervised learning framework.

Recent worldwide defense trends reflect an increasing investment in automating various battlefield functions [112–114]. Initial indicators suggest that related applications have already impacted the course of recent conflicts [115]. Computer communications and networks, paired with sensing and machine intelligence, are at the center of enabling technologies for these applications, making the military domain a potential key focus for the intelligent edge and Internet of Things (IoT) research. An important goal is to support performant and resilient sensor-to-decision loops at the tactical edge [116, 117]. For example, one might want to reduce latency in neutralizing threats in the battlefield. The Internet of Battlefield Things (IoBT) [118] is an operating environment for future cyber-physical battlefields, where physical and computational assets must collaborate to produce effects. Prior work articulated IoBT challenges in resilience [119, 120], distributed computing [121], Bayesian learning [122], uncertainty estimation [123] and intelligent data fusion [124–126], among other areas [127–129]. This work focuses on optimizing the efficiency and efficacy of the sensor-to-decision loop and offers an architecture, called IoBT-OS, to accomplish the optimization goals.

The need for IoBT-OS arises from the increasing complexity of networked computational resources in future battlefields. In general computing contexts, streamlining application development and operation necessitates the introduction of operating systems to address common performance and resilience challenges. Similarly, in a modern battlefield, where mission success depends in large part on computational artifacts, a new operating-system-like construct is in order. Its goal is to ensure that the execution of sensor-to-decision loops (that involve multiple intelligent devices and systems) meet the challenges arising from spatial distribution, accelerated mission-tempo, transient resources, and the potential presence of adversarial activity. IoBT provides the underlying support for ensuring performance and resilience of the networked computational and sensing substrate for the envisioned cyber-physical decision loops at the tactical edge.

6.2 Background: The Decision Loop

A central concept for organizing IoBT functions is the multi-domain operation effect loop, proposed in prior work [116]. As mentioned in [116], it breaks down the data processing workflow into stages; namely, (1) detect (targets or threats), (2) identify, (3) track, (4) aggregate (information), (5) distribute (to stakeholders), (6) decide, and (7) actuate. The execution of the loop starts with sensors that measure different signal modalities. It continues through communication components, and computational elements that execute appropriate machine analytics on sensor data. These analytics furnish information for decision making, such as detected target types and trajectories. The decisions usually involve selection of appropriate means of actuation (called *effects*) whose purpose is often to neutralize the identified threats. The success of IoBT support in executing this loop is measured against three key goals:

- *Tactical edge efficiency:* Recognizing that time is a decisive factor in gaining advantage, it is desired to accelerate the intelligent sensor-to-decision loops (ideally without loss of inference quality) and push key functionality towards the edge (to reduce dependency on large cloud-like infrastructure support). These advances allow pushing intelligent data analytics closer to the point of sensor data collection, thereby significantly shortening decision chains, while offering intelligent mission-informed data filtering as early as possible (at the edge) to reduce load on the possibly contested tactical network.
- *Edge resilience:* Recognizing that IoBT executes in adversarial settings, where new attacks on machine intelligence are possible (to disrupt decision loops and foil intelligent automation), it is important to develop foundations of resilience, especially in contexts where neural-network-based solutions are used for implementing the edge analytics. Such foundations must offer improved resilience to adversarial inputs, enhance risk analysis, speed-up sensor attack detection, and bound worst-case neural network outcomes in the presence of adversarial activity.
- *Tailored intelligence at the point of need:* Novel capabilities are needed at the tactical edge to take better advantage of heterogeneity, exploit multimodal sensing, and compute uncertainty. Examples include opportunistic exploitation of commodity radios as sensors, neural networks for inference in the frequency domain, distribution of machine learning models across heterogeneous edge systems, and sensor fusion to enhance target classification accuracy and reduce false positives (e.g., due to decoys) while meeting latency constraints.

Below, we focus primarily on edge efficiency. Issues of resilience and novel tailored intelligence services are beyond the scope of this work.

6.3   IoBT-OS

To support the above goals, the IoBT-OS architecture includes three types of modules: (1) an edge AI efficiency library, (2) mission-informed real-time data management algorithms, (3) digital twin support, and (4) offline training support.

The *edge AI efficiency library* comprises modules that encapsulate different functions for target/threat detection, identification, and tracking, using neural network components that have been compressed and optimized to execute in resource-scarce environments, while offering a controllable low latency to the application.

The *mission-informed real-time data management algorithms* segment and prioritize data processing by the supported real-time edge AI components. A key challenge is to allocate more resources to the processing of more critical stimuli, which requires a combination of (1) data segmentation by some notion of data importance or urgency, and (2) prioritized allocation of capacity to process more important/urgent data first. These algorithms should further maximize resource utilization.

Digital twin support features a set of value-added auxiliary capabilities implemented on a high-end computing platform thereby allowing the system to exploit additional resources when available (e.g., when a connection to the higher-end server can be established). A primary capability of the digital twin is to replicate key IoBT system state and environmental conditions for purposes of conducting various compute-intensive functions centrally, such as search for an optimal system configuration, global anomaly detection, root cause analysis, and model checking.

Offline training support features solutions that pre-train various models ahead of deployment. Pre-training includes execution-time profiling in order to develop accurate latency models of different system components, as well as neural network training, especially in regimes where accurate data labeling of these data is very time-consuming. A challenge is therefore to exploit mostly unlabeled data to train the neural networks involved in the sensor-to-decision loop.

IoBT-OS builds upon earlier frameworks by the authors that describe a vision of edge intelligence systems [130] and highlight challenges in applying machine intelligence to the IoT edge [131, 132].

6.4   The Case Study

6.4.1   Hardware Set-Up and Execution Loop

We use a vehicle detection and classification application to study the efficiency of IoBT-OS. In our setup, a Geophone and microphone are used to collect seismic and acoustic signals for preliminary vehicle detection and classification. The sensing devices consist of a Raspberry Shake (Raspberry Pi + one vertical-axis geophone)[4], a microphone array, and a portable power bank. A camera is also used to take pictures of the targets for the final visual confirmation on a separate edge server. We used a Jetson Nano as the server.

---

[4] https://raspberryshake.org/

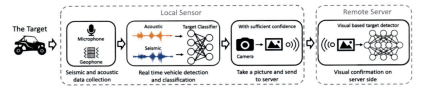

Fig. 2.23 The execution loop

The execution loop is shown in Fig. 2.23. The Geophone and microphone collect the seismic and acoustic signals from the environment. The seismic and acoustic data are then processed to detect and classify targets in real time. The target classifier is a DeepSense [54] neural network trained with our self-supervised contrastive learning framework. When the target classifier becomes confident enough in its prediction that a given target is found (*i.e.*, the confidence level is larger than a threshold), a message is sent to a camera that takes a picture of the target for confirmation. In order to reduce the transmission delay, the picture is first compressed by compressive offloading framework [133] and then sent to the edge server. A vision-based object detector (based on YOLO [21]) on the server processes the picture and confirm the target (or not). We use an object detector that was compressed using our DeepIoT [55] framework to reduce its inference delay and computational resource consumption.

6.4.2 Experimentation Results

To evaluate the performance of IoBT-OS, we deployed our devices on the grounds of the DEVCOM Army Research Laboratory Robotics Research Collaboration Campus (R2C2) and collected seismic and acoustic signals, while different ground vehicles were driven around the site. Data of three different targets: a Polaris all-terrain vehicle, a Chevrolet Silverado, and Warthog UGV were collected. Each target repeatedly passed by the sensors. The total length of the experiment was 115 minutes, spread roughly equally across the three targets. The seismic data was collected at 100 Hz and the acoustic data was collected at 16 kHz. To reduce overhead, the acoustic data was low-pass filtered at 400 Hz and high-pass filtered at 25 Hz, then down-sampled to 100 Hz. Based on the collected data, we study the decision accuracy and delay of IoBT-OS. A camera was employed to simultaneously record video of the target.

Decision Accuracy

We first studied the accuracy of target classification based on seismic and acoustic data. We applied the DeepSense neural network architecture [54] as the structure of the target classifier. Figure 2.24 illustrates its structure. Our previous work [37] has

2 Self-Supervised Learning from Unlabeled IoT Data

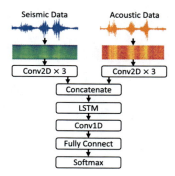

Fig. 2.24 The target classifier

Table 2.8 Decision accuracy

Three-layers CNN	Target classifier	Target classifier+SCL
73.6%	89.1%	91.2%

shown that physical sensing signals have sparser and more compact representations in the frequency domain. Thus, we used a spectrogram instead of time domain measurements as the input to the classifier. We trained the target classifier with our self-supervised contrastive learning framework.

We used roughly two thirds of the data for training and one third for testing. The seismic and acoustic data were cut into segments of 1 second each using a sliding window without overlap. We built a three-layers convolutional neural network (CNN) as a baseline and also compared the accuracy of our target classifier with and without self-supervised contrastive learning (SCL). The results are shown in Table 2.8. We observed that our target classifier has much higher recognition accuracy than the CNN baseline (89.1% vs 73.6%). And the that self-supervised contrastive learning further improves the accuracy from 89.1% to 91.2%.

Performance Under Different Data Lengths

In the experiment, the seismic and acoustic signals were continuously fed into the target classifier. Given N consecutive windows (*i.e.*, N seconds window of data) we took majority vote as the final classification result. We varied N from 1 s to 20 s and studied the classification accuracy for different data lengths to understand the trade-off between quality and delay. The results are shown in Fig. 2.25. We observed that our target classifier trained with self-supervised contrastive learning (the blue curve) always performs the best, and shows a 99% classification accuracy given 20 seconds of seismic and acoustic data.

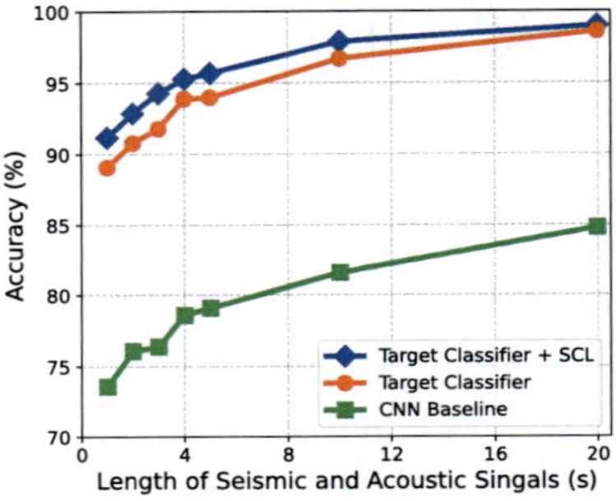

Fig. 2.25 Different data lengths

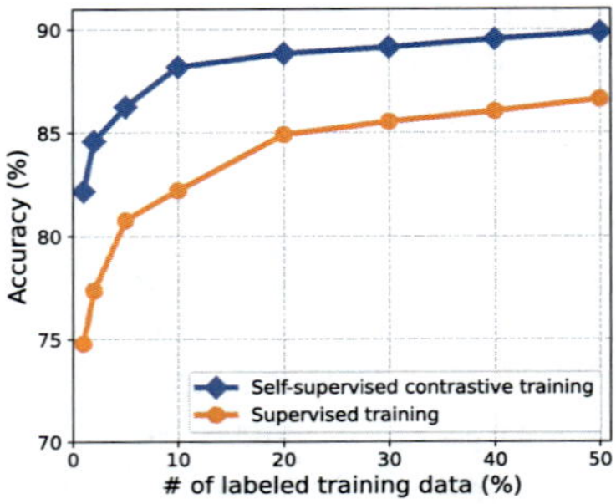

Fig. 2.26 Different # of labels

Performance Under Different Number of Labels

We also studied the effectiveness of the self-supervised contrastive learning framework at utilizing unlabeled data. Specifically, we assumed that 1%, 2%, 5%, 10%, 20%, 30%, 40%, and 50% of the training data has labels (and that the remaining training data is unlabeled). We then compared the performance of the target classifier trained with supervised learning (*i.e.*, with the labeled data only) and the self-supervised contrastive learning (trained with both labeled and unlabeled data) given the above ratios of labeled data. Figure 2.26 demonstrates the results. It can be seen that the target classifier trained with our self-supervised contrstive learning (the blue curve) performs much better than that trained in a supervised manner (the orange curve). This illustrates the efficiency of our self-supervised contrastive learning framework in utilizing unlabeled data.

Decision Delay

In this part, we study the performance of IoBT-OS on decision delay. The decision delay consists of different parts including the detection delay, the network transmission delay, as well as the delay of the virtual confirmation. Our self-supervised learning framework helps to reduce the detection delay, and thus we focus on the detection delay in this part. The detection delay comes from two parts: the time needed to get confident-enough detection results, and the inference time of

2 Self-Supervised Learning from Unlabeled IoT Data

Table 2.9 Inference time for the target classifier

	Three-layers CNN	Target classifier (SCL)
Raspberry Pi 3B+	10.6 ms	66.0 ms
Raspberry Pi 4	4.7 ms	17.7 ms

the classifier. According to the results in Fig. 2.25, our self-supervised learning framework as well as the DeepSense neural network reduce the time needed to get confident-enough detection results a lot compared with the CNN baseline. For the inference time of the target classifier based on acoustic and seismic sensors, the target classifier runs on the local sensors and continuously takes the seismic and acoustic data as input. It is required that the classifier's inference speed should be shorter than the interval at which seismic and acoustic data comes in. Otherwise, data will back up. Table 2.9 demonstrates the inference time (for a 1 second data segment as input) on different devices. We observed that the inference times of our target classifier are slightly larger than those of the three-layers CNN baseline. However, they are much smaller than 1 second. For example, on the Raspberry Pi 4, it takes only 17.7 ms to process the 1 second data window. This means that our classifier could generate its classification result in time.

7 Chapter Summary and Future Work

7.1 Summary

In this chapter, I studied the self-supervised learning techniques which demonstrated outstanding performance on dealing with the lack of training labels, and then customized those self-supervised learning techniques to the IoT applications. By carefully taking the unique characteristics of the IoT sensing signals into consideration, we can significantly improve the performance of self-supervised learning frameworks (originally designed for computer vision/NLP) in IoT application contexts. My customization thus provided an effective way to deal with the absence of sufficient labels in IoT applications. The work in this dissertation are summarized as follows.

7.1.1 Time-Domain Self-Supervised Contrastive Learning for IoT

We started the customization by directly borrowing the self-supervised learning framework from computer vision and applied it to IoT applications. In this part, we applied the popular self-supervised contrastive learning framework and built a model for IoT sensing signals from the time-domain. We studied the performance of the self-supervised contrastive learning framework on the radio modulation classification dataset, and the evaluation results demonstrated the effectiveness of the self-supervised contrastive learning technique on IoT applications.

7.1.2 Frequency-Domain Self-Supervised Contrastive Learning for IoT

We then re-designed the self-supervised contrastive learning framework by carefully studying the frequency-domain characteristics of the IoT sensing signals. We re-designed the encoder using STFNet, which is specially designed for dealing with IoT sensing signals, instead of the commonly used CNN or RNN models. Beside, we applied data augmentation from both time-domain and frequency-domain. According to our evaluation results, studying on the frequency-domain would largely improve the performance of self-supervised contrastive learning for IoT applications.

7.1.3 Semi-Supervised Contrastive Learning for IoT

After taking the frequency-domain characteristics into consideration, we then proposed a semi-supervised contrastive learning framework for IoT. Different from the original self-supervised contrastive learning, where only unlabeled data is used in the pre-training process, our semi-supervised framework uses both of the labeled and unlabeled data to train the encoder during the pre-training process. We designed a novel way to combine the labeled and unlabeled data together, to train the encoder. Our experiments showed obvious performance gain of our semi-supervised contrastive learning framework compared with the original one.

7.1.4 Spectrogram Masked AutoEncoder for IoT

The self-supervised contrastive learning demonstrated outstanding performance on IoT applications after our customization. However, data augmentation is an essential part of it and it requires a lot of human expertise to design a proper data augmentation strategy. Our experiments also demonstrated that the performance of the self-supervised contrastive learning framework would largely rely on the choice of data augmentation. To reduce the dependency on data augmentation, we carefully studied the masked autoencoder framework, which does not rely on the data augmentation, and customized it to IoT applications by re-designing the loss function. The evaluation results also verified the success of our customization.

7.1.5 A Case Study: Self-Supervised Learning on IoBT-OS

We finally applied our self-supervised learning framework to a real IoT system and studied its performance on IoBT-OS. IoBT-OS is an architecture focusing on optimizing the efficiency and efficacy of the sensor-to-decision loop. We used the seismic and acoustic based vehicle recognition application as an example to evaluate our self-supervised learning framework on IoBT-OS. We carried out a lot of experiments on IoBT-OS, and they demonstrated that our self-supervised learning framework can help to reduce the sensor-to-decision latency. This shows the effectiveness of our framework on real IoT systems.

7.2 Lessons

Data labeling is one of the key logistic costs of machine learning pipelines and this dissertation addressed the challenge of the lack of sufficient labeled data in the IoT space. By carefully customizing the self-supervised learning frameworks from computer vision/NLP to IoT applications, our self-supervised learning techniques achieved accuracy gains of up to $\times 1.5$ for IoT applications in regimes when labeled data was sparse. During the study of self-supervised learning and customizing them to IoT applications, we found some interesting and useful lessons.

7.2.1 Self-Supervised Contrastive Learning

Self-supervised contrastive learning is an effective and popular way to utilize the unlabeled data. If we have enough expertise with application-specific data transformations that do not change data labels (*i.e.*, data augmentation), contrastive learning can get good performance. Here, a good data augmentation strategy for sensing signals is to simulate the data-distorting behavior of sensors. For the same input, we can design multiple data augmentation strategies and different augmentation strategies were proved to lead to very different performance on self-supervised contrastive learning. Hence, the top challenge in applying contrastive learning is the need of human expertise to design a good data augmentation strategy. We also found that some data augmentations are generally not helpful. For example, adding Gaussion noise is generally a bad choice because it does not exploit the underlying structure of the input data. Another challenge is that the training process of contrastive learning is easy to overfit. We should design the optimization process very carefully in oder to avoid overfitting.

7.2.2 Masked Autoencoding

Masked autoencoding is another popular self-supervised learning architecture. Its success is based on the redundancy in the input data. If there is enough internal redundancy in the data input, masked autoencoding is generally a good choice for self-supervised learning framework. A key component of the masked autoencoding framework is the masking ratio. The mask ratio should be designed very carefully. If we use a very high mask ratio, then it would be very difficult to reconstruct the input and hence the encoder can only learn very little knowledge. If we use a very small mask ratio, then most of the input would be kept. This means that the reconstruction can be easily done by predicting the missing values according to their neighbors and the latent structure of the input could not be learned in this way. A proper mask ratio is another key point to make masked autoencoder model successful.

7.3 Future Work

In the end of this dissertation, we introduce some possible future research directions. In this dissertation, we studied the customization of self-supervised learning frameworks to IoT applications. For the future works, we cover three interesting related topics: (1) self-supervised learning frameworks for multi-modality inputs; (2) training/inferencing deep neural network models with noisy data; and (3) model compression for self-supervised learning frameworks.

7.3.1 Self-Supervised Learning Frameworks For Multi-Modality Inputs

In this dissertation, we customized the self-supervised learning frameworks to IoT applications by taking the frequency-domain characteristic into consideration. Beside learning in frequency-domain, the IoT applications have another very unique and important characteristic, that is multi-modalities. Different from the computer vision or NLP applications, the inputs of which are images, videos, or natural language, the inputs of IoT applications are usually from multiple sensors. For example, the accelerometer and gyroscope can be collaboratively used to recognized the human activities. It would be an interesting topic to study the self-supervised learning frameworks on multi-modality inputs. There exist close correlations among the different modalities. Combining such correlations with the design of the encoder in self-supervised learning would be a valuable research topic.

7.3.2 Training/Inferencing Deep Neural Network Models with Noisy Data

In this dissertation, we studied the self-supervised learning framework to address the challenge of the lack of sufficient labeled data in the IoT space. This is motivated by the fact that data labeling is very time-consuming and requires a lot of human labor work. It has been studied that 25% of time spent for machine learning project is allocated on data labeling.[5] And at the same time, another 25% of the time is spent on the data cleaning. Facing this observation, it would be an interesting topic

---

[5] https://medium.com/whattolabel/data-labeling-ais-human-bottleneck-24bd10136e52

to design neural network models which could deal with noisy data. Specifically, the study can be divided into two stages: (1) training neural network model with noisy data and (2) inferencing neural network models with noisy data.

7.3.3 Model Compression for Self-Supervised Learning Frameworks

Model compression is very important for deep neural network models in IoT applications. The IoT devices usually have limited memory and computation power, and cannot support the running of large and deep neural network models. A common way to solve this problem is to compress the neural network models and then deploy the compressed models on the IoT devices. However, the previous model compression techniques performed the compression along with the supervised training process. In the future, the self-supervised learning would be more and more popular and become the major training strategy. So, studying the model compression along with the self-supervised training would be also an interesting topic.

Besides the future work mentioned above, I believe that there are a lot of other interesting topics related to deep learning on IoT. By going deeper to the deep learning techniques and designing proper customizations on IoT applications, I believe that the area of IoT would get outstanding progress in the future.

References

1. N. Bui, A. Nguyen, P. Nguyen, H. Truong, A. Ashok, T. Dinh, R. Deterding, T. Vu, Pho2: smartphone based blood oxygen level measurement systems using near-ir and red wave-guided light, in *Proceedings of the 15th ACM Conference on Embedded Network Sensor Systems* (2017), pp. 1–14
2. J.M. Sorber, M. Shin, R. Peterson, D. Kotz, Plug-n-trust: practical trusted sensing for mhealth, in *Proceedings of the 10th International Conference on Mobile Systems, Applications, and Services* (2012), pp. 309–322
3. Y. Xiang, R. Piedrahita, R.P. Dick, M. Hannigan, Q. Lv, L. Shang, A hybrid sensor system for indoor air quality monitoring, in *2013 IEEE International Conference on Distributed Computing in Sensor Systems* (IEEE, 2013), pp. 96–104
4. A. Saeed, T. Ozcelebi, J. Lukkien, Multi-task self-supervised learning for human activity detection. Proc. ACM Interactive Mobile Wearable Ubiquit. Technol. **3**(2), 1–30 (2019)
5. C.I. Tang, I. Perez-Pozuelo, D. Spathis, C. Mascolo, Exploring contrastive learning in human activity recognition for healthcare. Preprint (2020). arXiv:2011.11542
6. H. Haresamudram, A. Beedu, V. Agrawal, P.L. Grady, I. Essa, J. Hoffman, T. Plötz, Masked reconstruction based self-supervision for human activity recognition, in *Proceedings of the 2020 International Symposium on Wearable Computers* (2020), pp. 45–49
7. E. Hoque, R.F. Dickerson, J.A. Stankovic, Vocal-diary: a voice command based ground truth collection system for activity recognition, in *Proceedings of the Wireless Health 2014 on National Institutes of Health* (2014), pp. 1–6
8. L. Capra, W. Emmerich, C. Mascolo, Carisma: context-aware reflective middleware system for mobile applications. IEEE Trans. Software Eng. **29**(10), 929–945 (2003)
9. S. Nirjon, R.F. Dickerson, Q. Li, P. Asare, J.A. Stankovic, D. Hong, B. Zhang, X. Jiang, G. Shen, F. Zhao, Musicalheart: a hearty way of listening to music, in *Proceedings of the 10th ACM Conference on Embedded Network Sensor Systems* (2012), pp. 43–56

10. A. Rowe, M. Berges, R. Rajkumar, Contactless sensing of appliance state transitions through variations in electromagnetic fields, in *Proceedings of the 2nd ACM Workshop on Embedded Sensing Systems for Energy-Efficiency in Building* (2010), pp. 19–24

11. C.-Y. Li, Y.-C. Chen, W.-J. Chen, P. Huang, H.-H. Chu, Sensor-embedded teeth for oral activity recognition, in *Proceedings of the 2013 International Symposium on Wearable Computers* (2013), pp. 41–44

12. E. Cho, K. Wong, O. Gnawali, M. Wicke, L. Guibas, Inferring mobile trajectories using a network of binary proximity sensors, in *2011 8th Annual IEEE Communications Society Conference on Sensor, Mesh and Ad Hoc Communications and Networks* (IEEE, 2011), pp. 188–196

13. B. Kusy, A. Ledeczi, X. Koutsoukos, Tracking mobile nodes using rf doppler shifts, in *Proceedings of the 5th International Conference on Embedded Networked Sensor Systems* (2007), pp. 29–42

14. J. Powar, C. Gao, R. Harle, Assessing the impact of multi-channel ble beacons on fingerprint-based positioning, in *2017 International Conference on Indoor Positioning and Indoor Navigation (IPIN)* (IEEE, 2017), pp. 1–8

15. P. Lazik, N. Rajagopal, O. Shih, B. Sinopoli, A. Rowe, Alps: a bluetooth and ultrasound platform for mapping and localization, in *Proceedings of the 13th ACM Conference on Embedded Networked Sensor Systems* (2015), pp. 73–84

16. K. Langendoen, N. Reijers, Distributed localization in wireless sensor networks: a quantitative comparison. Comput. Netw. **43**(4), 499–518 (2003)

17. M. Mirshekari, S. Pan, P. Zhang, H.Y. Noh, Characterizing wave propagation to improve indoor step-level person localization using floor vibration, in *Sensors and Smart Structures Technologies for Civil, Mechanical, and Aerospace Systems 2016*, vol. 9803 (SPIE, 2016), pp. 30–40

18. Y. LeCun, Y. Bengio, G. Hinton, Deep learning. Nature **521**(7553), 436–444 (2015)

19. J. Deng, W. Dong, R. Socher, L.-J. Li, K. Li, L. Fei-Fei, Imagenet: a large-scale hierarchical image database, in *2009 IEEE Conference on Computer Vision and Pattern Recognition* (IEEE, 2009), pp. 248–255

20. A. Krizhevsky, I. Sutskever, G.E. Hinton, Imagenet classification with deep convolutional neural networks, in *Advances in Neural Information Processing Systems* (2012), pp. 1097–1105

21. J. Redmon, S. Divvala, R. Girshick, A. Farhadi, You only look once: Unified, real-time object detection, in *Proceedings of the IEEE Conference on Computer Vision and Pattern Recognition* (2016), pp. 779–788

22. Z.-Q. Zhao, P. Zheng, S.-T. Xu, X. Wu, Object detection with deep learning: a review. IEEE Trans. Neural Netw. Learn. Syst. **30**(11), 3212–3232 (2019)

23. J. Devlin, M.-W. Chang, K. Lee, K. Toutanova, Bert: pre-training of deep bidirectional transformers for language understanding. Preprint (2018). arXiv:1810.04805

24. S. Suthaharan, Support vector machine, in *Machine Learning Models and Algorithms for Big Data Classification* (Springer, Berlin, 2016), pp. 207–235

25. N.D. Lane, P. Georgiev, L. Qendro, Deepear: robust smartphone audio sensing in unconstrained acoustic environments using deep learning, in *Proceedings of the 2015 ACM International Joint Conference on Pervasive and Ubiquitous Computing* (2015), pp. 283–294

26. T. Wang, S. Yao, S. Liu, J. Li, D. Liu, H. Shao, R. Wang, T. Abdelzaher, Audio keyword reconstruction from on-device motion sensor signals via neural frequency unfolding. Proc. ACM Interactive Mobile Wearable Ubiquit. Technol. **5**(3), 1–29 (2021)

27. A.A. Hammam, M.M. Soliman, A.E. Hassanein, Deeppet: a pet animal tracking system in internet of things using deep neural networks, in *2018 13th International Conference on Computer Engineering and Systems (ICCES)* (IEEE, 2018), pp. 38–43

28. B. El Boudani, L. Kanaris, A. Kokkinis, M. Kyriacou, C. Chrysoulas, S. Stavrou, T. Dagiuklas, Implementing deep learning techniques in 5g iot networks for 3d indoor positioning: delta (deep learning-based co-operative architecture). Sensors **20**(19), 5495 (2020)

2 Self-Supervised Learning from Unlabeled IoT Data

29. D. Liu, T. Abdelzaher, Semi-supervised contrastive learning for human activity recognition, in *2021 17th International Conference on Distributed Computing in Sensor Systems (DCOSS)* (IEEE, 2021), pp. 45–53
30. D. Liu, T. Wang, S. Liu, R. Wang, S. Yao, T. Abdelzaher, Contrastive self-supervised representation learning for sensing signals from the time-frequency perspective, in *2021 International Conference on Computer Communications and Networks (ICCCN)* (IEEE, 2021), pp. 1–10
31. S. Gidaris, P. Singh, N. Komodakis, Unsupervised representation learning by predicting image rotations. Preprint (2018). arXiv:1803.07728
32. R. Zhang, P. Isola, A.A. Efros, Colorful image colorization, in *European Conference on Computer Vision* (Springer, 2016), pp. 649–666
33. D. Pathak, P. Krahenbuhl, J. Donahue, T. Darrell, A.A. Efros, Context encoders: feature learning by inpainting, in *Proceedings of the IEEE Conference on Computer Vision and Pattern Recognition* (2016), pp. 2536–2544
34. T. Chen, S. Kornblith, M. Norouzi, G. Hinton, A simple framework for contrastive learning of visual representations. Preprint (2020). arXiv:2002.05709
35. W. Su, X. Zhu, Y. Cao, B. Li, L. Lu, F. Wei, J. Dai, Vl-bert: pre-training of generic visual-linguistic representations. Preprint (2019). arXiv:1908.08530
36. T.J. O'shea, N. West, Radio machine learning dataset generation with gnu radio, in *Proceedings of the GNU Radio Conference*, vol. 1, no. 1 (2016)
37. S. Yao, A. Piao, W. Jiang, Y. Zhao, H. Shao, S. Liu, D. Liu, J. Li, T. Wang, S. Hu et al., Stfnets: learning sensing signals from the time-frequency perspective with short-time fourier neural networks, in *The World Wide Web Conference* (2019), pp. 2192–2202
38. P. Khosla, P. Teterwak, C. Wang, A. Sarna, Y. Tian, P. Isola, A. Maschinot, C. Liu, D. Krishnan, Supervised contrastive learning. Preprint (2020), arXiv:2004.11362
39. K. He, X. Chen, S. Xie, Y. Li, P. Dollár, R. Girshick, Masked autoencoders are scalable vision learners. Preprint (2021), arXiv:2111.06377
40. C. Long, K. Chugg, A. Polydoros, Further results in likelihood classification of qam signals, in *Proceedings of MILCOM'94* (IEEE, 1994), pp. 57–61
41. N.E. Lay, A. Polydoros, Modulation classification of signals in unknown isi environments, in *Proceedings of MILCOM'95*, vol. 1 (IEEE, 1995), pp. 170–174
42. O.A. Dobre, A. Abdi, Y. Bar-Ness, W. Su, The classification of joint analog and digital modulations, in *MILCOM 2005–2005 IEEE Military Communications Conference* (IEEE, 2005), pp. 3010–3015
43. K. Ho, W. Prokopiw, Y. Chan, Modulation identification of digital signals by the wavelet transform. *IEE Proceedings-Radar, Sonar and Navigation*, vol. 147, no. 4 (2000), pp. 169–176
44. S. Huang, Y. Yao, Z. Wei, Z. Feng, P. Zhang, Automatic modulation classification of overlapped sources using multiple cumulants. IEEE Transa. Veh. Technol. **66**(7) (2016)
45. T.J. O'Shea, J. Corgan, T.C. Clancy, Convolutional radio modulation recognition networks, in *International Conference on Engineering Applications of Neural Networks* (Springer, 2016), pp. 213–226
46. T.J. O'Shea, T. Roy, T.C. Clancy, Over-the-air deep learning based radio signal classification. IEEE J. Sel. Top. Signal Process. **12**(1), 168–179 (2018)
47. S. Rajendran, W. Meert, D. Giustiniano, V. Lenders, S. Pollin, Deep learning models for wireless signal classification with distributed low-cost spectrum sensors. IEEE Trans. Cognitive Commun. Netw. **4**(3), 433–445 (2018)
48. J. Xu, C. Luo, G. Parr, Y. Luo, A spatiotemporal multi-channel learning framework for automatic modulation recognition. IEEE Wirel. Commun. Lett. **9**(10), 1629–1632 (2020)
49. L. Huang, W. Pan, Y. Zhang, L. Qian, N. Gao, Y. Wu, Data augmentation for deep learning-based radio modulation classification. IEEE Access **8**, 1498–1506 (2019)
50. Q. Zheng, P. Zhao, Y. Li, H. Wang, Y. Yang, Spectrum interference-based two-level data augmentation method in deep learning for automatic modulation classification. Neural Comput. Appl. 1–23 (2020)

51. P. Wang, M. Vindiola, Data augmentation for blind signal classification, in *MILCOM 2019–2019 IEEE Military Communications Conference (MILCOM)* (IEEE, 2019), pp. 305–310
52. D. Liu, P. Wang, T. Wang, T. Abdelzaher, Self-contrastive learning based semi-supervised radio modulation classification, in *MILCOM 2021–2021 IEEE Military Communications Conference (MILCOM)* (IEEE, 2021), pp. 777–782
53. K. Sohn, Improved deep metric learning with multi-class n-pair loss objective, in *Proceedings of the 30th International Conference on Neural Information Processing Systems* (2016), pp. 1857–1865
54. S. Yao, S. Hu, Y. Zhao, A. Zhang, T. Abdelzaher, Deepsense: a unified deep learning framework for time-series mobile sensing data processing, in *Proceedings of the 26th International Conference on World Wide Web* (2017), pp. 351–360
55. S. Yao, Y. Zhao, A. Zhang, L. Su, T. Abdelzaher, Deepiot: Compressing deep neural network structures for sensing systems with a compressor-critic framework, in *Proceedings of the 15th ACM Conference on Embedded Network Sensor Systems* (2017), pp. 1–14
56. X. Li, S. Liu, S. De Mello, X. Wang, J. Kautz, M.-H. Yang, Joint-task self-supervised learning for temporal correspondence, in *Advances in Neural Information Processing Systems* (2019), pp. 318–328
57. R. Collobert, J. Weston, L. Bottou, M. Karlen, K. Kavukcuoglu, P. Kuksa, Natural language processing (almost) from scratch. J. Mach. Learn. Res. **12**(ARTICLE), 2493–2537 (2011)
58. Z. Li, Z. Xiao, B. Wang, B.Y. Zhao, H. Zheng, Scaling deep learning models for spectrum anomaly detection, in *Proceedings of the Twentieth ACM International Symposium on Mobile Ad Hoc Networking and Computing* (2019), 291–300
59. D.C. Mohr, M. Zhang, S.M. Schueller, Personal sensing: understanding mental health using ubiquitous sensors and machine learning. Ann. Rev. Clin. Psychol. **13**, 23–47 (2017)
60. K. Han, D. Yu, I. Tashev, Speech emotion recognition using deep neural network and extreme learning machine, in *Fifteenth Annual Conference of the International Speech Communication Association* (2014)
61. Y. Guan, T. Plötz, Ensembles of deep lstm learners for activity recognition using wearables. *Proceedings of the ACM on Interactive, Mobile, Wearable and Ubiquitous Technologies* **1**(2), 1–28 (2017)
62. L. Peng, L. Chen, Z. Ye, Y. Zhang, Aroma: a deep multi-task learning based simple and complex human activity recognition method using wearable sensors. Proc. ACM Interactive Mobile Wearable Ubiquit. Technol. **2**(2), 1–16 (2018)
63. V. Radu, C. Tong, S. Bhattacharya, N.D. Lane, C. Mascolo, M.K. Marina, F. Kawsar, Multimodal deep learning for activity and context recognition. Proc. ACM Interactive Mobile Wearable Ubiquit. Technol. **1**(4), 1–27 (2018)
64. S. Bhattacharya, N.D. Lane, Sparsification and separation of deep learning layers for constrained resource inference on wearables, in *Proceedings of the 14th ACM Conference on Embedded Network Sensor Systems CD-ROM* (2016), pp. 176–189
65. S. Yao, Y. Zhao, H. Shao, S. Liu, D. Liu, L. Su, T. Abdelzaher, Fastdeepiot: towards understanding and optimizing neural network execution time on mobile and embedded devices, in *Proceedings of the 16th ACM Conference on Embedded Networked Sensor Systems* (2018), pp. 278–291
66. M. Noroozi, P. Favaro, Unsupervised learning of visual representations by solving jigsaw puzzles, in *European Conference on Computer Vision* (Springer, 2016), pp. 69–84
67. P. Vincent, H. Larochelle, Y. Bengio, P.-A. Manzagol, Extracting and composing robust features with denoising autoencoders, in *Proceedings of the 25th International Conference on Machine Learning* (2008), pp. 1096–1103
68. T. Mikolov, K. Chen, G. Corrado, J. Dean, Efficient estimation of word representations in vector space. Preprint (2013). arXiv:1301.3781
69. G. Zhong, L.-N. Wang, X. Ling, J. Dong, An overview on data representation learning: from traditional feature learning to recent deep learning. J. Finance Data Sci. **2**(4), 265–278 (2016)
70. Y. Bengio, A. Courville, P. Vincent, Representation learning: a review and new perspectives. IEEE Trans. Pattern Anal. Mach. Intell. **35**(8), 1798–1828 (2013)

71. K. Pearson, Liii. on lines and planes of closest fit to systems of points in space. Lond. Edinb. Dublin Philos. Mag. J. Sci. **2**(11), 559–572 (1901)
72. R.A. Fisher, The use of multiple measurements in taxonomic problems. Ann. Eugen. **7**(2), 179–188 (1936)
73. S.T. Roweis, L.K. Saul, Nonlinear dimensionality reduction by locally linear embedding. Science **290**(5500), 2323–2326 (2000)
74. C. Doersch, A. Gupta, A.A. Efros, Unsupervised visual representation learning by context prediction, in *Proceedings of the IEEE International Conference on Computer Vision* (2015), pp. 1422–1430
75. Y. Yao, C. Liu, D. Luo, Y. Zhou, Q. Ye, Video playback rate perception for self-supervised spatio-temporal representation learning, in *Proceedings of the IEEE/CVF Conference on Computer Vision and Pattern Recognition* (2020), pp. 6548–6557
76. J. Wang, J. Jiao, L. Bao, S. He, Y. Liu, W. Liu, Self-supervised spatio-temporal representation learning for videos by predicting motion and appearance statistics, in *Proceedings of the IEEE/CVF Conference on Computer Vision and Pattern Recognition* (2019), pp. 4006–4015
77. T. Han, W. Xie, A. Zisserman, Self-supervised co-training for video representation learning. Preprint (2020). arXiv:2010.09709
78. P. Morgado, N. Vasconcelos, T. Langlois, O. Wang, Self-supervised generation of spatial audio for 360 video. Preprint (2018). arXiv:1809.02587
79. B. Korbar, D. Tran, L. Torresani, Cooperative learning of audio and video models from self-supervised synchronization. Preprint (2018). arXiv:1807.00230
80. L. Kong, C.D.M. d'Autume, W. Ling, L. Yu, Z. Dai, D. Yogatama, A mutual information maximization perspective of language representation learning. Preprint (2019). arXiv:1910.08350
81. J.O. Smith, *Mathematics of the Discrete Fourier Transform (DFT): With Audio Applications* (Julius Smith, Stanford, 2007)
82. F. Yu, V. Koltun, Multi-scale context aggregation by dilated convolutions. Preprint (2015). arXiv:1511.07122
83. K. He, X. Zhang, S. Ren, J. Sun, Deep residual learning for image recognition, in *Proceedings of the IEEE Conference on Computer Vision and Pattern Recognition* (2016), pp. 770–778
84. O. Steven Eyobu, D.S. Han, Feature representation and data augmentation for human activity classification based on wearable imu sensor data using a deep lstm neural network. Sensors **18**(9), 2892 (2018)
85. J. Gao, X. Song, Q. Wen, P. Wang, L. Sun, H. Xu, Robusttad: robust time series anomaly detection via decomposition and convolutional neural networks. Preprint (2020). arXiv:2002.09545
86. A. Stisen, H. Blunck, S. Bhattacharya, T.S. Prentow, M.B. Kjærgaard, A. Dey, T. Sonne, M.M. Jensen, Smart devices are different: assessing and mitigatingmobile sensing heterogeneities for activity recognition, in *Proceedings of the 13th ACM Conference on Embedded Networked Sensor Systems* (2015), pp. 127–140
87. C. Chatzaki, M. Pediaditis, G. Vavoulas, M. Tsiknakis, Human daily activity and fall recognition using a smartphone's acceleration sensor, in *International Conference on Information and Communication Technologies for Ageing Well and e-Health* (Springer, 2016), pp. 100–118
88. M. Malekzadeh, R.G. Clegg, A. Cavallaro, H. Haddadi, Protecting sensory data against sensitive inferences, in *Proceedings of the 1st Workshop on Privacy by Design in Distributed Systems* (2018), pp. 1–6
89. D. Anguita, A. Ghio, L. Oneto, X. Parra, J.L. Reyes-Ortiz, A public domain dataset for human activity recognition using smartphones, in *Esann*, vol. 3 (2013), p. 3
90. J.R. Kwapisz, G.M. Weiss, S.A. Moore, Activity recognition using cell phone accelerometers. ACM SigKDD Explor. Newsl. **12**(2), 74–82 (2011)
91. W. Jiang, C. Miao, F. Ma, S. Yao, Y. Wang, Y. Yuan, H. Xue, C. Song, X. Ma, D. Koutsonikolas et al., Towards environment independent device free human activity recognition, in *Proceedings of the 24th Annual International Conference on Mobile Computing and Networking* (2018), pp. 289–304

92. D.P. Kingma, J. Ba, Adam: a method for stochastic optimization. Preprint (2014). arXiv:1412.6980
93. L. Van der Maaten, G. Hinton, Visualizing data using t-sne. J. Mach. Learn. Res. **9**(11) (2008)
94. J.S. Bauer, S. Consolvo, B. Greenstein, J. Schooler, E. Wu, N.F. Watson, J. Kientz, Shuteye: encouraging awareness of healthy sleep recommendations with a mobile, peripheral display, in *Proceedings of the SIGCHI Conference on Human Factors in Computing Systems* (2012), pp. 1401–1410
95. F.R. Bentley, Y.-Y. Chen, C. Holz, Reducing the stress of coordination: sharing travel time information between contacts on mobile phones, in *Proceedings of the 33rd Annual ACM Conference on Human Factors in Computing Systems* (2015), pp. 967–970
96. M. Faurholt-Jepsen, M. Vinberg, M. Frost, S. Debel, E. Margrethe Christensen, J.E. Bardram, L.V. Kessing, Behavioral activities collected through smartphones and the association with illness activity in bipolar disorder. Int. J. Methods Psychiatric Res. **25**(4), 309–323 (2016)
97. H. Zou, Z. Chen, H. Jiang, L. Xie, C. Spanos, Accurate indoor localization and tracking using mobile phone inertial sensors, wifi and ibeacon, in *2017 IEEE International Symposium on Inertial Sensors and Systems (INERTIAL)* (IEEE, 2017), pp. 1–4
98. E. Martin, O. Vinyals, G. Friedland, R. Bajcsy, Precise indoor localization using smart phones, in *Proceedings of the 18th ACM International Conference on Multimedia* (2010), pp. 787–790
99. N.Y. Hammerla, S. Halloran, T. Plötz, Deep, convolutional, and recurrent models for human activity recognition using wearables. Preprint (2016). arXiv:1604.08880
100. F.J.O. Morales, D. Roggen, Deep convolutional feature transfer across mobile activity recognition domains, sensor modalities and locations, in *Proceedings of the 2016 ACM International Symposium on Wearable Computers* (2016), pp. 92–99
101. J. Wang, Y. Chen, S. Hao, X. Peng, L. Hu, Deep learning for sensor-based activity recognition: a survey. Pattern Recognit. Lett. **119**, 3–11 (2019)
102. C.I. Tang, I. Perez-Pozuelo, D. Spathis, S. Brage, N. Wareham, C. Mascolo, Selfhar: improving human activity recognition through self-training with unlabeled data. Preprint (2021). arXiv:2102.06073
103. I. Misra, L.V.D. Maaten, Self-supervised learning of pretext-invariant representations, in *Proceedings of the IEEE/CVF Conference on Computer Vision and Pattern Recognition* (2020), pp. 6707–6717
104. R. Qian, T. Meng, B. Gong, M.-H. Yang, H. Wang, S. Belongie, Y. Cui, Spatiotemporal contrastive video representation learning. Preprint (2020). arXiv:2008.03800
105. D. Jiang, W. Li, M. Cao, R. Zhang, W. Zou, K. Han, X. Li, Speech simclr: combining contrastive and reconstruction objective for self-supervised speech representation learning. Preprint (2020). arXiv:2010.13991
106. G. Jin, B. Ye, Y. Wu, F. Qu, Vehicle classification based on seismic signatures using convolutional neural network. IEEE Geosci. Remote Sensing Lett. **16**(4), 628–632 (2018)
107. G.P. Mazarakis, J.N. Avaritsiotis, Vehicle classification in sensor networks using time-domain signal processing and neural networks. Microprocess. Microsyst. **31**(6), 381–392 (2007)
108. M. Kim, J. Tack, S.J. Hwang, Adversarial self-supervised contrastive learning. Adv. Neural Inf. Process. Syst. **33**, 2983–2994 (2020)
109. E.K. Antonsson, R.W. Mann, The frequency content of gait. J. Biomech. **18**(1), 39–47 (1985)
110. A. Dosovitskiy, L. Beyer, A. Kolesnikov, D. Weissenborn, X. Zhai, T. Unterthiner, M. Dehghani, M. Minderer, G. Heigold, S. Gelly et al., An image is worth 16x16 words: transformers for image recognition at scale. Preprint (2020). arXiv:2010.11929
111. I. Loshchilov, F. Hutter, Decoupled weight decay regularization. Preprint (2017). arXiv:1711.05101
112. A.B. Assessment, C.P. Brief, US military investments in autonomy and AI (2020)
113. S. Petrella, C. Miller, B. Cooper, Russia's artificial intelligence strategy: the role of state-owned firms. Orbis **65**(1), 75–100 (2021)
114. M.C. Horowitz, Artificial intelligence, international competition, and the balance of power. Texas Natl. Security Rev. **22** (2018)

115. J.F. Antal, *7 Seconds to Die: A Military Analysis of the Second Nagorno-Karabakh War and the Future of Warfighting* (Casemate, Philadelphia, 2022)
116. T. Abdelzaher, A. Taliaferro, P. Sullivan, S. Russell, The multi-domain operations effect loop: from future concepts to research challenges, in *Artificial Intelligence and Machine Learning for Multi-Domain Operations Applications II*, vol. 11413 (International Society for Optics and Photonics, Bellingham, 2020), p. 1141304
117. S. Russell, T. Abdelzaher, N. Suri, Multi-domain effects and the internet of battlefield things, in *MILCOM 2019–2019 IEEE Military Communications Conference (MILCOM)* (IEEE, 2019), pp. 724–730
118. S. Russell, T. Abdelzaher, The internet of battlefield things: the next generation of command, control, communications and intelligence (c3i) decision-making, in *MILCOM 2018–2018 IEEE Military Communications Conference (MILCOM)* (IEEE, 2018), pp. 737–742
119. T. Abdelzaher, N. Ayanian, T. Basar, S. Diggavi, J. Diesner, D. Ganesan, R. Govindan, S. Jha, T. Lepoint, B. Marlin et al., Toward an internet of battlefield things: a resilience perspective. Computer **51**(11), 24–36 (2018)
120. S. Liu, S. Yao, Y. Huang, D. Liu, H. Shao, Y. Zhao, J. Li, T. Wang, R. Wang, C. Yang et al., Handling missing sensors in topology-aware iot applications with gated graph neural network. Proc. ACM Interactive Mobile Wearable Ubiquit. Technol. **4**(3), 1–31 (2020)
121. T. Abdelzaher, N. Ayanian, T. Basar, S. Diggavi, J. Diesner, D. Ganesan, R. Govindan, S. Jha, T. Lepoint, B. Marlin et al., Will distributed computing revolutionize peace? the emergence of battlefield IoT, in *2018 IEEE 38th International Conference on Distributed Computing Systems (ICDCS)* (IEEE, 2018), pp. 1129–1138
122. A.D. Cobb, B.A. Jalaian, N.D. Bastian, S. Russell, Robust decision-making in the internet of battlefield things using bayesian neural networks, in *2021 Winter Simulation Conference (WSC)* (IEEE, 2021), pp. 1–12
123. B.M. Marlin, T. Abdelzaher, G. Ciocarlie, A.D. Cobb, M. Dennison, B. Jalaian, L. Kaplan, T. Raber, A. Raglin, P.K. Sharma et al., On uncertainty and robustness in large-scale intelligent data fusion systems, in *2020 IEEE Second International Conference on Cognitive Machine Intelligence (CogMI)* (IEEE, 2020), pp. 82–91
124. E. Blasch, T. Pham, C.-Y. Chong, W. Koch, H. Leung, D. Braines, T. Abdelzaher, Machine learning/artificial intelligence for sensor data fusion–opportunities and challenges. IEEE Aerospace Electron. Syst. Mag. **36**(7), 80–93 (2021)
125. S. Liu, S. Yao, J. Li, D. Liu, T. Wang, H. Shao, T. Abdelzaher, Giobalfusion: a global attentional deep learning framework for multisensor information fusion. Proc. ACM Interactive Mobile Wearable Ubiquit. Technol. **4**(1), 1–27 (2020)
126. S. Yao, Y. Zhao, H. Shao, D. Liu, S. Liu, Y. Hao, A. Piao, S. Hu, S. Lu, T.F. Abdelzaher, Sadeepsense: self-attention deep learning framework for heterogeneous on-device sensors in internet of things applications, in *IEEE INFOCOM 2019-IEEE Conference on Computer Communications* (IEEE, 2019), pp. 1243–1251
127. J. Huang, C. Samplawski, D. Ganesan, B. Marlin, H. Kwon, Clio: enabling automatic compilation of deep learning pipelines across iot and cloud, in *Proceedings of the 26th Annual International Conference on Mobile Computing and Networking* (2020), pp. 1–12
128. T. Li, J. Huang, E. Risinger, D. Ganesan, Low-latency speculative inference on distributed multi-modal data streams, in *Proceedings of the 19th Annual International Conference on Mobile Systems, Applications, and Services* (2021), pp. 67–80
129. D. Basu, D. Data, C. Karakus, S. Diggavi, Qsparse-local-sgd: distributed sgd with quantization, sparsification and local computations. Adv. Neural Inf. Process. Syst. **32** (2019)
130. S. Yao, Y. Hao, Y. Zhao, A. Piao, H. Shao, D. Liu, S. Liu, S. Hu, D. Weerakoon, K. Jayarajah et al., Eugene: towards deep intelligence as a service, in *2019 IEEE 39th International Conference on Distributed Computing Systems (ICDCS)* (IEEE, 2019), pp. 1630–1640
131. S. Yao, Y. Zhao, A. Zhang, S. Hu, H. Shao, C. Zhang, L. Su, T. Abdelzaher, Deep learning for the internet of things. Computer **51**(5), 32–41 (2018)

132. T. Abdelzaher, Y. Hao, K. Jayarajah, A. Misra, P. Skarin, S. Yao, D. Weerakoon, K.-E. Årzén, Five challenges in cloud-enabled intelligence and control. ACM Trans. Internet Technol. (TOIT) **20**(1), 1–19 (2020)
133. S. Yao, J. Li, D. Liu, T. Wang, S. Liu, H. Shao, T. Abdelzaher, Deep compressive offloading: speeding up neural network inference by trading edge computation for network latency, in *Proceedings of the International Conference on Embedded Networked Sensor Systems (SenSys)* (2020)

Chapter 3
On the Generalization Power of Overfitted Two-Layer Neural Tangent Kernel Models

Peizhong Ju, Xiaojun Lin, and Ness B. Shroff

Abstract In this chapter, we study the generalization performance of min ℓ_2-norm overfitting solutions for the neural tangent kernel (NTK) model of a two-layer neural network with ReLU activation that has no bias term. We show that, depending on the ground-truth function, the test error of overfitted NTK models exhibits characteristics that are different from the "double-descent" of other overparameterized linear models with simple Fourier or Gaussian features. Specifically, for a class of learnable functions, we derive a new upper bound of the generalization error that approaches a small limiting value, even when the number of neurons p approaches infinity. This limiting value further decreases with the number of training samples n. For functions outside of this class, we provide a lower bound on the generalization error that does not diminish to zero even when n and p are both large.

1 Introduction

Increasingly, AI problems are being solved by edge devices (e.g., [1–4]). This is especially true with the growth of internet of things (IoT) and other edge devices, which are fueling the growth of distributed AI. Among all learning models and methods that are deployed in edge devices, deep neural network (DNN) is considered as an important one and is widely used in many real-world applications (e.g., [5, 6]). A typical DNN usually consists of several layers that involve a large

---

P. Ju
Department of ECE, The Ohio State University, Columbus, OH, USA
e-mail: ju.171@osu.edu

X. Lin (✉)
Elmore Family School of Electrical and Computer Engineering, Purdue University, West Lafayette, IN, USA
e-mail: linx@purdue.edu

N. B. Shroff
Department of ECE and CSE, The Ohio State University, Columbus, OH, USA
e-mail: shroff.11@osu.edu

© The Author(s), under exclusive license to Springer Nature Switzerland AG 2023
M. Srivatsa et al. (eds.), *Artificial Intelligence for Edge Computing*,
https://doi.org/10.1007/978-3-031-40787-1_3

number of neurons (i.e., nonlinear activation units) and weights (i.e., parameters that needs to be trained), which endows the DNN with the power to fit a large number of training data and express a complicated model.

However, just because DNNs can be trained to fit training data from a complex ground-truth model, it does not mean that they will work well for unseen test data. In fact, with the growth of IoT and other wireless edge services, machine learning problems are going to experience a large variety of scenarios, some of them with more limited training data than others. Therefore, it is critical that the trained machine learning model can not only fit the training data, but also generalize to new information and data. Indeed, in classical statistical learning theory [7–12], the well-known *bias-variance tradeoff* suggests a U-shaped curve of generalization error. That is, when the model complexity increases, the error first decreases due to smaller bias, but then increases due to larger variance (i.e., overfitting). Therefore, it is considered important to control the model complexity carefully (e.g., using regularization), in order to avoid overfitting and attain good generalization power. However, DNNs seems to defy this understanding by not only overfitting the training data, but also generalizing well [13, 14]. Understanding why overfitting and overparameterization do not harm the generalization power of DNNs could have implications on the design of future network architecture and training methods.

Towards this direction, a recent line of study has focused on overparameterized linear models [15–21]. For linear models with simple features (e.g., Gaussian features and Fourier features) [15–20], an interesting "double-descent" phenomenon has been observed. Thus, there is a region where the number of model parameters (or linear features) is larger than the number of samples (and thus overfitting occurs), but the generalization error actually decreases with the number of features. However, linear models with these simple features are still quite different from nonlinear neural networks. Thus, although such results provide some hint why overparameterization and overfitting may be harmless, it is still unclear whether similar conclusions apply to neural networks.

In this chapter, we are interested in linear models based on the neural tangent kernel (NTK) [22], which can be viewed as a useful intermediate step towards modeling nonlinear neural networks. Essentially, NTK can be seen as a linear approximation of neural networks when the weights of the neurons do not change much. Indeed, [23, 24] have shown that, for a wide and fully-connected two-layer neural network, both the neuron weights and their activation patterns do not change much after gradient descent (GD) training with a sufficiently small step size. As a result, when there are a sufficient number of neurons, such a shallow and wide neural network is approximately linear in the weights and thus easier to analyse, which suggests the utility of the NTK model.

Despite its linearity, however, characterizing the double descent of such a NTK model remains elusive. In [21], the double-descent of a linear version of two-layer neural network is studied. The authors use the so-called "random-feature" model, where the bottom-layer weights are random and fixed, and only the top-layer weights are trained. (In comparison, the NTK model for such a two-layer neural network corresponds to training only the bottom-layer weights.) However, the setting there requires the number of neurons, the number of samples, and the data dimension to all grow proportionally to infinity. In contrast, we are interested in the setting where the number of samples is given, and the number of neurons is allowed to be much larger than the number of samples. As a consequence of the different setting, in [21] eventually only *linear* ground-truth functions can be learned. (Similar settings are also studied in [25].) In contrast, we will show that far more complex functions can be learned in our setting. In a related work, [26] shows that both the random-feature model and the NTK model can approximate highly *nonlinear* ground-truth functions with a sufficient number of neurons. However, [26] mainly studies the *expressiveness* of the models, and therefore does not explain why overfitting solutions can still generalize well. To the best of our knowledge, our work is the first to characterize the double-descent of overfitting solutions based on the NTK model.

Specifically, in this chapter we study the generalization error of the min ℓ_2-norm overfitting solution for a linear model based on the NTK of a two-layer neural network with ReLU activation that has no bias. Only the bottom-layer weights are trained. We focus on min ℓ_2-norm overfitting solutions, since gradient descent (GD) can be shown to converge to such solutions while driving the training error to zero [13] (see also Sect. 2). Given a class of ground truth functions (see details in Sect. 3), which we refer to as "learnable functions," our main result (Theorem 1) provides an upper bound on the generalization error of the min ℓ_2-norm overfitting solution for the two-layer NTK model with n samples and p neurons (for any finite p larger than a polynomial function of n). This upper bound confirms that the generalization error of the overfitting solution indeed exhibits descent in the overparameterized regime when p increases. Further, our upper bound can also account for the noise in the training samples.

There are several important insights revealed by our results. First, we find that the (double) descent of the overfitted two-layer NTK model is drastically different from that of linear models with simple Gaussian or Fourier features [15–19]. Specifically, for linear models with simple features, when the number of features p increases, the generalization error will eventually grow again and approach the so-called "null risk" [18], which is the error of a trivial model that predicts zero. In contrast, for the class of learnable functions described earlier, the generalization error of the overfitted NTK model will continue to descend as p grows to infinity, and will approach a limiting value that depends on the number of samples n. Further, when there is no noise, this limiting value will decrease to zero as the number of samples n increases. This difference is shown in Fig. 3.1a. As p increases, the test mean-square-error (MSE) of min-ℓ_1 and min-ℓ_2 overfitting solutions for Fourier features (blue and red curves) eventually grow back to the null risk (the black dashed line), even though they exhibit a descent at smaller p. In contrast, the error of the overfitted NTK model continues to descend to a much lower level.

The second important insight is that the aforementioned behavior critically depends on the ground-truth function belonging to the class of "learnable functions." Further, this class of learnable functions depend on the specific network architecture. For our NTK model (with RELU activation that has no bias), we precisely characterize this class of learnable functions. Specifically, for ground-truth functions that are outside the class of learnable functions, we show a lower bound on the generalization error that does not diminish to zero for any n and p (see Proposition 1 and Sect. 4). We use Fig. 3.1b to show this difference, where an almost identical setting as Fig. 3.1a is used, except that the ground-truth function is different. We can see in Fig. 3.1b that the test-error of the overfitted NTK model is always above the null risk and looks very different from that in Fig. 3.1a. We note that whether certain functions are learnable or not critically depends on the specific structure of the NTK model, such as the choice of the activation unit. Recently, [27] shows that all polynomials can be learned by 2-layer NTK model with ReLU activation that has a bias term, provided that the number of neurons p is sufficiently large. (See further discussions in Remark 2. However, [27] does not characterize the descent of generalization errors as p increases.) This difference in the class of learnable functions between the two settings (ReLU with or without bias) also turns out to be consistent with the difference in the expressiveness of the neural networks. That is, shallow networks with biased-ReLU are known to be universal function approximators [28], while those without bias can only approximate the sum of linear functions and even functions [26].

Our results are related to the work in [29], which characterizes the generalization performance of wide two-layer neural networks whose bottom-layer weights are trained by gradient descent (GD) to overfit the training samples. In particular, our class of learnable functions almost coincides with that of [29]. This is not surprising because, when the number of neurons is large, NTK becomes a close approximation of such two-layer neural networks. In that sense, the results in [29] are even more faithful in following the GD dynamics of the original two-layer network. However, the advantage of the NTK model is that it is easier to analyze. In particular, the results in this chapter can quantify how the generalization error descends with p. In contrast, the results in [29] provide only a generalization bound that is independent of p (provided that p is sufficiently large), but do not quantify the descent behavior as p increases. Our numerical results in Fig. 3.1a suggest that, over a wide range of p, the descent behavior of the NTK model (the green curve) matches well with that of two-layer neural networks trained by gradient descent (the cyan curve). Thus, we believe that our results also provide guidance for the latter model. The work in [30] studies a different neural network architecture with gated ReLU, whose NTK model turns out to be the same as ours. However, similar to [29], the result in [30] does not capture the speed of descent with respect to p either. Second, [29] only provides upper bounds on the generalization error. There is no corresponding lower bound to explain whether ground-truth functions outside a certain class are *not* learnable. Our result in Proposition 1 provides such a lower bound, and therefore more completely characterizes the class of learnable functions. (See further comparison in Remark 1 of Sect. 3 and Remark 3 of Sect. 5.) Another related work [31] also characterizes the class of learnable functions for two-layer and three-layer networks. However, [31] studies a training method that takes a new sample in every iteration, and thus does not overfit all training data. Finally, this chapter studies generalization of NTK

3 On the Generalization Power of Overfitted Two-Layer Neural Tangent... 115

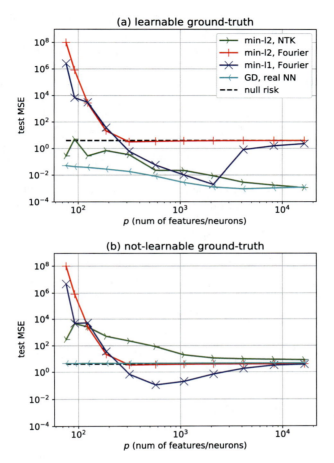

Fig. 3.1 The test mean-square-error(MSE) vs. the number of features/neurons p for **(a)** learnable function and **(b)** not-learnable function when $n = 60$, $d = 2$, $\|\epsilon\|_2^2 = 0.01$. The corresponding ground-truth are **(a)** $f(\theta) = \sum_{k \in \{0,1,2,4\}} (\sin(k\theta) + \cos(k\theta))$, and **(b)** $f(\theta) = \sum_{k \in \{3,5,7,9\}} (\sin(k\theta) + \cos(k\theta))$. (Note that in 2-dimension every input x on a unit circle can be represented by an angle $\theta \in [-\pi, \pi]$. See the end of Sect. 4.) Every curve is the average of 15 random simulation runs. For GD on the real neural network (NN), we use the step size $1/\sqrt{p}$ and the number of training epochs is fixed at 2000

models for the regression setting, which is different from the classification setting that assumes a separability condition, e.g., in [32].

2 Problem Setup

We consider the following data model $y = f(x) + \epsilon$, with the input $x \in \mathbb{R}^d$, the output $y \in \mathbb{R}$, the noise $\epsilon \in \mathbb{R}$, and $f : \mathbb{R}^d \mapsto \mathbb{R}$ denotes the ground-truth function.

Let $(\mathbf{X}_i, y_i)$, $i = 1, 2, \cdots, n$ denote n training samples. We collect them as $\mathbf{X} = [\mathbf{X}_1 \, \mathbf{X}_2 \, \cdots \, \mathbf{X}_n] \in \mathbb{R}^{d \times n}$, $\mathbf{y} = [y_1 \, y_2 \, \cdots \, y_n]^T \in \mathbb{R}^n$, $\boldsymbol{\epsilon} = [\epsilon_1 \, \epsilon_2 \, \cdots \, \epsilon_n]^T \in \mathbb{R}^n$, and $\mathbf{F}(\mathbf{X}) = [f(\mathbf{X}_1) \, f(\mathbf{X}_2) \, \cdots \, f(\mathbf{X}_n)]^T \in \mathbb{R}^n$. Then, the training samples can be written as $\mathbf{y} = \mathbf{F}(\mathbf{X}) + \boldsymbol{\epsilon}$. After training (to be described below), we use the function $\hat{f}$ to denote the trained model. Then, for any new test data $\mathbf{x}$, we will calculate the test error by $|\hat{f}(\mathbf{x}) - f(\mathbf{x})|$, and the mean squared error (MSE) by $\mathsf{E}_{\mathbf{x}}[\hat{f}(\mathbf{x}) - f(\mathbf{x})]^2$.

For training, consider a fully-connected two-layer neural network with p neurons. Let $\boldsymbol{w}_j \in \mathbb{R}$ and $\mathbf{V}_0[j] \in \mathbb{R}^d$ denote the top-layer and bottom-layer weights, respectively, of the j-th neuron, $j = 1, 2, \cdots, p$ (see Fig. 3.2). We collect them into $\boldsymbol{w} = [\boldsymbol{w}_1 \, \boldsymbol{w}_2 \, \cdots \, \boldsymbol{w}_p]^T \in \mathbb{R}^p$, and $\mathbf{V}_0 = [\mathbf{V}_0[1]^T \, \mathbf{V}_0[2]^T \, \cdots \, \mathbf{V}_0[p]^T]^T \in \mathbb{R}^{dp}$ (a column vector with dp elements). Note that with this notation, for any row or column vector $\boldsymbol{v}$ with dp elements, $\boldsymbol{v}[j]$ denotes a (row/column) vector that consists of the $(jd + 1)$-th to $(jd + d)$-th elements of $\boldsymbol{v}$. We choose ReLU as the activation function for all neurons. Throughout the chapter, we assume that there is no bias term in the ReLU activation function. We will discuss the implication of this assumption in Remark 2.

Now we are ready to introduce the NTK model [22]. We fix the top-layer weights $\boldsymbol{w}$, and we randomly choose the initial bottom-layer weights $\mathbf{V}_0$. We then train only the bottom-layer weights. Let $\mathbf{V}_0 + \overline{\Delta \mathbf{V}}$ denote the bottom-layer weights after training. Thus, the change of the output after training is

$$\sum_{j=1}^{p} \boldsymbol{w}_j \mathbf{1}_{\{\boldsymbol{x}^T(\mathbf{V}_0[j] + \overline{\Delta \mathbf{V}}[j]) > 0\}} \cdot (\mathbf{V}_0[j] + \overline{\Delta \mathbf{V}}[j])^T \boldsymbol{x}$$

$$- \sum_{j=1}^{p} \boldsymbol{w}_j \mathbf{1}_{\{\boldsymbol{x}^T \mathbf{V}_0[j] > 0\}} \cdot \mathbf{V}_0[j]^T \boldsymbol{x}.$$

In the NTK model, one assumes that $\overline{\Delta \mathbf{V}}$ is very small. As a result, for most $\boldsymbol{x}$, we have

$$\mathbf{1}_{\{\boldsymbol{x}^T(\mathbf{V}_0[j] + \overline{\Delta \mathbf{V}}[j]) > 0\}} = \mathbf{1}_{\{\boldsymbol{x}^T \mathbf{V}_0[j] > 0\}}.$$

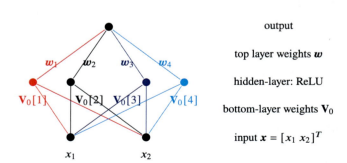

Fig. 3.2 A two-layer neural network where $d = 2$, $p = 4$

3 On the Generalization Power of Overfitted Two-Layer Neural Tangent...

Thus, the change of the output can be approximated by

$$\sum_{j=1}^{p} w_j \mathbf{1}_{\{x^T V_0[j]>0\}} \cdot \overline{\Delta V}[j]^T x = h_{V_0,x} \Delta V,$$

where $\Delta V \in \mathbb{R}^{dp}$ is given by $\Delta V[j] := w_j \overline{\Delta V}[j]$, $j = 1, 2, \cdots, p$, and $h_{V_0,x} \in \mathbb{R}^{1 \times (dp)}$ is given by

$$h_{V_0,x}[j] := \mathbf{1}_{\{x^T V_0[j]>0\}} \cdot x^T, \quad j = 1, 2, \cdots, p. \tag{3.1}$$

In the NTK model, we assume that the output of the trained model is exactly given by Eq. (3.1), i.e.,

$$\hat{f}_{\Delta V, V_0}(x) := h_{V_0,x} \Delta V. \tag{3.2}$$

In other words, the NTK model can be viewed as a linear approximation of the two-layer network when the change of the bottom-layer weights is small.

Define $H \in \mathbb{R}^{n \times (dp)}$ such that its i-th row is $H_i := h_{V_0, x_i}$. Throughout the paper, we will focus on the following min-ℓ_2-norm overfitting solution

$$\Delta V^{\ell_2} := \arg\min_{v} \|v\|_2, \text{ subject to } Hv = y.$$

Whenever ΔV^{ℓ_2} exists, it can be written in closed form as

$$\Delta V^{\ell_2} = H^T (HH^T)^{-1} y. \tag{3.3}$$

We are interested in ΔV^{ℓ_2} because gradient descent (GD) or stochastic gradient descent (SGD) for the NTK model in Eq. (3.2) is known to converge to ΔV^{ℓ_2} (proven in Supplementary Material, Appendix B in [33]).

Using Eqs. (3.2) and (3.3), the trained model is then

$$\hat{f}^{\ell_2}(x) := h_{V_0,x} \Delta V^{\ell_2}. \tag{3.4}$$

In the rest of the chapter, we will study the generalization error of Eq. (3.4).

In the rest of this section, we collect some assumptions as follows. Because $\hat{f}_{\Delta V, V_0}(ax) = a \cdot \hat{f}_{\Delta V, V_0}(x)$ for any $a \in \mathbb{R}$, we can always do preprocessing to normalize the input x. For simplicity, we focus on the simplest situation that the randomness for the inputs and the initial weights are uniform. Nonetheless, methods and results of this chapter can be readily generalized to other continuous random variable distributions, which we leave for future work. We thus make the following Assumption 1.

Assumption 1 *The input x are uniformly distributed in $\mathcal{S}^{d-1}$. The initial weights $\mathbf{V}_0[j]$'s are uniform in all directions. In other words, $\mu(\cdot)$ and $\tilde{\lambda}(\cdot)$ are both unif($\mathcal{S}^{d-1}$).* □

We study the overparameterized and overfitted setting, so in this chapter we always assume $p \geq n/d$, i.e., the number of parameters pd is larger than or equal to the number of training samples n. The situation of $d = 1$ is relatively trivial, so we only consider the case $d \geq 2$. We then make Assumption 2.

Assumption 2 $p \geq n/d$ *and* $d \geq 2$. □

If the input is a continuous random vector, then for any $i \neq j$, we have $\Pr\{\mathbf{X}_i = \mathbf{X}_j\} = 0$ and $\Pr\{\mathbf{X}_i = -\mathbf{X}_j\} = 0$ (because the probability that a continuous random variable equals to a given value is zero). Thus, $\Pr\{\mathbf{X}_i \parallel \mathbf{X}_j\} = 0$, and $\Pr\{\mathbf{X}_i \nparallel \mathbf{X}_j\} = 1$. Similarly, we can show that $\Pr\{\mathbf{V}_0[k] \nparallel \mathbf{V}_0[l]\} = 1$. We thus make Assumption 3.

Assumption 3 $\mathbf{X}_i \nparallel \mathbf{X}_j$ *for any* $i \neq j$, *and* $\mathbf{V}_0[k] \nparallel \mathbf{V}_0[l]$ *for any* $k \neq l$. □

With these assumptions, the following lemma says that when p is large enough, with high probability $\mathbf{H}$ has full row-rank (and thus $\Delta \mathbf{V}^{\ell_2}$ exists).

Lemma 1 $\lim_{p \to \infty} \Pr_{\mathbf{V}_0}\{\mathrm{rank}(\mathbf{H}) = n \mid \mathbf{X}\} = 1$.

Proof See Supplementary Material, Appendix E of [33]. □

3 Learnable Functions and Generalization Performance

We now show that the generalization performance of the overfitted NTK model in Eq. (3.4) crucially depends on the ground-truth function $f(\cdot)$. In other words, good generalization performance only occurs when the ground-truth function is "learnable." Below, we first describe a candidate class of ground-truth functions, and explain why they may correspond to the class of "learnable functions." Then, we will give an upper-bound on the generalization performance for this class of ground-truth functions. Finally, we will give a lower-bound on the generalization performance when the ground-truth functions are outside of this class.

We first define a set $\mathcal{F}^{\ell_2}$ of ground-truth functions.

Definition 1 $\mathcal{F}^{\ell_2} := \{f \overset{\text{a.e.}}{=} f_g \mid f_g(x) = \int_{\mathcal{S}^{d-1}} x^T z \frac{\pi - \arccos(x^T z)}{2\pi} g(z) d\mu(z), \|g\|_1 < \infty\}$.

We explain some notations used in Definition 1 as follows. The symbol $\overset{\text{a.e.}}{=}$ means two functions equals almost everywhere, and $\|g\|_1 := \int_{\mathcal{S}^{d-1}} |g(z)| d\mu(z)$. The function $g(z)$ may be any finite-value function in $L^1(\mathcal{S}^{d-1} \mapsto \mathbb{R})$. Further, we also allow $g(z)$ to contain (as components) Dirac δ-functions on $\mathcal{S}^{d-1}$. Note that a δ-function $\delta_{z_0}(z)$ has zero value for all $z \in \mathcal{S}^{d-1} \setminus \{z_0\}$, but $\|\delta_{z_0}\|_1 :=$

$\int_{S^{d-1}} \delta_{z_0}(z) d\mu(z) = 1$. Thus, the function $g(z)$ may contain any sum of δ-functions and finite-value L^1-functions.[1]

To see why $\mathcal{F}^{\ell_2}$ may correspond to the class of learnable functions, we can first examine what the learned function $\hat{f}^{\ell_2}$ in Eq. (3.4) should look like. Recall that $\mathbf{H}^T = [\mathbf{H}_1^T \cdots \mathbf{H}_n^T]$. Thus, $h_{\mathbf{V}_0,x}\mathbf{H}^T = \sum_{i=1}^n (h_{\mathbf{V}_0,x}\mathbf{H}_i^T)e_i^T$, where $e_i \in \mathbb{R}^n$ denotes the i-th standard basis. Combining Eqs. (3.3) and (3.4), we can see that the learned function in Eq. (3.4) is of the form

$$\hat{f}^{\ell_2}(x) = h_{\mathbf{V}_0,x}\mathbf{H}^T(\mathbf{HH}^T)^{-1}y$$

$$= \sum_{i=1}^n \left(\frac{1}{p}h_{\mathbf{V}_0,x}\mathbf{H}_i^T\right) pe_i^T(\mathbf{HH}^T)^{-1}y. \tag{3.5}$$

For all $x, z \in S^{d-1}$, define $C_{z,x}^{\mathbf{V}_0} := \{j \in \{1, 2, \cdots, p\} \mid z^T\mathbf{V}_0[j] > 0, x^T\mathbf{V}_0[j] > 0\}$, and its cardinality is given by

$$\left|C_{z,x}^{\mathbf{V}_0}\right| = \sum_{j=1}^p \mathbf{1}_{\{z^T\mathbf{V}_0[j]>0, \ x^T\mathbf{V}_0[j]>0\}}. \tag{3.6}$$

Then, using Eq. (3.1), we can show $\frac{1}{p}h_{\mathbf{V}_0,x}\mathbf{H}_i^T = x^T X_i \frac{|C_{X_i,x}^{\mathbf{V}_0}|}{p}$. It is not hard to show that

$$\frac{|C_{z,x}^{\mathbf{V}_0}|}{p} \xrightarrow{\text{P}} \frac{\pi - \arccos(x^T z)}{2\pi}, \quad \text{as } p \to \infty. \tag{3.7}$$

where $\xrightarrow{\text{P}}$ denotes converge in probability. (see Supplementary Material, Appendix D.5 of [33]). Thus, if we let

$$g(z) = \sum_{i=1}^n pe_i^T(\mathbf{HH}^T)^{-1}y\delta_{X_i}(z), \tag{3.8}$$

then as $p \to \infty$, Eq. (3.5) should approach a function in $\mathcal{F}^{\ell_2}$. This explains why $\mathcal{F}^{\ell_2}$ is a candidate class of "learnable functions." However, note that the above discussion only addresses the *expressiveness* of the model, and it does not provide any guarantee on the generalization performance of the trained model. In other words, it is still unclear whether any function in $\mathcal{F}^{\ell_2}$ can be learned with low generalization error. The following result provides the answer.

---

[1] Alternatively, we can also interpret $g(z)$ as a signed measure [34] on S^{d-1}. Then, δ-functions correspond to point masses, and the condition $\|g\|_1 < \infty$ implies that the corresponding unsigned version of the measure on S^{d-1} is bounded.

For some $m \in \left[1, \frac{\ln n}{\ln \frac{\pi}{2}}\right]$, define (recall that d is the dimension of $\boldsymbol{x}$)

$$J_m(n, d) := 2^{2d+5.5} d^{0.5d} n^{\left(2+\frac{1}{m}\right)(d-1)}. \tag{3.9}$$

Theorem 1 *Assume a ground-truth function $f \overset{a.e.}{=} f_g \in \mathcal{F}^{\ell_2}$ where $\|g\|_\infty < \infty^2$, $n \geq 2$, $m \in \left[1, \frac{\ln n}{\ln \frac{\pi}{2}}\right]$, $d \leq n^4$, and $p \geq 6 J_m(n, d) \ln \left(4 n^{1+\frac{1}{m}}\right)$. Then, for any $q \in [1, \infty)$ and for almost every $\boldsymbol{x} \in \mathcal{S}^{d-1}$, we must have[3]*

$$\Pr_{\boldsymbol{V}_0, \boldsymbol{X}} \left\{ |\hat{f}^{\ell_2}(\boldsymbol{x}) - f(\boldsymbol{x})| \geq \underbrace{n^{-\frac{1}{2}\left(1-\frac{1}{q}\right)}}_{\text{Term 1}} \right.$$

$$+ \underbrace{\left(1 + \sqrt{J_m(n, d)n}\right) p^{-\frac{1}{2}\left(1-\frac{1}{q}\right)}}_{\text{Term 2}} + \underbrace{\sqrt{J_m(n, d)n} \|\boldsymbol{\epsilon}\|_2}_{\text{Term 3}},$$

$$\left. \text{for all } \boldsymbol{\epsilon} \in \mathbb{R}^n \right\} \leq 2 e^2 \Bigg(\underbrace{\exp\left(-\frac{\sqrt[q]{n}}{8\|g\|_\infty^2}\right)}_{\text{Term 4}} \right.$$

$$+ \underbrace{\exp\left(-\frac{\sqrt[q]{p}}{8\|g\|_1^2}\right)}_{\text{Term 5}} + \underbrace{\exp\left(-\frac{\sqrt[q]{p}}{8n\|g\|_1^2}\right)}_{\text{Term 6}} \Bigg) + \underbrace{\frac{4}{\sqrt[m]{n}}}_{\text{Term 7}}. \tag{3.10}$$

We now interpret the upper bound shown in Theorem 1. We first focus on the noiseless case, where $\boldsymbol{\epsilon}$ and Term 3 are zero. If we fix n and let $p \to \infty$, then Terms 2, 5, and 6 all approach zero. We can then conclude that, in the noiseless and heavily overparameterized setting ($p \to \infty$), the generalization error will converge to a small limiting value (Term 1) that depends only on n. Further, this limiting value (Term 1) will converge to zero (so do Terms 4 and 7) as $n \to \infty$, i.e., when there are sufficiently many training samples. Finally, Theorem 1 holds even when there is noise.

When n and p are large, we can make Eq. (3.10) sharper by tuning the parameters q and m. For example, as we increase q, Term 1 will approach $n^{-0.5}$. Although a larger q makes Terms 4, 5, and 6 bigger, as long as n and p are sufficiently large,

---

[2] The requirement of $\|g\|_\infty < \infty$ can be relaxed. We show in Supplementary Material, Appendix L in [33] that, even when g is a δ-function (so $\|g\|_\infty = \infty$), we can still have a similar result of Eq. (3.10) but Term 1 will have a slower speed of decay $O(n^{-\frac{1}{2(d-1)}\left(1-\frac{1}{q}\right)})$ with respect to n instead of $O(n^{-\frac{1}{2}}(1 - \frac{1}{q}))$ shown in Eq. (3.10). Term 4 of Eq. (3.10) will also be different when g is a δ-function, but it still goes to zero when p and n are large.

[3] The notion $\Pr_M$ in Eq. (3.10) emphasizes that randomness is in M.

3 On the Generalization Power of Overfitted Two-Layer Neural Tangent... 121

those terms will still be close to 0. Similarly, if we increase m, then $J_m(n, d)$ will approach the order of $n^{2(d-1)}$. As a result, Term 3 approaches the order of $n^{2d-0.5}$ times $\|\boldsymbol{\epsilon}\|_2$ and the requirement $p \geq 6J_m(n, d) \ln\left(4n^{1+\frac{1}{m}}\right)$ approaches the order of $n^{2(d-1)} \ln n$.

Remark 1 We note that [29] shows that, for two-layer neural networks whose bottom-layer weights are trained by gradient descent, the generalization error for sufficiently large p has the following upper bound: for any $\zeta > 0$,

$$\Pr\left\{\mathsf{E}_x|\hat{f}(x) - f(x)| \leq \sqrt{\frac{2\boldsymbol{y}^T(\mathbf{H}^\infty)^{-1}\boldsymbol{y}}{n}} \right.$$
$$\left. + O\left(\sqrt{\frac{\log\frac{n}{\zeta \cdot \min \operatorname{eig}(\mathbf{H}^\infty)}}{n}}\right)\right\} \geq 1 - \zeta, \tag{3.11}$$

where $\mathbf{H}^\infty = \lim_{p \to \infty}(\mathbf{H}\mathbf{H}^T/p) \in \mathbb{R}^{n \times n}$. For certain class of learnable functions (we will compare them with our $\mathcal{F}^{\ell_2}$ in Sect. 4), the quantity $\boldsymbol{y}^T(\mathbf{H}^\infty)^{-1}\boldsymbol{y}$ is bounded. Thus, $\sqrt{\frac{2\boldsymbol{y}^T(\mathbf{H}^\infty)^{-1}\boldsymbol{y}}{n}}$ also decreases at the speed $1/\sqrt{n}$. The second $O(\cdot)$-term in Eq. (3.11) contains the minimum eigenvalue of $\mathbf{H}^\infty$, which decreases with n. (Indeed, we show that this minimum eigenvalue is upper bounded by $O(n^{-\frac{1}{d-1}})$ in Supplementary Material, Appendix G of [33].) Thus, Eq. (3.11) may decrease a little bit slower than $1/\sqrt{n}$. This decreasing speed is consistent with Term 1 in Eq. (3.10) (when q is large). Note that the term $2\boldsymbol{y}^T(\mathbf{H}^\infty)^{-1}\boldsymbol{y}$ in Eq. (3.11) captures how the complexity of the ground-truth function affects the generalization error. Similarly, the norm of $g(\cdot)$ also captures the impact[4] of the complexity of the ground-truth function in Eq. (3.10). However, we caution that the GD solution in [29] is based on the original neural network, which is usually different from our min ℓ_2-norm solution based on the NTK model (even though they are close for very large p). Thus, the two results may not be directly comparable.

Theorem 1 reveals several important insights on the generalization performance when the ground-truth function belongs to $\mathcal{F}^{\ell_2}$.

(i) **Descent in the overparameterized region:** When p increases, both sides of Eq. (3.10) decreases, which suggests that the test error of the overfitted NTK model decreases with p. In Fig. 3.1a, we choose a ground-truth function in $\mathcal{F}^{\ell_2}$ (we will explain why this function is in $\mathcal{F}^{\ell_2}$ later in Sect. 4). The test MSE of the aforementioned NTK model (green curve) confirms the overall trend[5] of descent in the overparameterized region. We note that while [29] provides a generalization error upper-bound for large p (i.e., Eq. (3.11)), the

---

[4] Although Term 1 in Eq. (3.10) in its current form does not depend on $g(\cdot)$, it is possible to modify our proof so that the norm of $g(\cdot)$ also enters Term 1.

upper bound there does not capture the dependency in p and thus does not predict this descent.

More importantly, we note a significant difference between the descent in Theorem 1 and that of min ℓ_2-norm overfitting solutions for linear models with simple features [15–19, 35, 36]. For example, for linear models with Gaussian features, we can obtain (see, e.g., Theorem 2 of [16]):

$$\text{MSE} = \|f\|_2^2 \left(1 - \frac{n}{p}\right) + \frac{\sigma^2 n}{p - n - 1}, \text{ for } p \geq n + 2 \quad (3.12)$$

where σ^2 denotes the variance of the noise. We illustrate this descent curve (in orange) in Fig. 3.3, and contrast it with the descent curve of min-ℓ_2 norm solution for NTK (in green). Note that for $p \geq n + 2$, Eq. (3.12) first descends to a small value at around $p = (n-1)/\left(1 - \frac{\sigma}{\|f\|_2}\right)$. It then quickly ascends again as p further increases (see Fig. 3.3). If we let $p \to \infty$ in Eq. (3.12), we can see that the MSE approaches $\|f\|_2^2$, which is referred to as the "null risk" [18], i.e., the MSE of a model that predicts zero. Note that the null-risk is at the level of the signal, and thus is quite large (see Fig. 3.3 again). In contrast, as $p \to \infty$, the test error of the NTK model converges to a value determined by n and ϵ (and is independent of the null risk). This difference is also confirmed in Fig. 3.1a, where the test MSE for the NTK model (green curve) is much lower than the null risk (the dashed line) when $p \to \infty$, while both the min ℓ_2-norm (the red curve) and the min ℓ_1-norm solutions (the blue curve) [20]

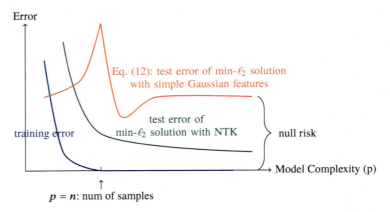

Fig. 3.3 Descent in the overparameterized region of the min ℓ_2-norm solution with NTK (compared with min ℓ_2-norm solutions of linear models with simple Gaussian features

---

[5] This curve oscillates at the early stage when p is small. We suspect it is because, at small p, the convergence in Eq. (3.7) has not occurred yet, and thus the randomness in $\mathbf{V}_0[j]$ makes the simulation results more volatile.

with Fourier features rise to the null risk when $p \to \infty$. Finally, note that the descent in Theorem 1 requires p to increase much faster than n. Specifically, to keep Term 2 in Eq. (3.10) small, it suffices to let p increase a little bit faster than $\Omega(n^{4d-1})$. This is again quite different from the descent shown in Eq. (3.12) and in other related work using Fourier and Gaussian features [35, 36], where p only needs to grow proportionally with n.

(ii) **Speed of the descent:** Since Theorem 1 holds for finite p, it also characterizes the speed of descent. In particular, Term 2 is proportional to $p^{-\frac{1}{2}\left(1-\frac{1}{q}\right)}$, which approaches $1/\sqrt{p}$ when q is large. Again, such a speed of descent is not captured in [29]. As we show in Fig. 3.1a, the test error of the gradient descent solution under the original neural network (cyan curve) is usually quite close to that of the NTK model (green curve). Therefore, our result in Theorem 1 provides useful guidance on how fast the generalization error descends with p for such neural networks.

(iii) **The effect of noise:** Term 3 in Eq. (3.10) characterizes the impact of the noise ϵ, which does not decrease or increase with p. Notice that this is again very different from Eq. (3.12), i.e., results of min ℓ_2-norm overfitting solutions for simple features, where the noise term $\frac{\sigma^2 n}{p-n-1} \to 0$ when $p \to \infty$. We use Fig. 3.4 to validate this insight. In Fig. 3.4, we fix $n = 60$ and plot curves of test MSE of NTK overfitting solution as p increases. We let the noise ϵ_i in the i-th training sample be $i.i.d.$ Gaussian with zero mean and variance σ^2. The green, red, and blue curves in Fig. 3.4 corresponds to the situation $\sigma^2 = 0$, $\sigma^2 = 0.09$, and $\sigma^2 = 0.25$, respectively. We can see that all three curves become flat when

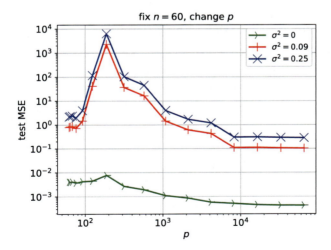

Fig. 3.4 The test MSE of the overfitted NTK model for the same ground-truth function as Fig. 3.1a. We fix $n = 60$ and increase p for different noise level σ^2. All data points in this figure are the average of 6 random simulation runs

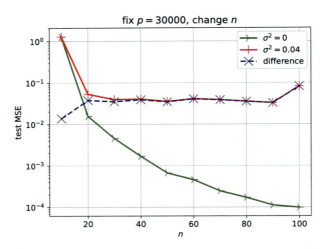

Fig. 3.5 The test MSE of the overfitted NTK model for the same ground-truth function as Fig. 3.1a. We fix $p = 30{,}000$ and increase n. All data points in this figure are the average of 6 random simulation runs

p is very large, and this phenomenon implies that the gap across different noise levels does not decrease when $p \to \infty$, which is in contrast to Eq. (3.12).

In Fig. 3.5, we instead fix $p = 30{,}000$, and increase n. We plot the test MSE both for the noiseless setting (green curve) and for $\sigma^2 = 0.04$ (red curve). The difference between the two curves (dashed blue curve) then captures the impact of noise, which is related to Term 3 in Eq. (3.10). Somewhat surprisingly, we find that the dashed blue curve is insensitive to n, which suggests that Term 3 in Eq. (3.10) may have room for improvement.

In summary, we have shown that any ground-truth function in $\mathcal{F}^{\ell_2}$ leads to low generalization error for overfitted NTK models when there are sufficiently many training samples. It is then natural to ask what happens if the ground-truth function is not in $\mathcal{F}^{\ell_2}$. Let $\overline{\mathcal{F}^{\ell_2}}$ denote the closure[6] of $\mathcal{F}^{\ell_2}$, and $D(f, \overline{\mathcal{F}^{\ell_2}})$ denotes the L^2-distance between f and $\overline{\mathcal{F}^{\ell_2}}$ (i.e., the infimum of the L^2-distance from f to every function in $\overline{\mathcal{F}^{\ell_2}}$).

Proposition 1

(i) *For any given* $(\mathbf{X}, \mathbf{y})$, *there exists a function* $\hat{f}_\infty^{\ell_2} \in \overline{\mathcal{F}^{\ell_2}}$ *such that, uniformly over all* $x \in \mathcal{S}^{d-1}$, $\hat{f}^{\ell_2}(x) \xrightarrow{P} \hat{f}_\infty^{\ell_2}(x)$ *as* $p \to \infty$.

---

[6] We consider the normed space of all functions in $L^2(\mathcal{S}^{d-1} \mapsto \mathbb{R})$. Notice that although $g(z)$ in Definition 1 may not be in L^2, f_g is always in L^2. Specifically, $f_g(x)$ is bounded for every $x \in \mathcal{S}^{d-1}$ when $\|g\|_1 < \infty$.

(ii) Consequently, if the ground-truth function $f \notin \overline{\mathcal{F}^{\ell_2}}$ (or equivalently, $D(f, \mathcal{F}^{\ell_2}) > 0$), then the MSE of $\hat{f}_\infty^{\ell_2}$ (with respect to the ground-truth function f) is at least $D(f, \mathcal{F}^{\ell_2})$.

Intuitively, Proposition 1 (proven in Supplementary Material Appendix J of [33]) suggests that, if a ground-truth function is outside the closure of $\mathcal{F}^{\ell_2}$, then no matter how large n is, the test error of a NTK model with infinitely many neurons cannot be small (regardless whether or not the training samples contain noise). We use Fig. 3.1b to validate this. We choose a ground-truth function outside $\overline{\mathcal{F}^{\ell_2}}$. The test MSE of NTK overfitting solutions (green curve) is above null risk (dashed black line) and thus is much higher compared with Fig. 3.1a. We also plot the test MSE of the GD solution of the real neural network (cyan curve), which seems to show the same trend.

Comparing Theorem 1 and Proposition 1, we can conclude that, all functions in $\mathcal{F}^{\ell_2}$ are learnable by the overfitted NTK model, and all functions not in $\overline{\mathcal{F}^{\ell_2}}$ are not.

4 What Exactly Are the Functions in $\mathcal{F}^{\ell_2}$?

Notice that the expression for learnable functions in Definition 1 is still in an indirect form, i.e., through the unknown function $g(\cdot)$. In [29], the authors show that all functions of the form $(x^T a)^l$, $l \in \{0, 1, 2, 4, 6, \cdots\}$ are learnable by GD (assuming large p and small step size), for a similar 2-layer network with ReLU activation that has no bias. In the following, we will show that our learnable functions in Definition 1 also have a similar form. Further, we can show that any functions of the form $(x^T a)^l$, $l \in \{3, 5, 7, \cdots\}$ are not learnable. Our characterization uses an interesting connection to harmonics and filtering on $\mathcal{S}^{d-1}$, which may be of independent interest.

Towards this end, we first note that the integral form in Definition 1 can be viewed as a convolution on $\mathcal{S}^{d-1}$ (denoted by $\circledast$). Specifically, for any $f_g \in \mathcal{F}^{\ell_2}$, we can rewrite it as

$$f_g(x) = g \circledast h(x) := \int_{\text{SO}(d)} g(\mathbf{S}e)h(\mathbf{S}^{-1}x)d\mathbf{S}, \tag{3.13}$$

$$h(x) := x^T e \frac{\pi - \arccos(x^T e)}{2\pi}, \tag{3.14}$$

where $e := [0\ 0\ \cdots\ 0\ 1]^T \in \mathbb{R}^d$, and $\mathbf{S}$ is a $d \times d$ orthogonal matrix that denotes a rotation in $\mathcal{S}^{d-1}$, chosen from the set $\text{SO}(d)$ of all rotations. An important property of the convolution Eq. (3.13) is that it corresponds to multiplication in the frequency domain, similar to Fourier coefficients. To define such a transformation to the frequency domain, we use a set of hyper-spherical harmonics $\Xi_{\mathbf{K}}^l$ [37, 38] when $d \geq 3$, which forms an orthonormal basis for functions on $\mathcal{S}^{d-1}$. These harmonics are indexed by l and $\mathbf{K}$, where $\mathbf{K} = (k_1, k_2, \cdots, k_{d-2})$ and $l = k_0 \geq k_1 \geq k_2 \geq \cdots \geq k_{d-2} \geq 0$ (those k_i's and l are all non-negative integers). Any function $f \in L^2(\mathcal{S}^{d-1} \mapsto \mathbb{R})$ (including even δ-functions [39]) can be decomposed uniquely into these harmonics, i.e., $f(\mathbf{x}) = \sum_l \sum_{\mathbf{K}} c_f(l, \mathbf{K}) \Xi_{\mathbf{K}}^l(\mathbf{x})$, where $c_f(\cdot, \cdot)$ are projections of f onto the basis function. In Eq. (3.13), let $c_g(\cdot, \cdot)$ and $c_h(\cdot, \cdot)$ denote the coefficients corresponding to the decompositions of g and h, respectively. Then, we must have [38]

$$c_{f_g}(l, \mathbf{K}) = \Lambda \cdot c_g(l, \mathbf{K}) c_h(l, \mathbf{0}), \tag{3.15}$$

where Λ is some normalization constant. Notice that in Eq. (3.15), the coefficient for h is $c_h(l, \mathbf{0})$ instead of $c_h(l, \mathbf{K})$. This is because of the intrinsic rotational symmetry of such convolution [38].

The above decomposition has an interesting "filtering" interpretation as follows. Specifically, we can regard the function h as a "filter" or "channel," while the function g as a transmitted "signal." Then, the function f_g in Eqs. (3.13) and (3.15) can be regarded as the received signal after the original signal g goes through the channel/filter h. Therefore, when coefficient $c_h(l, \mathbf{0})$ of h is non-zero, then the corresponding coefficient $c_{f_g}(l, \mathbf{K})$ for f_g can be any value (because we can arbitrarily choose g). In contrast, if a coefficient $c_h(l, \mathbf{0})$ of h is zero, then the corresponding coefficient $c_{f_g}(l, \mathbf{K})$ for f_g must also be zero for all $\mathbf{K}$.

Ideally, if the channel/filter h contains all "frequencies," i.e., all coefficients $c_h(l, \mathbf{0})$ are non-zero, then f_g can also contain all "frequencies," which means that $\mathcal{F}^{\ell_2}$ can contain almost all functions. Unfortunately, this is not true for the function h given in Eq. (3.14). Specifically, using the harmonics defined in [38], the basis $\Xi_{\mathbf{0}}^l$ for $(l, \mathbf{0})$ turns out to have the form

$$\Xi_{\mathbf{0}}^l(\mathbf{x}) = \sum_{k=0}^{\lfloor \frac{l}{2} \rfloor} (-1)^k \cdot a_{l,k} \cdot (\mathbf{x}^T \mathbf{e})^{l-2k}, \tag{3.16}$$

where $a_{l,k}$ are positive constants. Note that the expression Eq. (3.16) contains either only even powers of $\mathbf{x}^T \mathbf{e}$ (if l is even) or odd powers of $\mathbf{x}^T \mathbf{e}$ (if l is odd). Then, for the function h in Eq. (3.14), we have the following proposition (proven in Supplementary Material, Appendix K.4 of [33]). We note that [40] has a similar harmonics analysis, where the expression of $c_h(l, \mathbf{0})$ is given. However, it is not obvious that the expression of $c_h(l, \mathbf{0})$ for all $l = 0, 1, 2, 4, 6, \cdots$ given in [40] must be non-zero, which is made clear by Proposition 2 as follows.

Proposition 2 $c_h(l, \mathbf{0})$ *is zero for* $l = 3, 5, 7, \cdots$ *and is non-zero for* $l = 0, 1, 2, 4, 6, \cdots$.

We are now ready to characterize what functions are in $\mathcal{F}^{\ell_2}$. By the form of Eq. (3.16), for any non-negative integer k, any even power $(\mathbf{x}^T \mathbf{e})^{2k}$ is a linear combination of $\Xi_{\mathbf{0}}^0, \Xi_{\mathbf{0}}^2, \cdots, \Xi_{\mathbf{0}}^{2k}$, and any odd power $(\mathbf{x}^T \mathbf{e})^{2k+1}$ is a linear combination of $\Xi_0^1, \Xi_0^3, \cdots, \Xi_0^{2k+1}$. By Proposition 2, we thus conclude that any function $f_g(x) = (x^T e)^l$ where $l \in \{0, 1, 2, 4, 6, \cdots\}$ can be written in the form of Eq. (3.15) in the frequency domain, which implies that these functions are in $\mathcal{F}^{\ell_2}$. In contrast, any function $f(x) = (x^T e)^l$ where $l \in \{3, 5, 7, \cdots\}$ cannot be written in the form of Eq. (3.15), and are thus not in $\mathcal{F}^{\ell_2}$. Further, the ℓ_2-norm of any latter function will also be equal to its distance to $\mathcal{F}^\infty$. Therefore, the generalization-error lower-bound in Proposition 1 will apply (with $D(f, \mathcal{F}^{\ell_2}) = \|f\|_2$). Finally, by Eq. (3.13), $\mathcal{F}^{\ell_2}$ is invariant under rotation and finite linear summation. Therefore, any finite sum of $(x^T a)^l, l = 0, 1, 2, 4, 6, \cdots$ must also belong to $\mathcal{F}^{\ell_2}$.

Special Case of $d = 2$: The above analysis becomes much simpler for the special case of $d = 2$, where the input x corresponds to an angle $\theta \in [-\pi, \pi]$. The above-mentioned harmonics then become the Fourier series $\sin(k\theta)$ and $\cos(k\theta)$, $k = 0, 1, \cdots$. We can then explicitly derive the coefficients for the harmonics, which confirms that frequencies of $k \in \{0, 1, 2, 4, 6, \cdots\}$ are learnable (while others are not). These learnable and not-learnable functions are then used in Fig. 3.1. Details can be found in Sect. 4.1.

Remark 2 We caution that the above claim on non-learnable functions critically depends on the network architecture. That is, we assume throughout this chapter that the ReLU activation has no bias. From an expressiveness point of view, it is known that, using ReLU without bias, a shallow network can only approximate the sum of linear functions and even functions [26]. Thus, it is not surprising that other odd-power (but non-linear) polynomials cannot be learned. In contrast, by adding a bias, a shallow network using ReLU becomes a universal approximator [28]. The recent work of [27] shows that polynomials with all powers can be learned by the corresponding 2-layer NTK model. These results are consistent with ours because a ReLU activation function operating on $\tilde{x} \in \mathbb{R}^{d-1}$ with a bias can be equivalently viewed as one operating on a d-dimension input (with the last-dimension being fixed at $1/\sqrt{d}$) but with no bias. Even though only a subset of functions are learnable in the d-dimension space, when projected into a $(d - 1)$-dimension subspace, they may already span all functions. For example, one could write $(x^T a)^3$ as a linear combination of $\left(\begin{bmatrix} \tilde{x} \\ 1/\sqrt{d} \end{bmatrix}^T b_i\right)^{l_i}$, where $i \in \{1, 2, \cdots, 5\}$, $[l_1, \cdots, l_5] = [4, 4, 2, 1, 0]$, and $b_i \in \mathbb{R}^d$ depends only on a. (See Supplementary Material, Appendix K.6 in [33] for details.) It remains an interesting question whether similar difference arises for other network architectures (e.g., with more than 2 layers).

4.1 A Special Case: When $d = 2$

When $d = 2$, $\mathcal{S}^{d-1}$ denotes a unit circle. Therefore, every x corresponds to an angle $\varphi \in [-\pi, \pi]$ such that $x = [\cos\varphi \ \sin\varphi]^T$. In this situation, the hyper-spherical harmonics are the well-known Fourier series, i.e.,

1, $\cos(\theta)$, $\sin(\theta)$, $\cos(2\theta)$, $\sin(2\theta)$, $\cdots$. Thus, we can explicitly calculate all Fourier coefficients of h more easily.

Similar to Eq. (3.13), we first write down the convolution for $d = 2$, which is also in a simpler form. For any function $f_g \in \mathcal{F}^{\ell_2}$, we have

$$
\begin{aligned}
f_g(\varphi) &= \frac{1}{2\pi} \int_{\varphi-\pi}^{\varphi+\pi} \frac{\pi - |\theta - \varphi|}{2\pi} \cos(\theta - \varphi) g(\theta) d\theta \\
&= \frac{1}{2\pi} \int_{-\pi}^{\pi} \frac{\pi - |\theta|}{2\pi} \cos\theta \, g(\theta + \varphi) \, d\theta \ \text{(replace } \theta \text{ by } \theta - \varphi) \\
&= \frac{1}{2\pi} \int_{-\pi}^{\pi} \frac{\pi - |\theta|}{2\pi} \cos\theta \, g(\varphi - \theta) \, d\theta \ \text{(replace } \theta \text{ by } -\theta).
\end{aligned}
$$

Define $h(\theta) := \frac{\pi - |\theta|}{2\pi} \cos\theta$. We then have

$$
f_g(\varphi) = \frac{1}{2\pi} h(\varphi) \circledast g(\varphi),
$$

where $\circledast$ denotes (continuous) circular convolution. Let $c_{f_g}(k)$, $c_h(k)$ and $c_g(k)$ (where $k = \cdots, -1, 0, 1, \cdots$) denote the (complex) Fourier series coefficients for $f_g(\varphi)$, $h(\varphi)$, and $g(\varphi)$, correspondingly. Specifically, we have

$$
f_g(\varphi) = \sum_{k=-\infty}^{\infty} c_{f_g}(k) e^{ik\varphi}, \quad h(\varphi) = \sum_{k=-\infty}^{\infty} c_h(k) e^{ik\varphi}, \quad g(\varphi) = \sum_{k=-\infty}^{\infty} c_g(k) e^{ik\varphi}.
$$

Thus, we have

$$
c_{f_g}(k) = c_h(k) c_g(k). \tag{3.17}
$$

Next, we calculate $c_h(k)$, i.e., the Fourier decomposition of $h(\cdot)$. We have

$$
\begin{aligned}
c_h(k) &= \frac{1}{2\pi} \int_{-\pi}^{\pi} \frac{\pi - |\theta|}{2\pi} \cos\theta \, e^{-ik\theta} d\theta \\
&= \frac{1}{4\pi} \int_{-\pi}^{\pi} \left(1 - \frac{|\theta|}{\pi}\right) \frac{e^{-i(k+1)\theta} + e^{-i(k-1)\theta}}{2} d\theta \\
&= -\frac{1}{8\pi^2} \int_{-\pi}^{\pi} |\theta| \left(e^{-i(k+1)\theta} + e^{-i(k-1)\theta}\right) d\theta \\
&\quad + \frac{1}{8\pi} \int_{-\pi}^{\pi} \left(e^{-i(k+1)\theta} + e^{-i(k-1)\theta}\right) d\theta.
\end{aligned}
$$

We first consider $k = 1$ or -1. It is easy to verify that

3 On the Generalization Power of Overfitted Two-Layer Neural Tangent. . . 129

$$\int xe^{cx}\,dx = e^{cx}\left(\frac{cx-1}{c^2}\right), \quad \forall c \neq 0.$$

Thus, we have

$$
\begin{aligned}
c_h(1) &= -\frac{1}{8\pi^2}\int_{-\pi}^{\pi} |\theta|\left(e^{-i2\theta}+1\right)d\theta + \frac{1}{4} \\
&= -\frac{1}{8\pi^2}\left(\pi^2 - \int_{-\pi}^{0}\theta e^{-i2\theta}\,d\theta + \int_{0}^{\pi}\theta e^{-i2\theta}\,d\theta\right) + \frac{1}{4} \\
&= -\frac{1}{8\pi^2}\left(\pi^2 + \frac{i2\pi}{-4} + \frac{-i2\pi}{-4}\right) + \frac{1}{4} \\
&= -\frac{1}{8} + \frac{1}{4} \\
&= \frac{1}{8}.
\end{aligned}
$$

Similarly, we have

$$c_h(-1) = \frac{1}{8}.$$

Now we consider the situation of $k \neq \pm 1$. We have

$$
\begin{aligned}
\int_{-\pi}^{0} |\theta|e^{-i(k+1)\theta}\,d\theta &= -e^{-i(k+1)\theta}\cdot\frac{-i(k+1)\theta-1}{-(k+1)^2}\bigg|_{-\pi}^{0} \\
&= -\frac{1}{(k+1)^2} + \frac{1-i(k+1)\pi}{(k+1)^2}e^{i(k+1)\pi}, \\
\int_{0}^{\pi} |\theta|e^{-i(k+1)\theta}\,d\theta &= e^{-i(k+1)\theta}\cdot\frac{-i(k+1)\theta-1}{-(k+1)^2}\bigg|_{0}^{\pi} \\
&= -\frac{1}{(k+1)^2} + \frac{1+i(k+1)\pi}{(k+1)^2}e^{-i(k+1)\pi}.
\end{aligned}
$$

Notice that $e^{-i(k+1)\pi} = e^{-i(k+1)2\pi}e^{i(k+1)\pi} = e^{i(k+1)\pi}$. Therefore, we have

$$\int_{-\pi}^{\pi} |\theta|e^{-i(k+1)\theta}\,d\theta = \frac{2}{(k+1)^2}\left(e^{i(k+1)\pi}-1\right).$$

Similarly, we have

$$\int_{-\pi}^{\pi} |\theta|e^{-i(k-1)\theta}\,d\theta = \frac{2}{(k-1)^2}\left(e^{i(k-1)\pi}-1\right).$$

In summary, we have

$$
c_h(k) = \begin{cases} \frac{1}{8}, & k = \pm 1 \\ -\frac{1}{4\pi^2}\left(\frac{1}{(k+1)^2} + \frac{1}{(k-1)^2}\right)\left(e^{i(k+1)\pi} - 1\right), & \text{otherwise} \end{cases}
$$

$$
= \begin{cases} \frac{1}{8}, & k = \pm 1 \\ \frac{1}{2\pi^2}\left(\frac{1}{(k+1)^2} + \frac{1}{(k-1)^2}\right), & k = 0, \pm 2, \pm 4, \cdots \\ 0, & k = \pm 3, \pm 5, \cdots \end{cases}.
$$

By Eq. (3.17), we thus have

$$
c_{f_g}(k) = \begin{cases} \frac{1}{8}c_g(k), & k = \pm 1 \\ \frac{1}{2\pi^2}\left(\frac{1}{(k+1)^2} + \frac{1}{(k-1)^2}\right)c_g(k), & k = 0, \pm 2, \pm 4, \cdots \\ 0, & k = \pm 3, \pm 5, \cdots \end{cases}.
$$

In other words, when $d = 2$, functions in $\mathcal{F}^{\ell_2}$ can only contain frequencies $0, \theta, 2\theta, 4\theta, 6\theta, \cdots$, and cannot contain other frequencies $3\theta, 5\theta, 7\theta, \cdots$.

5 Proof Sketch of Theorem 1

In this section, we illustrate some key steps to prove Theorem 1. Starting from Eq. (3.3), we have

$$
\Delta\mathbf{V}^{\ell_2} = \mathbf{H}^T(\mathbf{H}\mathbf{H}^T)^{-1}(\mathbf{F}(\mathbf{X}) + \boldsymbol{\epsilon}). \tag{3.18}
$$

For the learned model $\hat{f}^{\ell_2}(\mathbf{x}) = \mathbf{h}_{\mathbf{V}_0, \mathbf{x}}\Delta\mathbf{V}^{\ell_2}$ given in Eq. (3.4), the error for any test input $\mathbf{x}$ is then

$$
\hat{f}^{\ell_2}(\mathbf{x}) - f(\mathbf{x}) = \left(\mathbf{h}_{\mathbf{V}_0, \mathbf{x}}\mathbf{H}^T(\mathbf{H}\mathbf{H}^T)^{-1}\mathbf{F}(\mathbf{X}) - f(\mathbf{x})\right)
$$
$$
+ \mathbf{h}_{\mathbf{V}_0, \mathbf{x}}\mathbf{H}^T(\mathbf{H}\mathbf{H}^T)^{-1}\boldsymbol{\epsilon}. \tag{3.19}
$$

In the classical "bias-variance" analysis with respect to MSE [41], the first term on the right-hand-side of Eq. (3.19) contributes to the bias and the second term contributes to the variance. We first quantify the second term (i.e., the variance) in the following proposition.

Proposition 3 *For any* $n \geq 2$, $m \in \left[1, \frac{\ln n}{\ln \frac{\pi}{2}}\right]$, $d \leq n^4$, *if* $p \geq 6J_m(n,d)\ln\left(4n^{1+\frac{1}{m}}\right)$, *we must have* $\Pr\limits_{\mathbf{X},\mathbf{V}_0}\left\{|\mathbf{h}_{\mathbf{v}_0,x}\mathbf{H}^T(\mathbf{H}\mathbf{H}^T)^{-1}\boldsymbol{\epsilon}| \leq \sqrt{J_m(n,d)n}\right.$ $\|\boldsymbol{\epsilon}\|_2$, *for all* $\boldsymbol{\epsilon} \in \mathbb{R}^n\right\} \geq 1 - \frac{2}{\sqrt[m]{n}}$.

The proof is in Supplementary Material Appendix F in [33]. Proposition 3 implies that, for fixed n and d, when $p \to \infty$, with high probability the variance will not exceed a certain factor of the noise $\|\boldsymbol{\epsilon}\|_2$. In other words, the variance will not go to infinity when $p \to \infty$. The main step in the proof is to lower bound $\min \text{eig}\left(\mathbf{H}\mathbf{H}^T\right)/p$, which is given by $1/(J_m(n,d)n)$. Note that this is the main place where we used the assumption that x is uniformly distributed. We expect that our main proof techniques can be generalized to other distributions (with a different expression of $J_m(n,d)$), which we leave for future work.

Remark 3 In the upper bound in [29] (i.e., Eq. (3.11)), any noise added to y will at least contribute to the generalization upper bound Eq. (3.11) by a positive term $\boldsymbol{\epsilon}^T(\mathbf{H}^\infty)^{-1}\boldsymbol{\epsilon}/n$. This implies that their upper bound may also grow as $\min \text{eig}(\mathbf{H}^\infty)$ decreases. One of the contribution of Proposition 3 is to characterize this minimum eigenvalue.

It remains to bound the bias part. We first study the class of ground-truth functions that can be learned with fixed $\mathbf{V}_0$. We refer to them as *pseudo ground-truth*, to differentiate them with the set $\mathcal{F}^{\ell_2}$ of learnable functions for random $\mathbf{V}_0$. They are defined with respect to the same $g(\cdot)$ function, so that we can later extend to the "real" ground-truth functions in $\mathcal{F}^{\ell_2}$ when considering the randomness of $\mathbf{V}_0$.

Definition 2 Given $\mathbf{V}_0$, for any learnable ground-truth function $f_g \in \mathcal{F}^{\ell_2}$ with the corresponding function $g(\cdot)$, define the corresponding **pseudo ground-truth** as

$$f_{\mathbf{V}_0}^g(x) := \int_{\mathcal{S}^{d-1}} x^T z \frac{|C_{z,x}^{\mathbf{V}_0}|}{p} g(z)d\mu(z).$$

The reason that this class of functions may be the learnable functions for fixed $\mathbf{V}_0$ is similar to the discussions in Eqs. (3.5) and (3.6). Indeed, using the same choice of $g(z)$ in Eq. (3.8), the learned function $\hat{f}^{\ell_2}$ in Eq. (3.5) at fixed $\mathbf{V}_0$ is always of the form in Definition 2.

The following proposition gives an upper bound of the generalization performance when the data model is based on the pseudo ground-truth and the NTK model uses exactly the same $\mathbf{V}_0$. In other words, we temporarily remove the randomness of $\mathbf{V}_0$.

Proposition 4 *Assume fixed* $\mathbf{V}_0$ *(thus p and d are also fixed), there is no noise. If the ground-truth function is* $f = f_{\mathbf{V}_0}^g$ *in Definition 2 and* $\|g\|_\infty < \infty$, *then for any* $x \in \mathcal{S}^{d-1}$ *and* $q \in [1, \infty)$, *we have* $\Pr_{\mathbf{X}}\left\{|\hat{f}^{\ell_2}(x) - f(x)| \leq n^{-\frac{1}{2}\left(1-\frac{1}{q}\right)}\right\} \geq 1 - 2e^2\exp\left(-\frac{\sqrt[q]{n}}{8\|g\|_\infty^2}\right)$.

Fig. 3.6 Geometric interpretation for the proof idea of Proposition 4

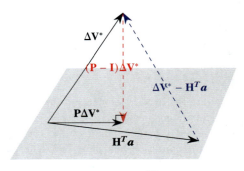

The proof is in Supplementary Material, Appendix H of [33]. Note that both the threshold of the probability event and the upper bound coincide with Term 1 and Term 4, respectively, in Eq. (3.10). Here we sketch the proof of Proposition 4. Based on the definition of the pseudo ground-truth, we can rewrite $f_{\mathbf{V}_0}^g$ as $f_{\mathbf{V}_0}^g(\mathbf{x}) = \mathbf{h}_{\mathbf{V}_0,\mathbf{x}}\Delta\mathbf{V}^*$, where $\Delta\mathbf{V}^* \in \mathbb{R}^{dp}$ is given by, for all $j \in \{1, 2, \cdots, p\}$, $\Delta\mathbf{V}^*[j] = \int_{\mathcal{S}^{d-1}} \mathbf{1}_{\{z^T\mathbf{V}_0[j]>0\}} z \frac{g(z)}{p} d\mu(z)$. From Eqs. (3.3) and (3.4), we can see that the learned model is $\hat{f}^{\ell_2}(\mathbf{x}) = \mathbf{h}_{\mathbf{V}_0,\mathbf{x}}\mathbf{P}\Delta\mathbf{V}^*$ where $\mathbf{P} := \mathbf{H}^T(\mathbf{H}\mathbf{H}^T)^{-1}\mathbf{H}$. Note that $\mathbf{P}$ is an orthogonal projection to the row-space of $\mathbf{H}$. Further, it is easy to show that $\|\mathbf{h}_{\mathbf{V}_0,\mathbf{x}}\|_2 \leq \sqrt{p}$. Thus, we have $|\hat{f}^{\ell_2}(\mathbf{x}) - f_{\mathbf{V}_0}^g(\mathbf{x})| = |\mathbf{h}_{\mathbf{V}_0,\mathbf{x}}(\mathbf{P}-\mathbf{I})\Delta\mathbf{V}^*| \leq \sqrt{p}\|(\mathbf{P}-\mathbf{I})\Delta\mathbf{V}^*\|_2$. The term $(\mathbf{P}-\mathbf{I})\Delta\mathbf{V}^*$ can be interpreted as the distance from $\Delta\mathbf{V}^*$ to the row-space of $\mathbf{H}$. (See Fig. 3.6 for an illustration of this distance to the projection of $\Delta\mathbf{V}^*$.) Note that this distance is no greater than the distance between $\Delta\mathbf{V}^*$ and any point in the row-space of $\mathbf{H}$. Thus, in order to get an upper bound on $\|(\mathbf{P}-\mathbf{I})\Delta\mathbf{V}^*\|_2$, we only need to find a vector $\mathbf{a} \in \mathbb{R}^n$ that makes $\|\Delta\mathbf{V}^* - \mathbf{H}^T\mathbf{a}\|_2$ as small as possible, especially when n is large. Our proof uses the vector $\mathbf{a}$ such that its i-th element is $\mathbf{a}_i := \frac{g(\mathbf{X}_i)}{np}$. See Supplementary Material, Appendix H in [33] for the rest of the details.

After Proposition 4, the final step to derive Theorem 1 is to allow $\mathbf{V}_0$ to be random. Given any random $\mathbf{V}_0$, any function $f_g \in \mathcal{F}^{\ell_2}$ can be viewed as the summation of a pseudo ground-truth function (with the same $g(\cdot)$) and a difference term. This difference can be viewed as a special form of "noise", and thus we can use Proposition 3 to quantify its impact. Further, the magnitude of this "noise" should decrease with p (because of Eq. (3.7)). Combining this argument with Proposition 4, we can then prove Theorem 1. See Supplementary Material, Appendix I in [33] for details.

6 Conclusions

In this chapter, we studied the generalization performance of the min ℓ_2-norm overfitting solution for a two-layer NTK model that uses ReLU without bias. We precisely characterize the learnable ground-truth functions for such models, by providing a generalization upper bound for all functions in $\mathcal{F}^{\ell_2}$, and a generalization lower bound for all functions not in $\overline{\mathcal{F}^{\ell_2}}$. We show that, while the test error of the overfitted NTK model also exhibits descent in the overparameterized regime, the descent behavior can be quite different from the double descent of linear models with simple features such as Gaussian or Fourier features.

For possible future work, there are several interesting directions. First, based on Fig. 3.5, our estimation of the effect of noise could be further improved. Second, it would be interesting to explore whether the methodology can be extended to NTK model for other neural networks, e.g., with different activation functions and with more than two layers.

Acknowledgments This work is partially supported by an NSF sub-award via Duke University (IIS-1932630), by NSF grants 2112471, CNS-2106932, CNS-2106933, CNS-1717493, CNS-1901057, and CNS-2007231, and by Office of Naval Research under Grant N00014-17-1-241. The authors would like to thank Professor R. Srikant at the University of Illinois at Urbana-Champaign for his insightful comments on our results.

References

1. J. Wang, H. Liang, G. Joshi, Overlap local-sgd: An algorithmic approach to hide communication delays in distributed sgd, in *ICASSP 2020-2020 IEEE International Conference on Acoustics, Speech and Signal Processing (ICASSP)* (IEEE, 2020), pp. 8871–8875
2. J. Wang, Q. Liu, H. Liang, G. Joshi, H.V. Poor, Tackling the objective inconsistency problem in heterogeneous federated optimization. Adv. Neural Inf. Process. Syst. **33**, 7611–7623 (2020)
3. K. Lee, M. Lam, R. Pedarsani, D. Papailiopoulos, K. Ramchandran, Speeding up distributed machine learning using codes. IEEE Trans. Inf. Theory **64**(3), 1514–1529 (2017)
4. R. Tandon, Q. Lei, A.G. Dimakis, N. Karampatziakis, Gradient coding: avoiding stragglers in distributed learning, in *International Conference on Machine Learning*. PMLR (2017), pp. 3368–3376
5. G. Li, S. K.S. Hari, M. Sullivan, T. Tsai, K. Pattabiraman, J. Emer, S.W. Keckler, Understanding error propagation in deep learning neural network (dnn) accelerators and applications, in *Proceedings of the International Conference for High Performance Computing, Networking, Storage and Analysis* (2017), pp. 1–12
6. B. Lin, Y. Huang, J. Zhang, J. Hu, X. Chen, J. Li, Cost-driven off-loading for dnn-based applications over cloud, edge, and end devices. IEEE Trans. Indust. Inform. **16**(8), 5456–5466 (2019)
7. C.M. Bishop, *Pattern Recognition and Machine Learning* (Springer, Berlin, 2006)
8. T. Hastie, R. Tibshirani, J. Friedman, *The Elements of Statistical Learning: Data Mining, Inference, and Prediction* (Springer Science & Business Media, New York, 2009)
9. C. Stein, Inadmissibility of the usual estimator for the mean of a multivariate normal distribution. Technical Report, Stanford University Stanford United States, 1956
10. W. James, C. Stein, Estimation with quadratic loss, in *Breakthroughs in Statistics* (Springer, Berlin, 1992), pp. 443–460
11. Y. LeCun, I. Kanter, S.A. Solla, Second order properties of error surfaces: Learning time and generalization, in *Advances in Neural Information Processing Systems* (1991)
12. A.N. Tikhonov, On the stability of inverse problems. Dokl. Akad. Nauk SSSR **39**, 195–198 (1943)

13. C. Zhang, S. Bengio, M. Hardt, B. Recht, O. Vinyals, Understanding deep learning requires rethinking generalization, in *5th International Conference on Learning Representations, ICLR 2017* (2017)
14. M.S. Advani, A.M. Saxe, H. Sompolinsky, High-dimensional dynamics of generalization error in neural networks. Neural Netw. **132**, 428–446 (2020)
15. M. Belkin, S. Ma, S. Mandal, To understand deep learning we need to understand kernel learning, in *International Conference on Machine Learning* (2018), pp. 541–549
16. M. Belkin, D. Hsu, J. Xu, Two models of double descent for weak features. Preprint (2019). arXiv:1903.07571
17. P.L. Bartlett, P.M. Long, G. Lugosi, A. Tsigler, Benign overfitting in linear regression. Proc. Natl. Acad. Sci. (2020)
18. T. Hastie, A. Montanari, S. Rosset, R.J. Tibshirani, Surprises in high-dimensional ridgeless least squares interpolation. Preprint (2019). arXiv:1903.08560
19. V. Muthukumar, K. Vodrahalli, A. Sahai, Harmless interpolation of noisy data in regression, in *2019 IEEE International Symposium on Information Theory (ISIT)* (IEEE, 2019), pp. 2299–2303
20. P. Ju, X. Lin, J. Liu, Overfitting can be harmless for basis pursuit, but only to a degree. Adv. Neural Inf. Process. Syst. **33** (2020)
21. S. Mei, A. Montanari, The generalization error of random features regression: precise asymptotics and double descent curve. Preprint (2019). arXiv:1908.05355
22. A. Jacot, F. Gabriel, C. Hongler, Neural tangent kernel: convergence and generalization in neural networks, in *Advances in Neural Information Processing Systems* (2018), pp. 8571–8580
23. Y. Li, Y. Liang, Learning overparameterized neural networks via stochastic gradient descent on structured data. Adv. Neural Inf. Process. Syst. **31**, 8157–8166 (2018)
24. S.S. Du, X. Zhai, B. Poczos, A. Singh, Gradient descent provably optimizes over-parameterized neural networks, in *International Conference on Learning Representations* (2018)
25. S. d'Ascoli, M. Refinetti, G. Biroli, F. Krzakala, Double trouble in double descent: bias and variance (s) in the lazy regime, in *International Conference on Machine Learning* (PMLR, 2020), pp. 2280–2290
26. B. Ghorbani, S. Mei, T. Misiakiewicz, A. Montanari, Linearized two-layers neural networks in high dimension. Preprint (2019). arXiv:1904.12191
27. S. Satpathi, R. Srikant, The dynamics of gradient descent for overparametrized neural networks, in *3rd Annual Learning for Dynamics and Control Conference (L4DC)* (2021)
28. Z. Ji, M. Telgarsky, R. Xian, Neural tangent kernels, transportation mappings, and universal approximation. Preprint (2019). arXiv:1910.06956
29. S. Arora, S. Du, W. Hu, Z. Li, R. Wang, Fine-grained analysis of optimization and generalization for overparameterized two-layer neural networks, in *International Conference on Machine Learning* (2019), pp. 322–332
30. J. Fiat, E. Malach, S. Shalev-Shwartz, Decoupling gating from linearity. Preprint (2019). arXiv:1906.05032
31. Z. Allen-Zhu, Y. Li, Y. Liang, Learning and generalization in overparameterized neural networks, going beyond two layers, in *Advances in Neural Information Processing Systems* (2019), pp. 6158–6169
32. Z. Ji, M. Telgarsky, Polylogarithmic width suffices for gradient descent to achieve arbitrarily small test error with shallow relu networks. Preprint (2019). arXiv:1909.12292
33. P. Ju, X. Lin, N. Shroff, On the generalization power of overfitted two-layer neural tangent kernel models, in *International Conference on Machine Learning*. PMLR (2021), pp. 5137–5147
34. K.B. Rao, M.B. Rao, *Theory of Charges: A Study of Finitely Additive Measures* (Academic Press, Amsterdam, 1983)
35. Z. Liao, R. Couillet, M.W. Mahoney, A random matrix analysis of random fourier features: beyond the gaussian kernel, a precise phase transition, and the corresponding double descent. Preprint (2020). arXiv:2006.05013

36. A. Jacot, B. Simsek, F. Spadaro, C. Hongler, F. Gabriel, Implicit regularization of random feature models, in *International Conference on Machine Learning*. PMLR (2020), pp. 4631–4640
37. N.Y. Vilenkin, Special functions and the theory of group representations. providence: American mathematical society, in *sftp* (1968)
38. I. Dokmanic, D. Petrinovic, Convolution on the n-sphere with application to pdf modeling. IEEE Trans. Signal Process. **58**(3), 1157–1170 (2009)
39. Y. Li, R. Wong, Integral and series representations of the dirac delta function. Preprint (2013). arXiv:1303.1943
40. R. Basri, D. Jacobs, Y. Kasten, S. Kritchman, The convergence rate of neural networks for learned functions of different frequencies. Preprint (2019). arXiv:1906.00425
41. M. Belkin, D. Hsu, S. Ma, S. Mandal, Reconciling modern machine learning and the bias-variance trade-off. Preprint (2018). arXiv:1812.11118

Chapter 4
Out of Distribution Detection

Wei-Han Lee and Mudhakar Srivatsa

Abstract Deep learning has gained tremendous success in transforming many data mining and machine learning tasks. Popular deep learning techniques are inapplicable to out of distribution detection (OOD) due to some unique characteristics of anomalies. OOD records are rare, heterogeneous, boundless, and prohibitively high costs for collecting large-scale OOD data. OOD records leads to false predictions for AI models. It reduces user confidence in AI products. However, recent studies in OOD detection either assume the existence of OOD records or require the retraining of the model, limiting their application in practice.

In this chapter, we introduce NeuralFP (Neuralfp: out-of-distribution detection using fingerprints of neural networks. In: 2020 25th International Conference on Pattern Recognition (ICPR), 2021) for OOD detection without requiring any access to OOD records by constructing non-linear fingerprints of neural network models. The key idea of NeuralFP is to exploit the different behavior in how the neural network model responds to normal data versus OOD data. Specifically, NeuralFP builds autoencoders for each layer of the neural network model and then analyzes the error distribution of the autocoders in reconstructing the training set to identify OOD records. We show the effectiveness of NeuralFP in detecting OOD records as well as its advantages over previous approaches through comprehensive experiments on multiple real-world datasets. For practical adoption of NeuralFP, we provide useful guidelines for parameter selection.

©2021 IEEE. Reprinted, with permission, from W. Lee, S. Millman, N. Desai, M. Srivatsa and C. Liu, "NeuralFP: Out-of-distribution Detection using Fingerprints of Neural Networks," 2020 25th International Conference on Pattern Recognition (ICPR), 2021.

W.-H. Lee (✉)
IBM Research, Yorktown Heights, NY, USA
e-mail: wei-han.lee1@ibm.com

M. Srivatsa
IBM Thomas J. Watson Research Center, New York, NY, USA
e-mail: msrivats@us.ibm.com

© The Author(s), under exclusive license to Springer Nature Switzerland AG 2023
M. Srivatsa et al. (eds.), *Artificial Intelligence for Edge Computing*,
https://doi.org/10.1007/978-3-031-40787-1_4

1 Introduction

Deep learning models learnt on the cloud serve as useful predictive functions for data at edge devices, which have been broadly applied in various application scenarios such as natural language processing, computer vision, etc. However, the deployed models may not be aware of new situations or new data due to due to data drift or adversarial design even if the prediction outputs are highly confident, which is the well-known out-of-distribution (OOD) problem [1–7]. Thus, the deep learning model needs to identify OOD records (statistically or adversarially) before obtaining meaningful prediction results.

OOD detection problem was first proposed by Hendrycks and Gimpel [1], which uses the predicted softmax sore to distinguish in-distribution (ID) data and OOD data. Following this research direction, later works have been proposed by implementing temperature scaling and input perturbation [3] or applying additional regularizers to the loss function [2]. However, these works either require the retraining of the neural network model or the prior knowledge of OOD records. Lee et al. proposed a novel OOD detection method [4], which explores the internal features of the convolutional neural networks (CNNs) by utilizing the Mahalanobis distance. Specifically, they leverage the principal component analysis (PCA) to remove the correlation between features of the last hidden layer. However, the detection capability might be restricted by the limited usage of the last hidden layer and the adopted linear transformation, which motivates us to consider the following two questions: (1) besides the linear transformation in [4], is it possible to utilize non-linear transformations to extract unique fingerprints of the training dataset in practice? and (2) what is the optimal strategy to integrate information of fingerprints across multiple layers (instead of only using the last hidden layer in [4]) for maximizing detection performance?

To examine the aforementioned questions, we propose a new OOD detection method (NeuralFP) which utilizes autoencoders to construct non-linear fingerprints of the training dataset on the deep learning model [8]. By carefully analyzing the fingerprints at each layer, NeuralFP designs an effective strategy to integrate the fingerprints across all the layers to further increase the detection accuracy. By experiments through multiple real-world datasets, we validate the effectiveness of NeuralFP in detecting OOD data which achieves better performance than the state-of-the-art approaches. In summary, our work has the following contributions:

1. We propose NeuralFP to fingerprint deep learning models, which can be applied to detect OOD data without requiring the knowledge of OOD data nor the retraining of the deep learning model.
2. Specifically, NeuralFP extracts non-linear fingerprints of the deep learning model by constructing autoencoders from the responses of the model applied to the training dataset, build on which NeuralFP further extracts the reconstruction errors of the training dataset for each layer. To achieve enhanced detection performance, NeuralFP proposes an effective integration strategy on the distributions of reconstruction errors across all the layers.

4 Out of Distribution Detection

3. We show the effectiveness of NeuralFP through extensive experiments on multiple real-world datasets including MNIST, Fashion-MNIST, SVHN and CIFAR-10. Furthermore, we demonstrate the superiority of NeuralTran over the state-of-the-art OOD methods and provide useful guidelines of selecting parameters for deploying NeuralFP in practice.

2 Related Work

Existing work of OOD detection using deep learning models can be generally categorized into two classes: 1) Threshold-based detectors [1–4]; 2) Predictive uncertainty estimator [5–7].

Threshold-Based Detectors Hendrycks and Gimpel [1] is the first to propose that the softmax score can be used to distinguish between ID records and OOD records. Lee et al. [2] proposed additional two regularization terms added to the original loss for detecting OOD records. Later on, Liang et al. [3] utilized temperature scaling and input perturbation to further improved the detection performance. However, these works either require the retraining of the deep learning model or assumes the prior knowledge of OOD data. Lee et al. [4] further utilized the Mahalanobis distance of class means (instead of the softmax score) to explore the internal feature maps of the CNNs by using the PCA to de-correlate the features of the last hidden layer.

NeuralFP belongs to this category which constructs model fingerprints to memorize the information of the training dataset on the deep learning models. Different from previous existing works, NeuralFP leverages the non-linear transformation to extract fingerprints of the training dataset and effectively integrate the information of fingerprints across all the layers. In Sect. 4.3, our experiment further demonstrates the advantages of NeuralFP over the state-of-the-art methods.

Predictive Uncertainty Estimator As another direction of research, predictive uncertainty estimation uses the bayesian probabilistic models [5, 6] or ensemble methods [7] to quantify uncertainty of input records that are far from the training dataset. However, these methods are extremely computationally expensive, limiting the practical applications [4]. It is worthy noting that the predictive uncertainty estimators can serve as complementary methods to NeuralFP for achieving better performance.

3 Our Method: NeuralFP

In this section, we will introduce the architecture of NeuralFP. Specifically, we will first show an motivating example and then detail the key components of NeuralFP.

3.1 Problem Statement

We consider a scenario where an edge device aims to detect whether its data are OOD or not by utilizing a deep learning model trained on the cloud. Here, we assume the training dataset on the cloud as $\mathcal{D} = \{(x_i, y_i)\}_{i=1}^{N}$, based on which a L-layer neural network model is trained. The model parameters for the l-th layer are denoted as θ_l and the overall model parameters are thus $\theta = \left[\theta_1^T, \cdots, \theta_L^T\right]^T$.

NeuralFP aims to construct fingerprints of the deep learning model applied to the training dataset, denoted by $\mathcal{FP}(\theta, \mathcal{D})$. For a given data record x^*, the edge device would communicate with the cloud and use $\theta, \mathcal{FP}(\theta, \mathcal{D})$ to detect whether the data is OOD or not.

3.2 Motivating Example

The key intuition of NeuralFP lies in the difference that the deep learning model responds to the ID data and the OOD data, which can be illustrated by the following motivating example. For a 7-layer deep learning model trained from the MNIST training dataset (detailed architecture is shown in Fig. 4.3a), we first feed the MNIST training data through each of the layers to obtain the corresponding activations, based on which we construct an autoencoder for each layer. Then, we pass the MNIST testing data (viewed as ID data) as well as the Fashion-MNIST testing data (viewed as OOD data) to the autoencoders and show the reconstruction error in Fig. 4.1.

Based on Fig. 4.1, we have the following observations: (1) the reconstruction errors are strong signals to distinguish OOD data (Fashion-MNIST testing data) from the ID data (MNIST testing data); (2) the reconstruction errors of OOD data may not always be smaller than that of the ID data (see the error distribution between Layer-6 and Layer-7); (3) different layers have various capability in distinguishing OOD data (compare the error distributions in Layer-1 and Layer-3 as an comparing example). These observations motivate the design of NeuralFP, where we carefully analyze the reconstruction errors across all the layers of the deep learning model to maximize the information that the model fingerprints contain about the original training data.

3.3 Design Details

In the following, we will describe the architecture of NeuralFP (as shown in Fig. 4.2), which is composed of two modes: fingerprinting in the cloud and OOD detection in the edge.

4 Out of Distribution Detection

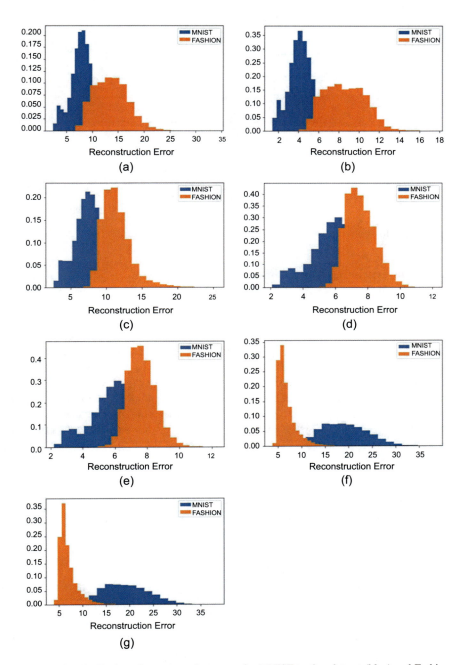

Fig. 4.1 The distribution of reconstruction errors for MNIST testing dataset (blue) and Fashion-MNIST testing dataset (orange) applied to the model fingerprints of MNIST training dataset. (**a**) Layer-1. (**b**) Layer-2. (**c**) Layer-3. (**d**) Layer-4. (**e**) Layer-5. (**f**) Layer-6. (**g**) Layer-7

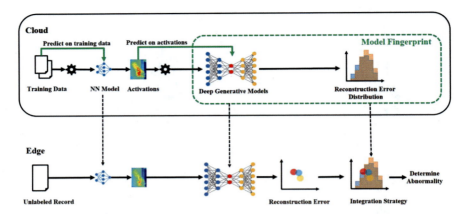

Fig. 4.2 The framework of NeuralFP

3.3.1 Fingerprinting on the Cloud

The fingerprints of NeuralFP in the cloud consists of two components: (1) deep generative models; and (2) the reconstruction error distributions of the training dataset. At a high level, NeuralFP first feeds all the training data over the deep learning model to collect activations, based on which deep generative models will be constructed. The deep generative models will then be used to collect latent information (e.g., the distribution of reconstruction errors) of the training data, which will serve as comparison benchmarks when a new unlabeled data is feeding to the deep learning model. Thus, we view NeuralFP as a *model fingerprinting* method where the fingerprints consist of the deep generative models and the reconstruction error distribution of the training dataset. In detail, the fingerprinting mode of NeuralFP contains the following key steps:

Step 1 NeuralFP first collects the activations by passing the training dataset through the deep learning model. Specifically, the activation in the l-th layer of the i-th data is denoted by $a_l(x_i)$, which can be computed as

$$
\begin{aligned}
a_1(x_i) &= g_l(\theta_1, x_i) \\
a_l(x_i) &= g_l(\theta_l, a_{l-1})
\end{aligned}
\quad (4.1)
$$

where g_l is an element-wise activation function such as a sigmoid function or a rectified linear unit [9], for the l-th layer of the deep learning model.

Step 2 For each layer of the deep learning model, the activations of all the training data would be used to construct the deep generative models f_l of the layer. Based on our analysis in Sect. 1, the non-linear fingerprints we considered here is the autoencoder model where the l-th autoencoder can be obtained by minimizing the following objective function.

$$f_l = \arg\min \sum_{i=1}^{N} \mathcal{L}(a_l(x_i), \hat{a}_l(x_i)) \qquad (4.2)$$

where $\hat{a}_l(x_i)$ is obtained by passing $a_l(x_i)$ through the autoencoder f_l. Then, the entire set of autoencoders for all the layers can be constructed as

$$f = [f_1^T, \cdots, f_L^T]^T \qquad (4.3)$$

Step 3 Finally, NeuralFP computes the reconstruction errors of the autoencoders applied to the training dataset, which can be viewed as the latent information of the training dataset and denoted by $s = [s_1, \cdots, s_L]$ where

$$s_l = \left[\mathcal{L}(a_l(x_1), \hat{a}_l(x_1)), \cdots, \mathcal{L}(a_l(x_N), \hat{a}_l(x_N)) \right] \qquad (4.4)$$

The deep learning model θ as well as its constructed fingerprints (the combination of the autoencoders for all the layers f and reconstruction errors of all the training data s) would be stored in the cloud for OOD detection.

3.3.2 OOD Detection in the Edge

For a given data x^*, it will pass through the deep learning model to compare with the constructed model fingerprints for determining whether it is OOD data or not. Based on the second observation in Sect. 3.2, the reconstruction error of an OOD data may not always be smaller than the ID data. Therefore, we consider data records with reconstruction errors locating outside of the error distribution of the training dataset as OOD data. Specifically, we define threshold parameters $\{\mu_l, \tau_l\}$ to bound the reconstruction error distribution of the training dataset at the l-th layer, and a data x^* would be classified as OOD if its reconstruction error $e_l^* = \mathcal{L}(a_l(x^*), \hat{a}_l(x^*))$ satisfies

$$e_l^* > \tau_l \quad \text{or} \quad e_l^* < \mu_l \qquad (4.5)$$

Next, we explore how to integrate the detection results across all the layers to enhance the OOD detection performance. Based on the third observation in Sect. 3.2, some layer may incorrectly recognize the OOD data as ID data. Therefore, we propose the *one-out* integration strategy where a data record would be labeled as *OOD* if it is recognized as OOD by at least one layer, i.e., we consider x^* as OOD data if

$$\exists l, \quad e_l^* > \tau_l \quad \text{or} \quad e_l^* < \mu_l \qquad (4.6)$$

It would be interesting to explore other integration strategies across all the layers, and a systematic comparison before and after adopting the *one-out* strategy will be presented in Sect. 4.2.3 to verify the effectiveness. In Sect. 4.4, we will also investigate how to select appropriate values of $\{\mu_l, \tau_l\}_{l=1}^{L}$ for practical adoption of NeuralFP.

4 Experiments

To show the effectiveness of NeuralFP, we evaluate our method across multiple real-world datasets. We will also show the advantages of NeuralFP by comparing with previous OOD detection methods. Finally, we will investigate the impact of threshold parameters in NeuralFP for practitioners to select useful values. All of our experiments are implemented on TensorFlow 2.2, and MacOS Catalina with 2.7 GHz Intel i7 4-cores processor.

4.1 Experimental Setup

4.1.1 Dataset and Model Architectures

Through our experiments, we use four image datasets listed in Table 4.1 with detailed information. MNIST dataset [10] represents handwritten digits and SVHN dataset [11] represents house numbers of street view, ranging from 0 to 9. Fashion-MNIST dataset [12] represents fashion objects, while CIFAR-10 dataset [13] represents different types of items (such as vehicles, animals, etc.).

For each dataset, the corresponding training dataset would be used to learn a deep learning model as well as constructing the model fingerprints according to Sect. 3.3.1. We will use its testing dataset together with the testing dataset of other datasets (serve as OOD data) for evaluating the OOD detection performance. Specifically, we construct deep learning models for MNIST and Fashion-MNIST (model structures are shown in Fig. 4.3) and the evaluation accuracy on the corresponding test dataset are 99.32% and 92.41%, respectively. At testing time, the test images from its own testing dataset can be viewed as the ID data, while

Table 4.1 Summary of datasets used in experiments

Dataset	# of training	# of testing	Category	Format	Description
MNIST	60,000	10,000	10	$28 \times 28 \times 1$	Hand-written digits
Fashion-MNIST	60,000	10,000	10	$28 \times 28 \times 1$	Fashion items
SVHN	73,257	26,032	10	$32 \times 32 \times 3$	Cropped digits from street view
CIFAR-10	50,000	10,000	10	$32 \times 32 \times 3$	Various objects

4 Out of Distribution Detection

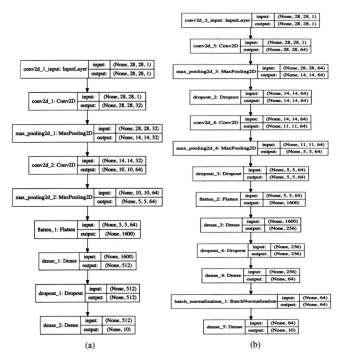

Fig. 4.3 Model structure for MNIST and fashion-MNIST dataset. (**a**) MNIST model structure. (**b**) FASHION model structure

testing images from the other three datasets would be treated as OOD data. Note that we have resized the 32 × 32 color images of SVHN and CIFAR-10 to 28 × 28 gray scale images for fair comparing with MNIST and Fashion-MNIST datasets.

4.1.2 Metrics

Following the convention, we consider OOD data as positive and the ID records as negative. Thus, we can compute *True Positive (TP)*: OOD data being correctly identified, *False Positive (FP)*: ID data being incorrectly recognized as OOD data, *True Negative (TN)*: ID data being correctly identified, *False Negative (FN)*: OOD data being incorrectly recognized as ID data, based on which we adopt the following three evaluation metrics to quantify the OOD detection performance of NeuralFP and its alternatives.

- Receiver operating characteristic (ROC) curve: measures the tradeoff between *True Positive Rate*: $\frac{TP}{TP+FN}$ and *False Positive Rate*: $\frac{FP}{TN+FP}$
- Area Under the ROC (AUC) [14]: quantifies the overall OOD detection performance under all values of the threshold

- Accuracy: $\frac{TP+TN}{TP+TN+FP+FN}$ for guiding selecting proper values of the threshold $\{\tau_l\}_{l=1}^{L}$ in practical adoption of NeuralFP

4.2 Detection Effectiveness

4.2.1 Detecting Statistical OOD Data

We evaluate various OOD dataset (Fashion-MNIST, MNIST, SVHN, CIFAR-10) on the deep learning model trained on MNIST dataset and Fashion-MNIST dataset, respectively. The ROC curves and AUC values are showed in Fig. 4.4 and Table 4.2. Based on the results, we have the following observations:

- NeuraFP can achieve good OOD detection performance since the TP rate can quickly increase to 1 under small values of FP rate for most scenarios.
- Practitioners of NeuralFP can select proper values of the threshold parameters so that a user-specified tradeoff between RP rate and FP rate can be achieved.
- MNIST dataset is more representative as compared to Fasion-MNIST dataset where the corresponding fingerprints constructed by NeuralFP are easier to be distinguished from OOD.

Therefore, we conclude that NeuralFP successfully captures the fingerprints of the training dataset on the deep learning model, and can achieve good OOD detection performance through a useful aggregation strategy of fingerprints across all the layers.

4.2.2 Detecting Adversarial OOD Data

4.2.3 Effectiveness of One-Out Integration Strategy

To evaluate the strategy of one-out integration in NeuralFP, we compute the AUC results of each individual layer of the deep learning model as shown in Table 4.3 (detecting using MNIST model) and Table 4.4 (detecting using Fashion-MNIST model), respectively. By comparing with Table 4.2, we know that combining fingerprints across all the layers can boost the overall detection performance which achieves significantly higher AUCs, thus validating the necessity and usefulness of the one-out integration strategy in NeuralFP.

4 Out of Distribution Detection

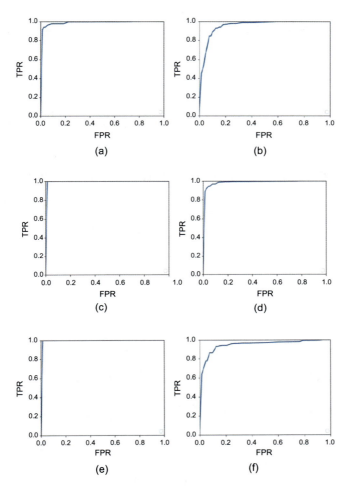

Fig. 4.4 ROC of detecting various OOD records. (**a**) Model:MNIST OOD:FASHION. (**b**) Model:FASHION OOD:MNIST. (**c**) Model:MNIST OOD:SVHN. (**d**) Model:FASHION OOD:SVHN. (**e**) Model:MNIST OOD:CIFAR10. (**f**) Model:FASHION OOD:CIFAR10

Table 4.2 AUCs under different experimental settings

Training	OOD			
	MNIST	FASHION	CIFAR10	SVHN
MNIST	N.A.	0.9826	0.9921	0.9921
FASHION	0.9498	N.A.	0.9442	0.9809

Table 4.3 AUCs for different layer under MNIST model

	FASHION	CIFAR10	SVHN
Layer-1	0.7298	0.8817	0.8821
Layer-2	0.8260	0.8848	0.8850
Layer-3	0.6893	0.8394	0.8634
Layer-4	0.4190	0.5340	0.7661
Layer-5	0.4298	0.4821	0.6927
Layer-6	0.8678	0.8902	0.8899
Layer-7	0.8741	0.8955	0.8944

Table 4.4 AUCs for different layer under Fashion-MNIST model

	MNIST	CIFAR10	SVHN
Layer-1	0.5280	0.6983	0.7499
Layer-2	0.5995	0.7178	0.7606
Layer-3	0.6013	0.7146	0.7546
Layer-4	0.8163	0.7851	0.7801
Layer-5	0.8236	0.8062	0.8447
Layer-6	0.7980	0.7868	0.8217
Layer-7	0.8009	0.7523	0.8064
Layer-8	0.6668	0.3737	0.3797
Layer-9	0.6446	0.2530	0.2418
Layer-10	0.1961	0.5216	0.5369
Layer-11	0.3826	0.5893	0.4637

4.3 Advantageous over Previous State-of-the-Arts

4.4 Guidelines for Parameter Selection

Next, we explore the impact of the thresholds on the overall OOD detection performance. In order to provide simple guidelines, we assume the thresholds are set to represent the same confidence interval across all the layers. Specifically, we set μ_l, τ_l $(l = 1, \cdots, L)$ to represent confidence intervals ranging from 1% to 99% and demonstrate the OOD detection performance in Fig. 4.5. We observe that setting the thresholds to the 95% confidence interval can achieve the highest OOD detection accuracy for most of the cases.

To further quantify the OOD detection accuracy under these selected thresholds, we construct a practical testing dataset with half ID data and half OOD data and show the OOD detection accuracy in Table 4.5. From these experimental results, we know that the selection of threshold to 95% confidence interval would result in good OOD detection accuracy in practice.

4 Out of Distribution Detection

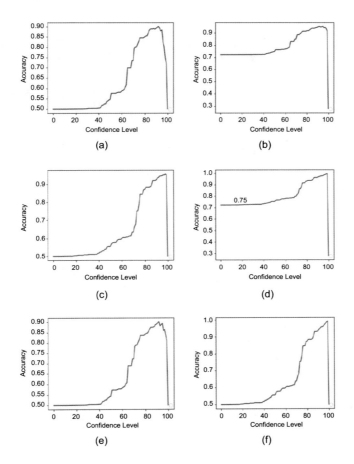

Fig. 4.5 Detection accuracy with varying threshold parameters. (**a**) Model:FASHION OOD:MNIST. (**b**) Model:FASHION OOD:SVHN. (**c**) Model:MNIST OOD:FASHION. (**d**) Model:MNIST OOD:SVHN. (**e**) Model:FASHION OOD:CIFAR10. (**f**) Model:MNIST OOD:CIFAR10

Table 4.5 Detection accuracy by setting threshold parameters to 95% confidence level

	OOD			
Training	MNIST	FASHION	CIFAR10	SVHN
MNIST	N.A.	95.15%	97.19%	98.43%
FASHION	88.36%	N.A.	89.38%	95.45%

4.5 Fingerprinting-Based Model Ranking

In addition to OOD detection, model fingerprinting is used to rank models by creating fingerprints on latent layers of deep learning classifiers. We create fingerprints of latent layers of image classifiers and calculate Silhouette scores of clusters

Table 4.6 Model ranking results using auto-encoder model fingerprinting

		C_{10}	C_{0-4}	C_{5-9}	E_{10}	E_{0-4}	E_{5-9}	F_{16}	F_{0-7}	F_{8-15}	P_{20}	P_{0-9}	P_{10-19}	ρ
CIFAR10	all SC	**0.012**	**−0.089**	**−0.039**	−0.142	−0.124	−0.142	−0.124	−0.119	−0.122	−0.094	−0.131	−0.173	0.524
	all Acc	**0.691**	**0.531**	**0.561**	0.443	0.45	0.45	0.444	0.43	0.445	0.479	0.48	0.456	
	0-4 SC	**0.105**	**0.119**	0.0	−0.071	−0.062	−0.084	−0.053	−0.061	−0.081	**0.003**	−0.019	−0.092	0.783
	0-4 Acc	**0.761**	**0.731**	**0.654**	0.583	0.584	0.582	0.58	0.577	0.586	0.608	0.613	0.576	
	5-9 SC	**0.2**	−0.005	**0.158**	−0.062	−0.068	−0.097	−0.066	−0.062	−0.058	**0.01**	−0.019	−0.064	0.879
	5-9 Acc	**0.852**	**0.72**	**0.85**	0.666	0.666	0.659	0.654	0.663	0.668	0.697	0.7	0.694	
ImageNet (Elec)	all SC	**−0.134**	−0.154	−0.141	−0.14	−0.162	−0.165	−0.166	**−0.137**	−0.167	**−0.109**	−0.158	−0.191	0.392
	all Acc	**0.352**	0.296	0.336	**0.397**	**0.343**	0.287	0.296	0.315	0.304	0.306	0.332	0.325	
	0-4 SC	−0.079	−0.101	−0.119	−0.097	**−0.073**	−0.111	−0.102	**−0.07**	−0.093	**−0.077**	−0.105	−0.133	0.203
	0-4 Acc	0.404	0.365	0.398	**0.445**	**0.466**	0.395	0.335	0.374	0.368	0.401	**0.409**	0.392	
	5-9 SC	−0.079	−0.102	−0.04	**0.022**	−0.117	**−0.005**	−0.095	−0.079	−0.123	**−0.013**	−0.082	−0.064	0.819
	5-9 Acc	0.541	0.454	0.549	**0.611**	0.535	**0.592**	0.527	0.519	0.519	0.554	0.514	**0.563**	
ImageNet (Fish)	all SC	−0.212	−0.236	**−0.179**	−0.216	−0.245	−0.25	**−0.174**	−0.212	−0.206	**−0.168**	−0.2	−0.227	0.344
	all Acc	**0.29**	**0.293**	0.285	0.265	0.283	0.257	**0.33**	0.248	0.257	0.287	0.276	0.261	
	0-7 SC	−0.154	−0.156	−0.157	**−0.143**	−0.205	−0.149	**−0.144**	**−0.116**	−0.174	−0.155	−0.176	−0.178	0.455
	0-7 Acc	**0.367**	0.366	0.33	0.342	0.338	0.323	**0.407**	**0.416**	0.328	0.349	0.355	0.333	
	8-15 SC	−0.152	−0.173	−0.098	−0.159	−0.141	−0.2	**−0.047**	−0.142	**−0.051**	**−0.081**	−0.119	−0.174	0.218
	8-15 Acc	**0.435**	0.411	0.43	0.399	0.397	0.404	**0.434**	0.38	**0.48**	0.394	0.404	0.411	
VOC12	all SC	−0.221	−0.231	−0.213	−0.228	−0.215	−0.218	−0.241	**−0.15**	−0.232	**−0.085**	−0.251	**−0.169**	0.294
	all Acc	**0.413**	0.376	**0.397**	0.365	0.367	0.364	0.358	0.356	0.354	**0.444**	0.395	0.379	
	0-9 SC	−0.183	−0.213	−0.187	−0.207	−0.181	−0.177	−0.187	**−0.131**	−0.18	**−0.056**	**−0.163**	−0.211	0.011
	0-9 Acc	**0.456**	0.41	0.424	0.38	0.377	0.372	0.367	0.365	0.379	**0.472**	**0.473**	0.414	
	10-19 SC	−0.177	−0.153	−0.163	−0.163	−0.178	−0.169	−0.217	**−0.113**	−0.208	**0.006**	−0.207	**0.053**	0.439
	10-19 Acc	**0.615**	0.583	0.601	0.584	0.583	0.58	0.578	0.573	0.58	**0.63**	0.595	**0.624**	

4 Out of Distribution Detection

created by new incoming test data. With higher Silhouette scores, we assume higher performance of a given model and rank them higher than the model with lower Silhouette score. Table 4.6 shows an experiment that compares the model rank scores (SC) and the accuracy (Acc) of 12 image classifiers. CIFAR10, ImageNet (categories of Electronics and Fish) and VOC12 datasets were used. Fingerprints were generated on the latent layer of each model using the training data and test data (rows) were used to calculate accuracy and Silhouette scores against each model (columns). The top three scores in each category are boldfaced and the correlation between accuracy (Acc) and model ranking score (SC) is clearly visible with the top three score. The overall correlation (Spearman's ρ) is provided in the last column. The higher correlation suggests that the fingerprinting-based model-ranking scores coincide well with the actual accuracy of the models.

5 Conclusion

In this chapter, we introduce NeuralFP, a practical and effective OOD detection method, which extracts representative information of the training dataset by constructing fingerprints of deep learning models. Through integrating the fingerprints across all the layers, NeuralFP can successfully distinguish the differences between ID data and OOD data. We have verified the effectiveness of NeuralFP through experiments on multiple real-world datasets, showed its advantages over existing OOD detection methods, and provided useful guidelines for parameter selecting in practice. To sum up, NeuralFP can serve as key technique in detecting OOD data in practical edge computing scenarios.

References

1. D. Hendrycks, K. Gimpel, A baseline for detecting misclassified and out-of-distribution examples in neural networks, in *In 5th International Conference on Learning Representations, ICLR* (2016)
2. K. Lee, H. Lee, K. Lee, J. Shin, Training confidence-calibrated classifiers for detecting out-of-distribution samples, in *In 7th International Conference on Learning Representations, ICLR* (2018)
3. S. Liang, Y. Li, R. Srikant, Enhancing the reliability of out-of-distribution image detection in neural networks, in *In 7th International Conference on Learning Representations, ICLR* (2018)
4. K. Lee, K. Lee, H. Lee, J. Shin, A simple unified framework for detecting out-of-distribution samples and adversarial attacks, in *Advances in Neural Information Processing Systems* (2018), pp. 7167–7177
5. Y. Li, Y. Gal, Dropout inference in bayesian neural networks with alpha-divergences, in *Proceedings of the 34th International Conference on Machine Learning-Volume 70. JMLR. org* (2017), pp. 2052–2061
6. C. Louizos, M. Welling, Multiplicative normalizing flows for variational bayesian neural networks, in *Proceedings of the 34th International Conference on Machine Learning-Volume 70. JMLR. org* (2017), pp. 2218–2227

7. B. Lakshminarayanan, A. Pritzel, C. Blundell, Simple and scalable predictive uncertainty estimation using deep ensembles, in *Advances in Neural Information Processing Systems* (2017), pp. 6402–6413
8. W.-H. Lee, S. Millman, N. Desai, M. Srivatsa, C. Liu, Neuralfp: out-of-distribution detection using fingerprints of neural networks, in *2020 25th International Conference on Pattern Recognition (ICPR)* (2021), pp. 9561–9568
9. Y. LeCun, Y. Bengio, G. Hinton, Deep learning. Nature **521**(7553), 436–444 (2015)
10. Y. LeCun, L. Bottou, Y. Bengio, P. Haffner, Gradient-based learning applied to document recognition. Proc. IEEE **86**(11), 2278–2324 (1998)
11. Y. Netzer, T. Wang, A. Coates, A. Bissacco, B. Wu, A.Y. Ng, Reading digits in natural images with unsupervised feature learning, in *Proceedings of the NIPS Workshop on Deep Learning and Unsupervised Feature Learning* (2011)
12. H. Xiao, K. Rasul, R. Vollgraf, Fashion-mnist: a novel image dataset for benchmarking machine learning algorithms (2017). arXiv preprint. arXiv:1708.07747
13. A. Krizhevsky, Learning multiple layers of features from tiny images. Master's thesis, University of Tront (2009)
14. C.X. Ling, J. Huang, H. Zhang, et al., Auc: a statistically consistent and more discriminating measure than accuracy, in *Ijcai*, vol. 3 (2003), pp. 519–524

Chapter 5
Model Compression for Edge Computing

Shuochao Yao and Tarek Abdelzaher

Abstract We are now ready to address what's arguably the primary inference challenge in Edge AI applications, namely, compressing an AI model to fit the limited resources of edge devices. Many traditional AI models are designed for large-scale cloud environments with ample GPUs. The computational environment at the edge is substantially different. Specifically, it is much more resource-constrained. Fortunately, often edge applications are also more restricted to a data sub-domain. For example, a vision application used for edge security (e.g., detecting intruders) might only need to recognize a relatively small number of object categories compared to a full-fledged general-purpose vision agent. Can model compression be driven by inference quality needs of only the subset of relevant data categories? Such an approach was used in the design of DeepIoT, an application-aware compression framework for neural network architecture. The chapter discusses DeepIoT and its extensions. It shows that dramatic improvements in resource footprint are possible that take advantage of the target data domain without sacrificing inference quality in that domain.

1 Introduction

Resource constraints of IoT devices remain an important impediment towards deploying deep learning models. A key question is therefore whether or not it is possible to compress deep neural networks, such as those described in the previous section, to a point where they fit comfortably on low-end embedded devices, enabling real-time "intelligent" interactions with their environment. Can a unified approach compress commonly used deep learning structures, including

---

S. Yao (✉)
George Mason University, Fairfax, VA, USA
e-mail: shuochao@gmu.edu

T. Abdelzaher
University of Illinois at Urbana Champaign, Urbana, IL, USA
e-mail: zaher@illinois.edu

© The Author(s), under exclusive license to Springer Nature Switzerland AG 2023
M. Srivatsa et al. (eds.), *Artificial Intelligence for Edge Computing*,
https://doi.org/10.1007/978-3-031-40787-1_5

fully-connected, convolutional, and recurrent neural networks, as well as their combinations? To what degree does the resulting compression reduce energy, execution time, and memory needs in practice?

We prosed such a compression framework, called DeepIoT [1], that compresses commonly used deep neural network structures for sensing applications through deciding the minimum number of elements in each layer. Previous illuminating studies on neural network compression sparsify large dense parameter matrices into large sparse matrices [2]. In contrast, DeepIoT minimizes the number of elements in each layer, which results in converting parameters into a set of small dense matrices. A small dense matrix does not require additional storage for element indices and is efficiently optimized for processing. DeepIoT greatly reduces the effort of designing efficient neural structures for sensing applications by deciding the number of elements in each layer in a manner informed by the topology of the neural network.

DeepIoT borrows the idea of dropping hidden elements from a widely-used deep learning regularization method called *dropout*. The dropout operation gives each hidden element a dropout probability. During the dropout process, hidden elements can be pruned according to their dropout probabilities. A "thinned" network structure can thus be generated. The challenge is to set these dropout probabilities in an informed manner to generate the optimal slim network structure that preserves the accuracy of sensing applications while maximally reducing their resource consumption. An important purpose of DeepIoT is thus to find the optimal dropout probability for each hidden element in the neural network.

To obtain the optimal dropout probabilities for nodes in the neural network, DeepIoT exploits the network parameters themselves. From the perspective of model compression, an element that is more redundant should have a higher probability to be dropped. A contribution of DeepIoT lies in exploiting a novel compressor neural network to solve this problem. It takes model parameters of each layer as input, learns parameter redundancies, and generates the dropout probabilities accordingly. The compressor neural network is optimized jointly with the original neural network to be compressed in an iterative manner that tries to minimize the loss function of the original IoT application.

DeepIoT greatly reduces the size of model parameters, and speeds up the execution time by getting rid of the inefficient sparse matrix multiplication. However, a formal way to explore the neural network structure design and underlying system efficiency is still unclear. Most manually designed time-efficient neural network structures for mobile devices use parameter size or FLOPs (floating point operations) as the indicator of model execution time [3–5]. Even the official TensorFlow website recommends to use the total number of floating number operations (FLOPs) of neural networks "to make rule-of-thumb estimates of how fast they will run on different devices".[1] However, in practice, counting the number of neural network parameters and the total FLOPs does not lead to good estimates of execution time

---

[1] https://www.tensorflow.org/versions/r1.5/mobile/optimizing.

because the relation between these predictors and execution time is not proportional. We therefore design FastDeepIoT [6], showing how a better understanding of the non-linear relation between neural network structure and performance can further improve execution time and energy consumption without impacting accuracy.

2 The Design of DeepIoT Framework

Without loss of generality, before introducing the technical details, we first use an example of compressing a 3-layer fully-connected neural network structure to illustrate the overall pipeline of DeepIoT. The detailed illustration is shown in Fig. 5.1. The basic steps of compressing neural network structures for sensing applications with DeepIoT can be summarized as follows.

1. Insert operations that randomly zeroing out hidden elements with probabilities $\mathbf{p}^{(l)}$ called dropout (red boxes in Fig. 5.1) into internal layers of the original neural network. The internal layers exclude input layers and output layers that have the fixed dimension for a sensing application. This step will be detailed in Sect. 2.1.
2. Construct the compressor neural network. It takes the weight matrices $\mathbf{W}^{(l)}$ (green boxes in Fig. 5.1) from the layers to be compressed in the original neural network as inputs, learns and shares the parameter redundancies among different layers, and generates optimal dropout probabilities $\mathbf{p}^{(l)}$, which is then fed back to the dropout operations in the original neural network. This step will be detailed in Sect. 2.2.
3. Iteratively optimize the compressor neural network and the original neural network with the compressor-critic framework. The compressor neural network is optimized to produce better dropout probabilities that can generate a more efficient network structure for the original neural network. The original neural

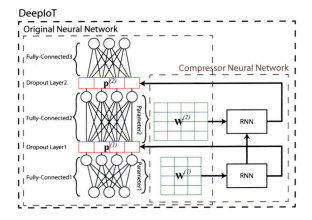

Fig. 5.1 Overall DeepIoT system framework. Orange boxes represent dropout operations. Green boxes represent parameters of the original neural network

network is optimized to achieve a better performance with the more efficient structure for a sensing application. This step will be detailed in Sect. 2.3.

2.1 Dropout Operations in the Original Neural Network

Dropout is commonly used as a regularization method that prevents feature co-adapting and model overfitting. The term "dropout" refers to dropping out units (hidden and visible) in a neural network. Since DeepIoT is a structure compression framework, we focus mainly on dropping out hidden units. The definitions of hidden units are distinct in different types of neural networks, and we will describe them in detail. The basic idea is that we regard neural networks with dropout operations as bayesian neural networks with Bernoulli variational distributions [7–9].

For the fully-connected neural networks, the fully-connected operation with dropout can be formulated as

$$
\begin{aligned}
\mathbf{z}_{[j]}^{(l)} &\sim \text{Bernoulli}(\mathbf{p}_{[j]}^{(l)}), \\
\tilde{\mathbf{W}}^{(l)} &= \mathbf{W}^{(l)}\text{diag}(\mathbf{z}^{(l)}), \\
\mathbf{Y}^{(l)} &= \mathbf{X}^{(l)}\tilde{\mathbf{W}}^{(l)} + \mathbf{b}^{(l)}, \\
\mathbf{X}^{(l+1)} &= f(\mathbf{Y}^{(l)}).
\end{aligned}
\tag{5.1}
$$

Refer to (5.1). The notation $l = 1, \cdots, L$ is the layer number in the fully-connected neural network. For any layer l, the weight matrix is denoted as $\mathbf{W}^{(l)} \in \mathbb{R}^{d^{(l-1)} \times d^{(l)}}$; the bias vector is denoted as $\mathbf{b}^{(l)} \in \mathbb{R}^{d^{(l)}}$; and the input is denoted as $\mathbf{X}^{(l)} \in \mathbb{R}^{1 \times d^{(l-1)}}$. In addition, $f(\cdot)$ is a nonlinear activation function.

As shown in (5.1), each hidden unit is controlled by a Bernoulli random variable. In the original dropout method, the success probabilities of $\mathbf{p}_{[j]}^{(l)}$ can be set to the same constant p for all hidden units [9], but DeepIoT uses the Bernoulli random variable with individual success probabilities for different hidden units in order to compress the neural network structure in a finer granularity.

For the convolutional neural networks, the basic fully-connected operation is replaced by the convolution operation [7]. However, the convolution can be reformulated as a linear operation as shown in (5.1). For any layer l, we denote $\mathcal{K}^{(l)} = \{\mathbf{K}_k^{(l)}\}$ for $k = 1, \cdots, c^{(l)}$ as the set of convolutional neural network (CNN)'s kernels, where $\mathbf{K}_k^{(l)} \in \mathbb{R}^{h^{(l)} \times w^{(l)} \times c^{(l-1)}}$ is the kernel of CNN with height $h^{(l)}$, width $w^{(l)}$, and channel $c^{(l-1)}$. The input tensor of layer l is denoted as $\hat{\mathbf{X}}^{(l)} \in \mathbb{R}^{\hat{h}^{(l-1)} \times \hat{w}^{(l-1)} \times c^{(l-1)}}$ with height $\hat{h}^{(l-1)}$, width $\hat{w}^{(l-1)}$, and channel $c^{(l-1)}$.

Next, we convert convolving the kernels with the input into performing matrix product. We extract $h^{(l)} \times w^{(l)} \times c^{(l-1)}$ dimensional patches from the input $\hat{\mathbf{X}}^{(l)}$ with stride s and vectorize them. Collect these vectorized n patches to be the rows of our

5 Model Compression for Edge Computing
157

new input representation $\mathbf{X}^{(l)} \in \mathbb{R}^{n \times (h^{(l)} w^{(l)} c^{(l-1)})}$. The vectorized kernels form the columns of the weight matrix $\mathbf{W}^{(l)} \in \mathbb{R}^{(h^{(l)} w^{(l)} c^{(l-1)}) \times c^{(l)}}$.

With this transformation, dropout operations can be applied to convolutional neural networks according to (5.1). The composition of pooling and activation functions can be regarded as the nonlinear function $f(\cdot)$ in (5.1). Instead of dropping out hidden elements in each layer, we drop out convolutional kernels in each layer. From the perspective of structure compression, DeepIoT tries to prune the number of kernels used in the convolutional neural networks.

For the recurrent neural network, we take a multi-layer Long Short Term Memory network (LSTM) as an example. The LSTM operation with dropout can be formulated as

$$
\mathbf{z}_{[j]}^{(l)} \sim \text{Bernoulli}(\mathbf{p}_{[j]}^{(l)}),
$$

$$
\begin{pmatrix} \mathbf{i} \\ \mathbf{f} \\ \mathbf{o} \\ \mathbf{g} \end{pmatrix} = \begin{pmatrix} \text{sigm} \\ \text{sigm} \\ \text{sigm} \\ \text{tanh} \end{pmatrix} \mathbf{W}^{(l)} \begin{pmatrix} \mathbf{h}_t^{(l-1)} \odot \mathbf{z}^{(l-1)} \\ \mathbf{h}_{t-1}^{(l)} \odot \mathbf{z}^{(l)} \end{pmatrix}, \tag{5.2}
$$

$$
\mathbf{c}_t^{(l)} = \mathbf{f} \odot \mathbf{c}_{t-1}^{(l)} + \mathbf{i} \odot \mathbf{g},
$$

$$
\mathbf{h}_t^{(l)} = \mathbf{o} \odot \tanh(\mathbf{c}_t^{(l)}).
$$

The notation $l = 1, \cdots, L$ is the layer number and $t = 1, \cdots, T$ is the step number in the recurrent neural network. Element-wise multiplication is denoted by $\odot$. Operators $sigm$ and $tanh$ denote sigmoid function and hyperbolic tangent respectively. The vector $\mathbf{h}_t^{(l)} \in \mathbb{R}^{n^{(l)}}$ is the output of step t at layer l. The vector $\mathbf{h}_t^{(0)} = \mathbf{x}_t$ is the input for the whole neural network at step t. The matrix $\mathbf{W}^{(l)} \in \mathbb{R}^{4n^{(l)} \times (n^{(l-1)}+n^{(l)})}$ is the weight matrix at layer l. We let $\mathbf{p}_{[j]}^{(0)} = 1$, since DeepIoT only drops hidden elements.

As shown in (5.2), DeepIoT uses the same vector of Bernoulli random variables $\mathbf{z}^{(l)}$ to control the dropping operations among different time steps in each layer, while individual Bernoulli random variables are used for different steps in the original LSTM dropout [10]. From the perspective of structure compression, DeepIoT tries to prune the number of hidden dimensions used in LSTM blocks. The dropout operation of other recurrent neural network architectures, such as Gated Recurrent Unit (GRU), can be designed similarly.

2.2 Compressor Neural Network

Now we introduce the architecture of the compressor neural network. A hidden element in the original neural network that is connected to redundant model parameters should have a higher probability to be dropped. Therefore we design the compressor neural network to take the weights of an original neural network $\{\mathbf{W}^{(l)}\}$ as inputs, learn the redundancies among these weights, and generate dropout probabilities $\{\mathbf{p}^{(l)}\}$ for hidden elements that can be eventually used to compress the original neural network structure.

A straightforward solution is to train an individual fully-connected neural network for each layer in the original neural network. However, since there are interconnections among weight redundancies in different layers, DeepIoT uses a variant LSTM as the structure of compressor to share and use the parameter redundancy information among different layers.

According to the description in Sect. 2.1, the weight in layer l of fully-connected, convolutional, or recurrent neural network can all be represented as a single matrix $\mathbf{W}^{(l)} \in \mathbb{R}^{d_f^{(l)} \times d_{\mathrm{drop}}^{(l)}}$, where $d_{\mathrm{drop}}^{(l)}$ denotes the dimension that dropout operation is applied and $d_f^{(l)}$ denotes the dimension of features within each dropout element. Here, we need to notice that the weight matrix of LSTM at layer l can be reshaped as $\mathbf{W}^{(l)} \in \mathbb{R}^{4 \cdot (n^{(l-1)} + n^{(l)}) \times n^{(l)}}$, where $d_{\mathrm{drop}}^{(l)} = n^{(l)}$ and $d_f^{(l)} = 4 \cdot (n^{(l-1)} + n^{(l)})$. Hence, we take weights from the original network layer by layer, $\mathcal{W} = \{\mathbf{W}^{(l)}\}$ with $l = 1, \cdots, L$, as the input of the compressor neural network. Instead of using a vanilla LSTM as the structure of compressor, we apply a variant l-step LSTM model shown as

$$
\begin{pmatrix} \mathbf{v}_i^{\mathsf{T}} \\ \mathbf{v}_f^{\mathsf{T}} \\ \mathbf{v}_o^{\mathsf{T}} \\ \mathbf{v}_g^{\mathsf{T}} \end{pmatrix} = \mathbf{W}_c^{(l)} \mathbf{W}^{(l)} \mathbf{W}_i^{(l)}, \qquad \begin{pmatrix} \mathbf{u}_i \\ \mathbf{u}_f \\ \mathbf{u}_o \\ \mathbf{u}_g \end{pmatrix} = \mathbf{W}_h \mathbf{h}_{l-1},
$$

$$
\begin{pmatrix} \mathbf{i} \\ \mathbf{f} \\ \mathbf{o} \\ \mathbf{g} \end{pmatrix} = \begin{pmatrix} \mathrm{sigm} \\ \mathrm{sigm} \\ \mathrm{sigm} \\ \tanh \end{pmatrix} \left(\begin{pmatrix} \mathbf{v}_i \\ \mathbf{v}_f \\ \mathbf{v}_o \\ \mathbf{v}_g \end{pmatrix} + \begin{pmatrix} \mathbf{u}_i \\ \mathbf{u}_f \\ \mathbf{u}_o \\ \mathbf{u}_g \end{pmatrix} \right), \qquad (5.3)
$$

$$
\mathbf{c}_l = \mathbf{f} \odot \mathbf{c}_{l-1} + \mathbf{i} \odot \mathbf{g},
$$

$$
\mathbf{h}_l = \mathbf{o} \odot \tanh(\mathbf{c}_l),
$$

$$
\mathbf{p}^{(l)} = \mathbf{p}_t = \mathrm{sigm}(\mathbf{W}_o^{(l)} \mathbf{h}_l),
$$

$$
\mathbf{z}_{[j]}^{(l)} \sim \mathrm{Bernoulli}(\mathbf{p}_{[j]}^{(l)}).
$$

Refer to (5.3), we denote d_c as the dimension of the variant LSTM hidden state. Then $\mathbf{W}^{(l)} \in \mathbb{R}^{d_f^{(l)} \times d_{\mathrm{drop}}^{(l)}}$, $\mathbf{W}_c^{(l)} \in \mathbb{R}^{4 \times d_f^{(l)}}$, $\mathbf{W}_i^{(l)} \in \mathbb{R}^{d_{\mathrm{drop}}^{(l)} \times d_c}$, $\mathbf{W}_h \in \mathbb{R}^{4d_c \times d_c}$, and $\mathbf{W}_o^{(l)} \in \mathbb{R}^{d_{\mathrm{drop}}^{(l)} \times d_c}$. The set of training parameters of the compressor neural network is denoted as ϕ, where $\phi = \{\mathbf{W}_c^{(l)}, \mathbf{W}_i^{(l)}, \mathbf{W}_h, \mathbf{W}_o^{(l)}\}$. The matrix $\mathbf{W}^{(l)}$ is the input

5 Model Compression for Edge Computing | 159

matrix for step l in the compressor neural network, which is also the lth layer's parameters of the original neural network in (5.1) or (5.2).

Compared with the vanilla LSTM that requires vectorizing the original weight matrix as inputs, the variant LSTM model preserves the structure of original weight matrix and uses less learning parameters to extract the redundancy information among the dropout elements. In addition, $\mathbf{W}_c^{(l)}$ and $\mathbf{W}_i^{(l)}$ convert original weight matrix $\mathbf{W}^{(l)}$ with different sizes into fixed-size representations. The binary vector $\mathbf{z}^{(l)}$ is the dropout mask and probability $\mathbf{p}^{(l)}$ is the dropout probabilities for the lth layer in the original neural network used in (5.1) and (5.2), which is also the stochastic dropout policy learnt through observing the weight redundancies of the original neural network.

2.3 Compressor-Critic Framework

In Sects. 2.1 and 2.2, we have introduced customized dropout operations applied on the original neural networks that need to be compressed and the structure of compressor neural network used to learn dropout probabilities based on parameter redundancies. In this subsection, we will discuss the detail of compressor-critic compressing process. It optimizes the original neural network and the compressor neural network in an iterative manner and enables the compressor neural network to gradually compress the original neural network with soft deletion.

We denote the original neural network as $F_{\mathcal{W}}(\mathbf{x}|\mathbf{z})$, and we call it critic. It takes $\mathbf{x}$ as inputs and generates predictions based on binary dropout masks $\mathbf{z}$ and model parameters $\mathcal{W}$ that refer to a set of weights $\mathcal{W} = \{\mathbf{W}^{(l)}\}$. We assume that $F_{\mathcal{W}}(\mathbf{x}|\mathbf{z})$ is a pre-trained model. We denote the compressor neural network by $\mathbf{z} \sim \mu_\phi(\mathcal{W})$. It takes the weights of the critic as inputs and generates the probability distribution of the mask vector $\mathbf{z}$ based on its own parameters ϕ. In order to optimize the compressor to drop out hidden elements in the critic, DeepIoT follows the objective function

$$\begin{aligned} \mathcal{L} &= \mathbb{E}_{\mathbf{z} \sim \mu_\phi}\big[L\big(\mathbf{y}, F_{\mathcal{W}}(\mathbf{x}|\mathbf{z})\big)\big] \\ &= \sum_{\mathbf{z} \sim \{0,1\}^{|\mathbf{z}|}} \mu_\phi(\mathbf{W}) \cdot L\big(\mathbf{y}, F_{\mathcal{W}}(\mathbf{x}|\mathbf{z})\big), \end{aligned} \tag{5.4}$$

where $L(\cdot, \cdot)$ is the objective function of the critic. The objective function can be interpreted as the expected loss of the original neural network over the dropout probabilities generated by the compressor.

DeepIoT optimizes the compressor and critic in an iterative manner. It reduces the expected loss as defined in (5.4) by applying the gradient descent method on compressor and critic iteratively. However, since there are discrete sampling operations, *i.e.*, dropout operations, within the computational graph, backpropagation is not directly applicable. Therefore we apply an unbiased likelihood-ratio estimator to calculate the gradient over ϕ [11, 12]:

$$\nabla_\phi \mathcal{L} = \sum_{\mathbf{z}} \nabla_\phi \mu_\phi(\mathcal{W}) \cdot L(\mathbf{y}, F_{\mathcal{W}}(\mathbf{x}|\mathbf{z}))$$

$$= \sum_{\mathbf{z}} \mu_\phi(\mathcal{W}) \nabla_\phi \log \mu_\phi(\mathcal{W}) \cdot L(\mathbf{y}, F_{\mathcal{W}}(\mathbf{x}|\mathbf{z})) \qquad (5.5)$$

$$= \mathbb{E}_{\mathbf{z} \sim \mu_\phi} [\nabla_\phi \log \mu_\phi(\mathcal{W}) \cdot L(\mathbf{y}, F_{\mathcal{W}}(\mathbf{x}|\mathbf{z}))].$$

Therefore an unbiased estimator for (5.5) can be

$$\widehat{\nabla_\phi \mathcal{L}} = \nabla_\phi \log \mu_\phi(\mathcal{W}) \cdot L(\mathbf{y}, F_{\mathcal{W}}(\mathbf{x}|\mathbf{z})) \quad \mathbf{z} \sim \mu_\phi. \qquad (5.6)$$

The gradient over $\mathbf{W}^{(l)} \in \mathcal{W}$ is

$$\nabla_{\mathbf{W}^{(l)}} \mathcal{L} = \sum_{\mathbf{z}} \mu_\phi(\mathcal{W}) \cdot \nabla_{\mathbf{W}^{(l)}} L(\mathbf{y}, F_{\mathcal{W}}(\mathbf{x}|\mathbf{z}))$$

$$= \mathbb{E}_{\mathbf{z} \sim \mu_\phi} [\nabla_{\mathbf{W}^{(l)}} L(\mathbf{y}, F_{\mathcal{W}}(\mathbf{x}|\mathbf{z}))]. \qquad (5.7)$$

Similarly, an unbiased estimator for (5.7) can be

$$\widehat{\nabla_{\mathbf{W}^{(l)}} \mathcal{L}} = \nabla_{\mathbf{W}^{(l)}} L(\mathbf{y}, F_{\mathcal{W}}(\mathbf{x}|\mathbf{z})) \quad \mathbf{z} \sim \mu_\phi. \qquad (5.8)$$

Now we provide more details of $\widehat{\nabla_\phi \mathcal{L}}$ in (5.6). Although the estimator (5.6) is an unbiased estimator, it tends to have a higher variance. A higher variance of estimator can make the convergence slower. Therefore, variance reduction techniques are typically required to make the optimization feasible in practice [13, 14].

One variance reduction technique is to subtract a constant c from learning signal $L(\mathbf{y}, F_{\mathcal{W}}(\mathbf{x}|\mathbf{z}))$ in (5.5), which still keeps the expectation of the gradient unchanged [13]. Therefore, we keep track of the moving average of the learning signal $L(\mathbf{y}, F_{\mathcal{W}}(\mathbf{x}|\mathbf{z}))$ denoted by c, and subtract c from the gradient estimator (5.6).

The other variance reduction technique is keeping track of the moving average of the signal variance v, and divides the learning signal by $\max(1, \sqrt{v})$ [14].

Combing the aforementioned two variance reduction techniques, the final estimator (5.6) for gradient over ϕ becomes

$$\widehat{\nabla_\phi \mathcal{L}} = \nabla_\phi \log \mu_\phi(\mathcal{W}) \cdot \frac{L(\mathbf{y}, F_{\mathcal{W}}(\mathbf{x}|\mathbf{z})) - c}{\max(1, \sqrt{v})} \quad \mathbf{z} \sim \mu_\phi, \qquad (5.9)$$

where c and v are the moving average of mean and the moving average of variance of learning signal $L(\mathbf{y}, F_{\mathcal{W}}(\mathbf{x}|\mathbf{z}))$ respectively.

After introducing the basic optimization process in DeepIoT, now we are ready to deliver the details of the compressing process. Compared with previous compressing algorithms that gradually delete weights without rehabilitation [15],

5 Model Compression for Edge Computing

DeepIoT applies "soft" deletion by gradually suppressing the dropout probabilities of hidden elements with a decay factor $\gamma \in (0, 1)$. During the experiments in Sect. 3, we set γ as the default value 0.5. Since it is impossible to make the optimal compression decisions from the beginning, suppressing the dropout probabilities instead of deleting the hidden elements directly can provide the "deleted" hidden elements changes to recover. This less aggressive compression process reduces the potential risk of irretrievable network damage and learning inefficiency.

Algorithm 1 Compressor-predictor compressing process

1: **Input:** pre-trained predictor $F_{\mathcal{W}}(\mathbf{x}|\mathbf{z})$
2: **Initialize:** compressor $\mu_\phi(\mathcal{W})$ with parameter ϕ, moving average c, moving average of variance v
3: **while** $\mu_\phi(\mathcal{W})$ is not convergent **do**
4:     $\mathbf{z} \sim \mu_\phi(\mathcal{W})$
5:     $c \leftarrow \text{movingAvg}\big(L\big(\mathbf{y}, F_{\mathcal{W}}(\mathbf{x}|\mathbf{z})\big)\big)$
6:     $v \leftarrow \text{movingVar}\big(L\big(\mathbf{y}, F_{\mathcal{W}}(\mathbf{x}|\mathbf{z})\big)\big)$
7:     $\phi \leftarrow \phi - \beta \cdot \nabla_\phi \log \mu_\phi(\mathcal{W}) \cdot \big(L\big(\mathbf{y}, F_{\mathcal{W}}(\mathbf{x}|\mathbf{z})\big) - c\big)/\max(1, \sqrt{v})$
8: **end while**
9: $\tau = 0$
10: **while** the percentage of left number of parameters in $F_{\mathcal{W}}(\mathbf{x}|\mathbf{z})$ is larger than α **do**
11:     $\mathbf{z} \sim \mu_\phi(\mathcal{W})$
12:     $c \leftarrow \text{movingAvg}\big(L\big(\mathbf{y}, F_{\mathcal{W}}(\mathbf{x}|\mathbf{z})\big)\big)$
13:     $v \leftarrow \text{movingVar}\big(L\big(\mathbf{y}, F_{\mathcal{W}}(\mathbf{x}|\mathbf{z})\big)\big)$
14:     $\phi \leftarrow \phi - \beta \cdot \nabla_\phi \log \mu_\phi(\mathcal{W}) \cdot \big(L\big(\mathbf{y}, F_{\mathcal{W}}(\mathbf{x}|\mathbf{z})\big) - c\big)/\max(1, \sqrt{v})$
15:     $\mathcal{W} \leftarrow \mathcal{W} - \beta \cdot \nabla_{\mathcal{W}} L\big(\mathbf{y}, F_{\mathcal{W}}(\mathbf{x}|\mathbf{z})\big)$
16:     update threshold τ: $\tau \leftarrow \tau + \Delta$ for every T rounds
17: **end while**
18: $\hat{\mathbf{z}}_{[j]}^{(l)} = \mathbb{1}\mathbf{p}_{[j]}^{(l)} > \tau$
19: **while** $F_{\mathcal{W}}(\mathbf{x}|\hat{\mathbf{z}})$ is not convergent **do**
20:     $\mathcal{W} \leftarrow \mathcal{W} - \beta \cdot \nabla_{\mathcal{W}} L\big(\mathbf{y}, F_{\mathcal{W}}(\mathbf{x}|\hat{\mathbf{z}})\big)$
21: **end while**

During the compressing process, DeepIoT gradually increases the threshold of dropout probability τ from 0 with step Δ. The hidden elements with dropout probability, $\mathbf{p}_{[j]}^{(l)}$ that is less than the threshold τ will be given decay on dropout probability, *i.e.*, $\hat{\mathbf{p}}_{[j]}^{(l)} \leftarrow \gamma \cdot \mathbf{p}_{[j]}^{(l)}$. Therefore, the operation in compressor (5.3) can be updated as

$$\mathbf{z}_{[j]}^{(l)} \sim \text{Bernoulli}\left(\mathbf{p}_{[j]}^{(l)} \cdot \gamma^{\mathbb{1}\mathbf{p}_{[j]}^{(l)} \leq \tau}\right), \tag{5.10}$$

where $\mathbb{1}$ is the indicator function; $\gamma \in (0, 1)$ is the decay factor; and $\tau \in [0, 1)$ is the threshold. Since the operation of suppressing dropout probability with the predefined decay factor γ is differentiable, we can still optimize the original and the compressor neural network through (5.8) and (5.9). The compression process will stop when the percentage of left number of parameters in $F_\mathcal{W}(\mathbf{x}|\mathbf{z})$ is smaller than a user-defined value $\alpha \in (0, 1)$.

After the compression, DeepIoT fine-tunes the compressed model $F_\mathcal{W}(\mathbf{x}|\hat{\mathbf{z}})$, with a fixed mask $\hat{\mathbf{z}}$, which is decided by the previous threshold τ. Therefore the mask generation step in (5.10) will be updated as

$$\hat{\mathbf{z}}_{[j]}^{(l)} = \mathbb{1}\mathbf{p}_{[j]}^{(l)} > \tau. \tag{5.11}$$

We summarize the compressor-critic compressing process of DeepIoT in Algorithm 1.

The algorithm consists of three parts. In the first part (Line 3 to Line 8), DeepIoT freezes the critic $F_\mathcal{W}(\mathbf{x}|\mathbf{z})$ and initializes the compressor $\mu_\phi(\mathcal{W})$ according to (5.9). In the second part (Line 9 to Line 17), DeepIoT optimizes the critic and compressor jointly with the gradients calculated by (5.8) and (5.9). At the same time, DeepIoT gradually compresses the predictor by suppressing dropout probabilities according to (5.10). In the final part (Line 18 to Line 21), DeepIoT fine-tunes the critic with the gradient calculated by (5.8) and a deterministic dropout mask is generated according to (5.11). After these three phases, DeepIoT generates a binary dropout mask $\hat{\mathbf{z}}$ and the fine-tuning parameters of the critic $\mathcal{W}$. With these two results, we can easily obtain the compressed model of the original neural network.

3 The Evaluation of DeepIoT

In this section, we evaluate DeepIoT through three representative sensing tasks. The first set is motivated by the prospect of enabling future smarter embedded "things" (physical objects) to interact with humans using user-friendly modalities such as visual cues, handwritten text, and speech commands, while the second evaluates human-centric context sensing, such as human activity recognition and user identification. In the following subsections, we first describe the comparison baselines that are current state of the art deep neural network compression techniques. We then present the first set of experiments that demonstrate accuracy and resource demands observed if IoT-style smart objects interacted with users via natural human-centric modalities thanks to deep neural networks compressed, for the resource-constrained hardware, with the help of our DeepIoT framework. Finally, we present the second set of experiments that demonstrate accuracy and resource demands when applying DeepIoT to compress deep neural networks trained for human-centric context sensing applications.

3.1 Evaluation Platforms

Our hardware is based on Intel Edison computing platform [16]. The Intel Edison computing platform is powered by the Intel Atom SoC dual-core CPU at 500 MHz

5   Model Compression for Edge Computing   163

and is equipped with 1GB memory and 4GB flash storage. For fairness, all neural network models are run solely on CPU during experiments.

All the original neural networks for all sensing applications are trained on the workstation with NVIDIA GeForce GTX Titan X. For all baseline algorithms mentioned in Sect. 3.2, the compressing processes are also conducted on the workstation. The compressed models are exported and loaded into the flash storage on Intel Edison for experiments.

We installed the Ubilinux operation system on Intel Edison computing platform [17]. Far fairness, all compressed deep learning models are run through Theano [18] with only CPU device on Intel Edison. The matrix multiplication operations and sparse matrix multiplication operations are optimized by BLAS and Sparse BLAS respectively during the implementation. No additional run-time optimization is applied for any compressed model and in all experiments.

3.2   Baseline Algorithms

We compare DeepIoT with other three baseline algorithms:

1. **DyNS:** This is a magnitude-based network pruning algorithm [2]. The algorithm prunes weights in convolutional kernels and fully-connected layer based on the magnitude. It retrains the network connections after each pruning step and has the ability to recover the pruned weights. For convolutional and fully-connected layers, DyNS searches the optimal thresholds separately.
2. **SparseSep:** This is a sparse-coding and factorization based algorithm [19]. The algorithm simplifies the fully-connected layer by finding the optimal code-book and code based on a sparse coding technique. For the convolutional layer, the algorithm compresses the model with matrix factorization methods. We greedily search for the optimal code-book and factorization number from the bottom to the top layer.
3. **DyNS-Ext:** The previous two algorithms mainly focus on compressing convolutional and fully-connected layers. Therefore we further enhance and extend the magnitude-based method used in DyNS to recurrent layers and call this algorithm DyNS-Ext. Just like DeepIoT, DyNS-Ext can be applied to all commonly used deep network modules, including fully-connected layers, convolutional layers, and recurrent layers. If the network structure does not contain recurrent layers, we apply DyNS instead of DyNS-Ext.

For magnitude-based pruning algorithms, DyNS and DyNS-Ext, hidden elements with zero input connections or zero output connections will be pruned to further compress the network structure. In addition, all models use 32-bit floats without any quantization.

Table 5.1 LeNet5 on MNIST dataset

Layer	Hidden units	Params	DeepIoT (Hidden units/ Params)		DyNS	SparseSep
conv1 (5 × 5)	20	0.5K	10	50.0%	24.2%	84%
conv2 (5 × 5)	50	25K	20	20.0%	20.7%	91%
fc1	500	400K	10	0.8%	1.0%	78.75%
fc2	10	5K	10	2.0%	16.34%	70.28%
Total		431K		1.98%	2.35%	72.39%
Test error	0.85%		0.85%		0.85%	1.05%

3.2.1 Handwritten Digits Recognition with LeNet5

The first human interaction modality is recognizing handwritten text. In this experiment, we consider a meaningful subset of that; namely recognizing handwritten digits from visual inputs. An example application that uses this capability might be a smart wallet equipped with a camera and a tip calculator. We use MNIST[2] as our training and testing dataset. The MNIST is a dataset of handwritten digits that is commonly used for training various image processing systems. It has a training set of 60,000 examples, and a test set of 10,000 examples.

We test our algorithms and baselines on the LeNet-5 neural network model. The corresponding network structure is shown in Table 5.1. Notice that we omit all the polling layers in Table 5.1 for simplicity, because they do not contain training parameters.

The first column of Table 5.1 represents the network structure of LeNet-5, where "convX" represents the convolutional layer and "fcY" represents the fully-connected layer. The second column represents the number of hidden units or convolutional kernels we used in each layer. The third column represents the number of parameters used in each layer and in total. The original LeNet-5 is trained and achieves an error rate of 0.85% in the test dataset.

We then apply DeepIoT and two other baseline algorithms, DyNS and SparseSep, to compress LeNet-5. Note that, we do not use DyNS-Ext because the network does not contain a recurrent layer. The network statistics of the compressed model are shown in Table 5.1. DeepIoT is designed to prune the number of hidden units for a more efficient network structure. Therefore, we illustrate both the remaining number of hidden units and the proportion of the remaining number of parameters in Table 5.1. Both DeepIoT and DyNS can significantly compress the network without hurting the final performance. SparseSep shows an acceptable drop of performance. This is because SparseSep is designed without fine-tuning. It has the benefit of not fine-tuning the model, but it suffers the loss in the final performance at the same time.

The detailed tradeoff between testing accuracy and memory consumption by the model is illustrated in Fig. 5.2. We compress the original neural network with

---

[2] http://yann.lecun.com/exdb/mnist/.

5 Model Compression for Edge Computing

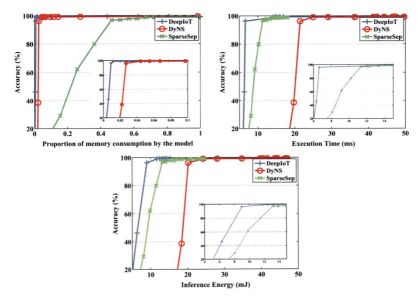

Fig. 5.2 System performance tradeoff for LeNet5 on MNIST dataset

different compression ratios and recode the final testing accuracy. In the zoom-in illustration, DeepIoT achieves at least ×2 better tradeoff compared with the two baseline methods. This is mainly due to two reasons. One is that the compressor neural network in DeepIoT obtains a global view of parameter redundancies and is therefore better capable of eliminating them. The other is that DeepIoT prunes the hidden units directly, which enables us to represent the compressed model parameters with a small dense matrix instead of a large sparse matrix. The sparse matrix consumes more memory for the indices of matrix elements. Algorithms such as DyNS generate models represented by sparse matrices that cause larger memory consumption.

The evaluation results on execution time of compressed models on Intel Edison, are illustrated in Fig. 5.2. We run each compressed model on Intel Edison for 5000 times and use the mean value for generating the tradeoff curves.

DeepIoT still achieves the best tradeoff compared with other two baselines by a significant margin. DeepIoT takes 14.2 ms to make a single inference, which reduces execution time by 71.4% compared with the original network without loss of accuracy. However SparseSep takes less execution time compared with DyNS at the cost of acceptable performance degradation (around 0.2% degradation on test error). The main reason for this observation is that, even though fully-connected layers occupy the most model parameters, most execution time is used by the convolution operations. SparseSep uses a matrix factorization method to covert the 2d convolutional kernel into two 1d convolutional kernels on two different dimensions. Although this method makes low-rank assumption on convolutional kernel, it can speed up convolution operations if the size of convolutional kernel is large (5 × 5 in this experiment). It can sometimes speed up the operation even when two 1d kernels have more parameters in total compared with the original 2d kernel. However DyNS applies a magnitude-based method that prunes most of the parameters in fully-connected layers. For convolutional layers, DyNS does not reduce the number of convolutional operations effectively, and sparse matrix multiplication is less efficient compared with regular matrix with the same number of elements. DeepIoT directly reduces the number of convolutional kernels in each layer, which reduces the number of operations in convolutional layers without making the low-rank assumption that can hurt the network performance.

The evaluation of energy consumption on Intel Edison is illustrated in Fig. 5.2. For each compressed model, we run it for 5000 times and measure the total energy consumption by a power meter. Then, we calculate the expected energy consumption for one-time execution and use the one-time energy consumption to generate the tradeoff curves in Fig. 5.2.

Not surprisingly, DeepIoT still achieves the best tradeoff in the evaluation on energy consumption by a significant margin. It reduces energy consumption by 73.7% compared with the original network without loss of accuracy. Being similar as the evaluation on execution time, energy consumption focuses more on the number of operations than the model size. Therefore, SparseSep can take less energy consumption compared with DyNS at the cost of acceptable loss on performance.

3.3 Image Recognition with VGGNet

The task image recognition through low-resolution camera. During this experiment, we use CIFAR10[3] as our training and testing dataset. The CIFAR-10 dataset consists of 60,000 32 × 32 colour images in 10 classes, with 6000 images per class. There are 50,000 training images and 10,000 test images. It is a standard testing benchmark dataset for the image recognition tasks. While not necessarily representative of seeing objects in the wild, it offers a more controlled environment for an apples-to-apples comparison.

During this evaluation, we use the VGGNet structure as our original network structure. It is a huge network with millions of parameters. VGGNet is chosen to show that DeepIoT is able to compress relative deep and large network structure. The detailed structure is shown Table 5.2.

In Table 5.2, we illustrate the detailed statistics of best compressed model that keeps the original testing accuracy for three algorithms. We clearly see that DeepIoT beats the other two baseline algorithms by a significant margin. This shows that the compressor in DeepIoT can handle networks with relatively deep structure. The compressor uses a variant of the LSTM architecture to share the redundancy

---

[3] https://www.kaggle.com/c/cifar-10.

5 Model Compression for Edge Computing   167

Table 5.2 VGGNet on CIFAR-10 dataset

Layer	Hidden units	Params	DeepIoT (Hidden units/ Params)		DyNS	SparseSep
conv1 (3 × 3)	64	1.7K	27	42.2%	53.9%	93.1%
conv2 (3 × 3)	64	36.9K	47	31.0%	40.1%	57.3%
conv3 (3 × 3)	128	73.7K	53	30.4%	52.3%	85.1%
conv4 (3 × 3)	128	147.5K	68	22.0%	67.0%	56.8%
conv5 (3 × 3)	256	294.9K	104	21.6%	71.2%	85.1%
conv6 (3 × 3)	256	589.8K	97	15.4%	65.0%	56.8%
conv7 (3 × 3)	256	589.8K	89	13.2%	61.2%	56.8%
conv8 (3 × 3)	512	1.179M	122	8.3%	36.5%	85.2%
conv9 (3 × 3)	512	2.359M	95	4.4%	10.6%	56.8%
conv10 (3 × 3)	512	2.359M	64	2.3%	3.9%	56.8%
conv11 (2 × 2)	512	1.049M	128	3.1%	3.0%	85.2%
conv12 (2 × 2)	512	1.049M	112	5.5%	1.7%	85.2%
conv13 (2 × 2)	512	1.049M	149	6.4%	2.4%	85.2%
fc1	4096	2.097M	27	0.19%	2.2%	95.8%
fc2	4096	16.777M	371	0.06%	0.39%	135%
fc3	10	41K	10	9.1%	18.5%	90.2%
Total		29.7M		2.44%	7.05%	112%
Test accuracy	90.6%		90.6%		90.6%	87.1%

information among different layers. Compared with other baselines considering only local information within each layer, sharing the global information among layers helps us learn about the parameter redundancy and compress the network structure. In addition, we observe performance loss in the compressed network generated by SparseSep. It is mainly due to the fact that SparseSep avoids the fine-tuning step. This experiment shows that fine-tuning (Line 18 to Line 21 in Algorithm 1) is important for model compression.

Figure 5.3 shows the tradeoff between testing accuracy and memory consumption for different models. DeepIoT achieves a better performance by even a larger margin, because the model generated by DeepIoT can still be represented by a standard matrix, while other methods that use a sparse matrix representation require more memory consumption.

Figure 5.3 shows the tradeoff between testing accuracy and execution time for different models. DeepIoT still achieves the best tradeoff. DeepIoT takes 82.2 ms for a prediction, which reduces 94.5% execution time without the loss of accuracy. DyNS uses less execution time compared with SparseSep in this experiment. There are two reasons for this. One is that VGGNet use smaller convolutional kernel compared with LeNet-5. Therefore factorizing 2d kernel into two 1d kernel helps less on reducing computation time. The other point is that SparseSep fails to compress the original network into a small size while keeping the original performance, because SparseSep avoids the fine-tuning.

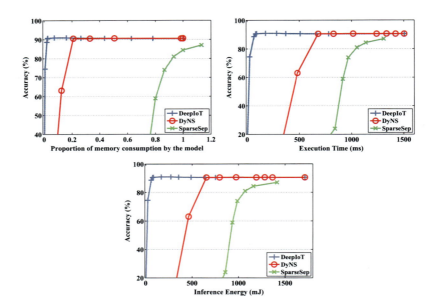

Fig. 5.3 System performance tradeoff for VGGNet on CIFAR-10 dataset

Figure 5.3 shows the tradeoff between testing accuracy and energy consumption for different models. DeepIoT reduces energy consumption by 95.7% compared with the original VGGNet without loss of accuracy. It greatly helps us to develop a long-standing application with deep neural network in energy-constrained embedded devices.

3.4 Speech Recognition with Deep Bidirectional LSTM

The task is about speech. The sensing system can take the voices of users from the microphone and automatically convert what users said into text. The previous experiment focus on the network structure with convolutional layers and fully-connected layers. We see how DeepIoT and the baseline algorithms work on the recurrent neural network in this section.

In this experiment, we use LibriSpeech ASR corpus [20] as our training and testing dataset. The LibriSpeech ASR corpus is a large-scale corpus of read English speech. It consists of 460-hour training data and 2-hour testing data.

We choose deep bidirectional LSTM as the original model [21] in this experiment. It takes mel frequency cepstral coefficient (MFCC) features of voices as inputs, and uses two 5-layer long short-term memory (LSTM) in both forward and backward direction. The output of two LSTM are jointly used to predict the spoken text. The detailed network structure is shown in the first column of Table 5.3, where

5 Model Compression for Edge Computing

Table 5.3 Deep bidirectional LSTM on LibriSpeech ASR corpus

Layer		Hidden unit		Params		DeepIoT (Hidden units/ Params)				DyNS-Ext	
LSTMf1	LSTMb1	512	512	1.090M	1.090M	55	20	10.74%	3.91%	34.9%	18.2%
LSTMf2	LSTMb2	512	512	2.097M	2.097M	192	71	4.03%	0.54%	37.2%	23.1%
LSTMf3	LSTMb3	512	512	2.097M	2.097M	240	76	17.58%	2.06%	43.1%	27.9%
LSTMf4	LSTMb4	512	512	2.097M	2.097M	258	81	23.62%	2.35%	52.3%	40.2%
LSTMf5	LSTMb5	512	512	2.097M	2.097M	294	90	28.93%	2.78%	72.6%	61.8%
fc1		29		59.3K		29		37.5%		69.0%	
Total				19.016M				9.98%		37.1%	
Word error rate (WER)		9.31				9.20				9.62	

"LSTMf" denotes the LSTM in forward direction and "LSTMb" denotes the LSTM in backward direction.

Two baseline algorithms are not applicable to the recurrent neural network, so we compared DeepIoT only with SyNS-Ext in this experiment. The word error rate (WER), defined as the edit distance between the true word sequence and the most probable word sequence predicted by the neural network, is used as the evaluation metric for this experiment.

We show the detailed statistics of best compressed model that keeps the original WER in Table 5.3. DeepIoT achieves a significantly better compression rate compared with DyNS-Ext, and the model generated by DeepIoT even has a little improvement on WER. However, compared with the previous two examples on convolutional neural network, DeepIoT fails to compress the model to less than 5% of the original parameters in the recurrent neural network case (still a 20-fold reduction though). The main reason is that compressing recurrent networks needs to prune both the output dimension and the hidden dimension. It has been shown that dropping hidden dimension can harm the network performance [10]. However DeepIoT is still successful in compressing network to less than 10% of parameters.

Figure 5.4 shows the tradeoff between word error rate and memory consumption by compressed models. DeepIoT achieves around $\times 7$ better tradeoff compared with magnitude-based method, DyNS-Ext. This means compressing recurrent neural networks requires more information about parameter redundancies within and among each layer. Compression using only local information, such as magnitude information, will cause degradation in the final performance.

Figure 5.4 shows the tradeoff between word error rate and execution time. DeepIoT reduces execution time by 86.4% without degradation on WER compared with the original network. With the evaluation on Intel Edision, the original network requires 71.15 seconds in average to recognize one human speak voice example with the average length of 7.43 seconds. The compressed structure generated by DeepIoT reduces the average execution time to 9.68 seconds without performance loss, which improves responsiveness of human voice recognition.

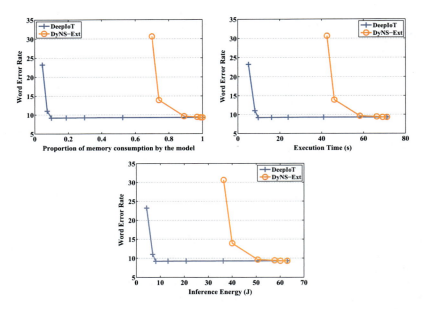

Fig. 5.4 System performance tradeoff for deep bidirectional LSTM on LibriSpeech ASR corpus

Figure 5.4 shows the tradeoff between word error rate and energy consumption. DeepIoT reduces energy by 87% compared with the original network. It performs better than DyNS-Ext by a large margin.

3.5 Supporting Human-Centric Context Sensing

In addition to experiments about supporting basic human-centric interaction modalities, we evaluate DeepIoT on one human-centric context sensing application. We compress the state-of-the-art deep learning model, DeepSense, [22] for these problems and evaluate the accuracy and other system performance for the compressed networks. DeepSense contains all commonly used modules, including convolutional, recurrent, and fully-connected layers, which is also a good example to test the performance of compression algorithms on the combination of different types of neural network modules.

The human-centric context sensing tasks we consider is heterogeneous human activity recognition (HHAR). The HHAR task recognizes human activities with motion sensors, accelerometer and gyroscope. "Heterogeneous" means that the task is focus on the generalization ability with human who has not appeared in the training set (Fig. 5.5; Table 5.4).

In this evaluation section, we use the dataset collected by Allan et al. [23]. This dataset contains readings from two motion sensors (accelerometer and gyroscope).

5 Model Compression for Edge Computing

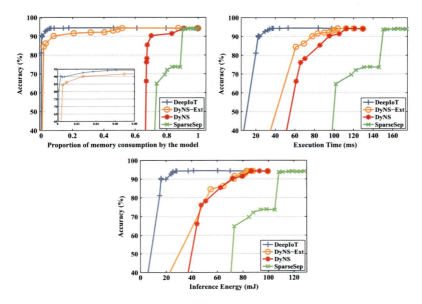

Fig. 5.5 System performance tradeoff for heterogeneous human activity recognition

Readings were recorded when users execute activities scripted in no specific order, while carrying smartwatches and smartphones. The dataset contains 9 users, 6 activities (biking, sitting, standing, walking, climbStairup, and climbStair-down), and 6 types of mobile devices.

The original network structure of DeepSense is shown in the first two columns of Table 5.12. HHAR uses a unified neural network structure DeepSense [22]. The structure contains both convolutional and recurrent layers. Since SparseSep and DyNS are not directly applicable to recurrent layers, we keep the recurrent layers unchanged while using them. In addition, we also compare DeepIoT with DyNS-Ext in this experiment.

Table 5.12 illustrates the statistics of final pruned network generated by four algorithms that have no or acceptable degradation on testing accuracy. DeepIoT is the best-performing algorithm considering the remaining number of network parameters. This is mainly due to the design of compressor network and compressor-critic framework that jointly reduce the redundancies among parameters while maintaining a global view across different layers. DyNS and SparseSep are two algorithms that can be only applied to the fully-connected and convolutional layers in the original structure. Therefore there exists a lower bound of the left proportion of parameters, *i.e.*, the number of parameters in recurrent layers. This lower bound is around 66%.

The detailed tradeoffs between testing accuracy and memory consumption by the models are illustrated in Fig. 5.12. DeepIoT still achieves the best tradeoff for sensing applications. Other than the compressor neural network providing global

Table 5.4 Heterogeneous human activity recognition

Layer		Hidden unit		Params		DeepIoT (Hidden units/Params)		DyNS-Ext		DyNS		SparseSep	
conv1a	conv1b (2×9)	64	64	1.1K	1.1K	20	19	31.25%	29.69%	92%	95.7%	50.3%	60.0%	100%	100%
conv2a	conv2b (1×3)	64	64	12.3K	12.3K	20	14	9.76%	6.49%	70.1%	77.7%	25.3%	40.5%	114%	114%
conv3a	conv3b (1×3)	64	64	12.3K	12.3K	23	23	11.23%	7.86%	69.9%	66.2%	32.1%	35.4%	114%	114%
conv4 (2×8)		64		65.5K		10		5.61%		40.3%		20.4%		53.7%	
conv5 (1×6)		64		24.6K		12		2.93%		27.2%		18.3%		100%	
conv6 (1×4)		64		16.4K		17		4.98%		24.6%		12.0%		100%	
gru1		120		227.5K		27		5.8%		1.2%		100%		100%	
gru2		120		86.4K		31		6.24%		3.6%		100%		100%	
fc1		6		0.7K		6		25.83%		98.6%		99%		70%	
Total				472.5K				6.16%		17.1%		74.5%		95.3%	
Test accuracy		94.6%				94.7%				94.6%		94.6%		93.7%	

S. Yao and T. Abdelzaher

parameter redundancies, directly pruning hidden elements in each layer also enables DeepIoT to obtain more concise representations in matrix form, which results in less memory consumption.

The tradeoffs between system execution time and testing accuracy are shown in Fig. 5.12. DeepIoT uses the least execution time when achieving the same testing accuracy compared with three baselines. It takes 36.7 ms for a single prediction, which reduces execution time by around 71.4% without loss of accuracy. DyNS and DyNS-Ext achieve better performance on time compared with SparseSep. As shown in Table 5.12, the original network uses 1-d filters in its structure. The matrix factorization based kernel compressing method used in SparseSep cannot help to reduce or even increase the parameter redundancies and the number of operations involved. Therefore, there are constraints on the network structure when applying matrix factorization based compression algorithm. In addition, SparseSep cannot be applied to the recurrent layers in the network, which consumes a large proportion of operations during running the neural network.

The tradeoffs between energy consumption and testing accuracy are shown in Fig. 5.12. DeepIoT is the best-performing algorithm for energy consumption. It reduces energy by around 72.2% without loss of accuracy. Due to the aforementioned problem of SparseSep on 1-d filter, redundant factorization causes more execution time and energy consumption in the experiment.

4 The Design of FastDeepIoT

In this section, we show how a better understanding of the non-linear relation between neural network structure and performance can further improve execution time and energy consumption without impacting accuracy.

4.1 Nonlinearities: Evidence and Exploitation

In practice, counting the number of neural network parameters and the total FLOPs does not lead to good estimates of execution time because the relation between these predictors and execution time is not proportional. On one hand, the fully-connected layer usually has more parameters but takes much less time to run compared to the convolutional layer [24]. On the other hand, one can easily find examples, where increasing the total FLOPs does not translate into added execution time. Caching effects, memory accesses, and compiler optimizations complicate the translation. Table 5.5 shows that CNN2 takes around $\times 2.6$ the execution time of CNN1, while both have *the same* total FLOPs. Moreover, CNN3 takes *longer* to run compared to CNN4 despite having *fewer* FLOPs. These observations indicate that current rules-of-thumb for estimating neural network execution time are not the best approximations.

Table 5.5 Execution time of convolutional layers with 3 × 3 kernel size, stride 1, same padding, and 224 × 224 input image size on the Nexus 5 phone

	in_channel	out_channel	FLOPs	Time (ms)
CNN1	8	32	452.4 M	114.9
CNN2	32	8	452.4 M	300.2
CNN3	66	32	3732.3 M	908.3
CNN4	43	64	4863.3 M	751.7

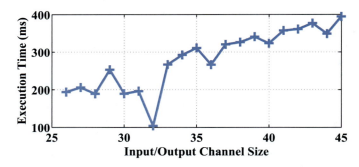

Fig. 5.6 The non-linearity of neural network execution time over input/output channel

FastDeepIoT answers two key questions to better parameterize neural network implementations for efficient execution on mobile and embedded platforms:

1. What are the main factors that affect the execution time of neural networks on mobile and embedded devices?
2. How to guide existing structure compression algorithms to minimize the neural network execution time properly?

FastDeepIoT consists of two main modules to tackle these two challenging problems, respectively.

Profiling Due to different code-level optimizations for different network structures within the deep learning library, the execution time of neural network layers can be extremely nonlinear over the structure configuration space. A simple illustration is shown in Fig. 5.6, where we plot the execution time of convolutional layers when changing the size of input and output channels simultaneously. The plot reveals non-monotonic effects, featuring periodic dips in execution time as network size increases.

A simple regression model over the entire space will thus not be a good approximation. Instead, we propose a tree-structured linear regression model. Specifically, we automatically detect key conditions at which linearity is violated and arrange them into a tree structure that splits the overall modeling space into piecewise linear regions. Within each region (tree branch), we use linear regression to convert input structure information into some key explanatory variables, predictive of execution time. The splitting of the overall space and the fitting of subspaces to predictive models are done jointly, which improves both model interpretability and accuracy.

The aforementioned modeling is done without specific knowledge of underlying hardware and deep learning library.

Compression Using the results of profiling, we then propose a compression steering module that guides existing neural network structure compression methods to better minimize execution time. The execution time model leads compression algorithms to focus more on the layer that takes longer to run instead of treating all layers equally or concentrating on inaccurate total metrics. It is also better able at exploiting non-monotonicity of execution time with respect to network structure size to reduce the former without hurting application-level accuracy metrics.

4.2 Profiling Module

We separate this module into two parts. The first part generates diverse training structures for profiling. The second part builds an accurate and interpretable model predicting the execution time of deep learning components for the corresponding structure information.

4.2.1 Neural Network Profiling

We introduce the basic system settings and the procedure of generating training structures for profiling here.

FastDeepIoT utilizes TensorFlow benchmark tool [25] to profile the execution time of all deep learning components on the target device. In order to make the profiling results fully reflect the changes on the neural network structures, we fix the frequencies of phone CPUs (processors) to be constants and stop all the power management services that can affect the processor frequency on target devices, such as fixing *mpdecision* on Qualcomm chips.

The next step is to generate diverse neural network structures for time profiling. As a deep learning component, such as a convolutional layer and recurrent layer, the combinations of its structure design choices can form an extremely huge structure configuration space. Therefore, we can only select a small proportion of structure configurations during our time profiling. The scope of our structure configuration is shown in Table 5.6, from which the network generation code chooses a random combination. Notice that we do not contain the activation function as the profiling choice, because it only occupies around 1–2% execution time of a deep learning component through empirical observations. By eliminating this insignificant configuration, *i.e.*, activation_function $\in$ {ReLU, Tanh, sigmoid}, we can save the number of profiling components by the factor of 3. Except for some pre-defined cases, such as sigmoid activation function for gate outputs in recurrent layers, we set all activation functions to be ReLU, which is one of the most widely

Table 5.6 The scope of our structure configuration for fully-connected (FC), convolutional (CNN), and recurrent (RNN) layers

Type	Structure configuration scope
FC	$in\_dim \in [1, 4096]$ $out\_dim \in [1, 4096]$
CNN	$in\_height \in [24, 225]$ $in\_width \in [24, 225]$ $kernel\_height \times kernel\_width \in \{2 \times 2, 3 \times 3, 4 \times 4, 5 \times 5, 2 \times 3\}$ $in\_channel \in [1, 256]$ $out\_channel \in [1, 256]$ $padding \in \{valid, same\}$ $stride \in \{1, 2\}$
RNN	$in\_dim \in [1, 512]$ $out\_dim \in [1, 512]$ $step \in \{8, 10, 15, 20\}$

used activation functions. In addition, the order of deep learning components in the network has little impact on their execution time empirically.

In our profiling module, for each target device, we profile around 120 neural networks with about 1300 deep learning components in total. These time profiling results form a time profiling dataset, $\mathcal{D} = \{S_i, y_i\}$, where S_i is the structure configuration and y_i the execution time.

4.2.2 Execution Time Model Building

Due to the code-level optimization for different component configuration choices in the deep learning library, execution-time non-linearity appears over the structure configuration space as shown in Fig. 5.6. The main challenge here is to build a model that can automatically figure out the conditions that cause the execution-time non-linearity without specific knowledge of underlying library and hardware.

In order to maintain both the accuracy and interpretability, we propose a tree-structure linear regression model. The model can recursively partition the structure configuration space such that the time profiling samples fitting the same linear relationship are grouped together. The intuition behind this model is that the execution time of deep learning component under each particular code-level optimization can be formulated with a linear relationship given a set of well-designed explanatory variables. In addition, different deep learning components, *i.e.*, fully connected, convolutional, and recurrent layer, learn their own execution time models.

Each time profiling data is composed of three elements. The feature vector **f**, used for identifying the condition that causes the execution-time non-linearity; the execution time y; and the explanatory variable vector **x**, used for fitting the execution time y.

The basic idea of tree-structure linear regression is to find out the most significant condition causing the execution-time non-linearity within the current dataset recursively. These conditions will form a binary tree structure. In order to figure out key conditions causing the execution-time non-linearity, we take two conditioning functions into account.

5 Model Compression for Edge Computing 177

1. *Range condition* $C_1(\mathbf{f}[j], \tau) := f[j] \leq \tau$: identifies execution-time non-linearity caused by cache and memory hit as well as specific implementation for a certain feature range.
2. *Integer multiple condition* $C_2(\mathbf{f}[j], \tau) := f[j] \equiv 0\tau$: identifies execution-time non-linearity caused by loop unrolling, data alignment, and parallelized operations.

Assume that we are generating node m in the binary tree with dataset $\mathcal{D}_m$. The model creates a set of conditions $\{\phi\}$. Each of them can partition the dataset into two subsets $\mathcal{D}_m^{(l)}(\phi)$ and $\mathcal{D}_m^{(r)}(\phi)$. Each condition ϕ consists of three elements, $\phi = \{\mathbf{f}[j], \tau_m, k\}$, where $k \in \{1, 2\}$ is the conditioning function type.

$$
\begin{aligned}
\mathcal{D}_m^{(l)}(\phi) &= \mathcal{D}_m | C_k(x_j, \tau_m), \\
\mathcal{D}_m^{(r)}(\phi) &= \mathcal{D}_m \backslash \mathcal{D}_m^{(l)}(\phi).
\end{aligned}
\tag{5.12}
$$

Node m selects the most significant condition ϕ^* by minimizing the impurity function $G(\mathcal{D}_m, \phi)$,

$$
\phi^* = {}_\phi G(\mathcal{D}_m, \phi),
\tag{5.13}
$$

$$
G(\mathcal{D}_m, \phi) = \frac{|\mathcal{D}_m^{(l)}(\phi)|}{|\mathcal{D}_m|} H\big(\mathcal{D}_m^{(l)}(\phi)\big) + \frac{|\mathcal{D}_m^{(r)}(\phi)|}{|\mathcal{D}_m|} H\big(\mathcal{D}_m^{(r)}(\phi)\big),
\tag{5.14}
$$

$$
H(\mathcal{D}) = \min_{\mathbf{w},b} \frac{1}{|\mathcal{D}|} \sum_{(\mathbf{x},y)\in\mathcal{D}} (\mathbf{w}^\mathsf{T}\mathbf{x} + b - y)^2 \quad \text{s.t.} \mathbf{w}, b \geq 0.
\tag{5.15}
$$

The impurity function is designed as the weighted mean square errors of linear regressions over two sub-datasets partitioned by the condition ϕ.

Next, we describe the feature vector $\mathbf{f}$. Our choice of feature vector $\mathbf{f}$ contains three parts: the structure features, the memory features, and the parameter feature. The structure features refer to *in_dim* and *out_dim* for fully-connected and recurrent layers as well as *in_channel* and *out_channel* for convolutional layers. The memory features include the memory size of input, *mem_in*, the memory size of output, *mem_out*, and the memory size of internal representations, *mem_inter*. The parameter feature refers to the size of parameters, *param_size*. The detailed definitions of memory and parameter features are shown in Table 5.7. All notations in Table 5.7 are consistent with the notations of structure configurations in Table 5.6, except for the height and width of output image, *out_height* and *out_width*, in the convolutional layer. However, we can easily calculate these two values based on other structure information, *i.e.*, *in_height*, *in_width*, *kernel_height*, *kernel_width*, *stride*, and *padding*.[4]

Last, we discuss about our explanatory variable vector $\mathbf{x}$ for linear regression. In this dissertation, we build an intuitive performance model that the execution time of

---

[4] https://www.tensorflow.org/api_guides/python/nn#Convolution.

Algorithm 2 Execution time model building

1: **Input:** time profiling dataset $\mathcal{D}_z$, feature vector $\mathbf{f}$, two conditioning functions $f[j] \leq \tau$ and $f[j] \equiv 0\tau$, and explanatory variable $\mathbf{x}_z$.
2: Fit $\mathcal{D}_z$ with $\mathbf{w}_r$ and b_r according to (5.15).
3: Save root node $r = [\emptyset, \mathbf{w}_r, b_r]$
4: **Initialize:** que $= \big[[r, \mathcal{D}_z]\big]$.
5: **while** len(que) > 0 **do**
6:     $q, \mathcal{D}_q =$ que.deque()
7:     **if** $\mathcal{D}_q$ meets stoping condition **then**
8:         Continue
9:     **end if**
10:     Search for the optimal partition $\phi^* = \{f[j^*], \tau^*, k^*\}$ according to (5.13) (5.14) (5.15).
11:     Generate partitioned dataset $\mathcal{D}_q^{(l)}(\phi^*)$ and $\mathcal{D}_m^{(r)}(\phi^*)$ according to (5.12).
12:     Fit $\mathcal{D}_q^{(l)}(\phi^*)$ with $\mathbf{w}_q^{(l)}$ and $b_q^{(l)}$ according to (5.15).
13:     Fit $\mathcal{D}_q^{(r)}(\phi^*)$ with $\mathbf{w}_q^{(r)}$ and $b_q^{(r)}$ according to (5.15).
14:     Save q's left child node $q^{(l)} = \big[[\text{True}, \phi^*], \mathbf{w}_q^{(l)}, b_q^{(l)}\big]$
15:     Save q's right child node $q^{(r)} = \big[[\text{False}, \phi^*], \mathbf{w}_q^{(r)}, b_q^{(r)}\big]$
16:     que.enque$\big([q^{(l)}, \mathcal{D}_q^{(l)}(\phi^*)]\big)$
17:     que.enque$\big([q^{(r)}, \mathcal{D}_q^{(r)}(\phi^*)]\big)$
18: **end while**

Table 5.7 The definition of parameter and memory information for Fully-Connected layer (FC), Convolutional layer (CNN), Gated Recurrent Unit (GRU), and Long Short Term Memory (LSTM)

Type	mem_in	mem_out	mem_inter
FC	in_dim	out_dim	0
CNN	in_height × in_width×	out_height × out_width×	out_height × out_width × kernel_height×
	in_channel	out_channel	kernel_width × in_channel
GRU	step × in_dim	step × out_dim	3 × step × out_dim
LSTM	2 × step × in_dim	2 × step × out_dim	4 × step × out_dim

Type	param_size
FC	in_dim × out_dim + out_dim
CNN	kernel_height × kernel_width× in_channel × out_channel +1
GRU	3 × out_dim× (in_dim + out_dim + 1)
LSTM	4 × out_dim× (in_dim + out_dim + 1)

a program is contributed by three parts, CPU operations, memory operations, and disk I/O operations. For a deep learning component, these parts refer to FLOPs, memory size, and parameter size,

$$\mathbf{x} = [\text{FLOPs}, mem, param\_size]. \tag{5.16}$$

5 Model Compression for Edge Computing 179

where $mem = mem\_in + mem\_out + mem\_inter$.

With the weight vector $\mathbf{w}$ and the bias term b, the overall execution time of a deep learning component, y, can be modelled as $y = \mathbf{w}^\mathsf{T}\mathbf{x} + b$. Since every term should have a positive contribution to the execution time, we add an additional constraint, $\mathbf{w}, b \geq 0$, as shown in (5.15).

The tree-structure linear regression model builds a binary tree that gradually picks out conditions that cause execution-time non-linearity and breaks the dataset into subsets that contain more "linearity". Our designed explanatory variable vector $\mathbf{x}$ is able to fit the dataset with linear relationships better level by level, especially for fully-connected and convolutional layer.

The recurrent layers, however, still have flaws. We analyze the error and find out that recurrent layers have a constant initialization overhead or set-up time for each step. Therefore, we update explanatory variable vector $\mathbf{x}$,

$$\mathbf{x}_{fc} = \mathbf{x}_{cnn} = [\text{FLOPs}, mem, param\_size],$$
$$\mathbf{x}_{rnn} = [\text{FLOPs}, mem, param\_size, step]. \tag{5.17}$$

We summarize our execution time model building process in Algorithm 2. There is a stopping condition in Line 7 that keeps tree-structure linear regression from growing infinitely. In our case, the stopping condition occurs when a linear regression can fit the current dataset $\mathcal{D}_q$ with a mean absolute percentage error less than 5% or when the size of current dataset is smaller than 15, $|\mathcal{D}_q| < 15$.

4.2.3 Execution Time Model with Statistical Analysis

In this part, we provide an illustration of the FastDeepIoT profiling module on Nexus 5 phone with statistical analysis. The module first profiles and generates the execution time profiling dataset. Then, the module builds an execution time model for each deep learning component based on the tree-structure linear regression in Algorithm 2. Additional evaluations on the execution time model will be shown in Sect. 5.2.

For fully-connected layers and recurrent layers, including GRU and LSTM, their execution time has a perfect linear relationship with our explanatory variable vector $\mathbf{x}_{fc}$ and $\mathbf{x}_{rnn}$. However, the execution time model of convolutional layers reflects a strong non-linearity over the structure configuration space. As shown in Fig. 5.7, the execution time of convolutional layer has local minima when $in\_channel$ or $out\_channel$ is a multiple of 4.

Then we calculate the p-values to evaluate the mathematical relationship between each explanatory variable and the execution time. The p-value for each explanatory variable tests the null hypothesis that the variable has no correlation with the execution time. Results are shown in Table 5.8. The p-values of explanatory variables, FLOPs, mem, and $step$, are less than the significance level (0.05) for all deep learning components (Fig. 5.8).

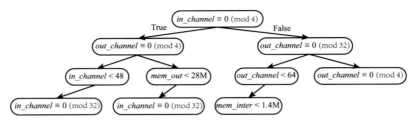

Fig. 5.7 The execution time model of convolutional layers on Nexus 5

Table 5.8 The p-values of explanatory variables

Type	FLOPs	mem	param_size	step
FC	0.000	0.000	1.000	
CNN	0.037	0.009	1.000	
GRU	0.000	0.000	1.000	0.000
LSTM	0.000	0.000	1.000	0.000

Fig. 5.8 Simplified execution time model of convolutional layers on Nexus 5

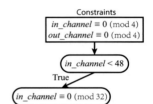

So our empirical time profiling data provides enough evidence that the correlation between these explanatory variables and the execution time are statistically significant. However, the p-values for *param_size* is high for all cases, which shows that the number of parameters has limited correlation with the execution time. This experiment, again, highlights the importance of proposing a compression algorithm targeting on minimizing the execution time instead the number of parameters.

4.3 Compression Steering Module

Profiling and modelling deep learning execution time is not enough for speeding up the model execution. In this section, we introduce the compression steering module that is designed to empower existing deep learning structure compression algorithms to minimize model execution time properly.

We assume that $\mathcal{S} = \{\mathbf{s}_l\}$ and $\mathcal{W} = \{\mathbf{W}_l\}$ for $l = 1, \cdots, L$ is structure information and weight matrix of a neural network from layer 1 to layer L respectively. We denote our execution time model as $t_l = \mathcal{T}(\mathbf{s}_l)$, which takes the structure information $\mathbf{s}_l$ as input and predicts the component execution time t_l. For a general neural network structure compression algorithm, we denote the original

5 Model Compression for Edge Computing

Algorithm 3 Layer structure expansion and local minima searching

---

1: **Input:** the execution time model $\mathcal{T}()$ with root node r and the layer structure $\{\mathbf{f}, \mathbf{x}\}$.
2: Set node $= r$, condL $= []$.
3: **while** $\neg$(node.left $==$ None&node.right $==$ None) **do**
4:     **if** node.cond is a range condition **then**
5:         condL.append(node.cond)
6:         **if f** obeys node.cond **then**
7:             node $=$ node.left
8:         **else**
9:             node $=$ node.right
10:         **end if**
11:     **else**
12:         $\hat{\mathbf{f}} = \mathbf{f}$ and $\hat{\mathbf{x}} = \mathbf{x}$.
13:         $\hat{\mathbf{f}}[\text{node}.j] = \text{node}.\tau \times [\,]\mathbf{f}[\text{node}.j]/\text{node}.\tau$
14:         Update $\hat{\mathbf{x}}$ according to $\hat{\mathbf{f}}$.
15:         **if** node.$\mathbf{w}_T^\mathsf{T}\hat{\mathbf{x}} + $ node.$b_T > $ node.$\mathbf{w}_F^\mathsf{T}\mathbf{x} + $ node.b_F &$\hat{\mathbf{f}}$ obeys condL **then**
16:             $\mathbf{f} = \hat{\mathbf{f}}$ and $\mathbf{x} = \hat{\mathbf{x}}$.
17:             node $=$ node.left
18:         **else**
19:             node $=$ node.right
20:         **end if**
21:     **end if**
22: **end while**
23: **Return: f**

---

compression process as,

$$\min_{\mathcal{S},\mathcal{W}} \mathcal{L}_\theta(\mathcal{S}, \mathcal{W}), \tag{5.18}$$

where the compression algorithm minimizes a loss function, concerning prediction error or parameter size, with either the gradient descend or searching based optimization method.

In order to enable the compression algorithm to minimize the execution time, our first step is to incorporate the execution time model into the original objective function (5.18),

$$\min_{\mathcal{S},\mathcal{W}} \mathcal{L}_\theta(\mathcal{S}, \mathcal{W}) + \lambda \sum_{l=1}^{L} \mathcal{T}(\mathbf{s}_l), \tag{5.19}$$

where λ is a hyper-parameter that make the tradeoff between minimizing training loss and minimizing execution time.

Adding execution time to the compression objective function can encourage the compression algorithm to concentrate more on the layers with higher execution time, which helps to speed up the whole neural network.

However, due to the existence of execution-time local minima, compressing neural network structure is not always the optimal choice for minimizing the execution time. As shown in Fig. 5.6, enlarging neural network structure can find a nearby execution-time local minimum that reduces the execution time. Notice that enlarging structure is a lossless operation. We can at least enlarge weight matrices with zeros that keeps performance the same.

In general, utilizing execution-time local minima for speeding up involves two steps:

1. Identifying an expanded structure configuration that can trigger a nearby execution-time local minimum.
2. Deciding whether the expanded structure can speed up the execution time.

For an execution time model trained with a complex method, such as neural networks, identifying a nearby execution-time local minimum can be almost impossible by blindly searching a large configuration space. However, our tree-structure linear regression can easily identify a nearby local minimum speeding up the neural network execution.

Local extrema, *i.e.*, maxima and minima, are identified by the integer multiple condition, $f[j] \equiv 0\tau$, in our tree-structure linear regression model. Our compression steering module searches for the nearby local maxima by gradually expanding the structure that fits the integer multiple conditions from root node to leaf node in the execution time model.

Assume that node m is under the condition $\mathbf{f}[j_m] \equiv 0\tau_m$ with two sets of linear regression parameters $\{\mathbf{w}_T, b_T\}$ and $\{\mathbf{w}_F, b_F\}$ used for fitting the dataset that obeying and against the condition respectively. A deep learning layer is denoted with the feature vector $\mathbf{f}_l$ and the explanatory variable vector $\mathbf{x}_l$. The compression steering module generates an expanded layer with feature vector $\hat{\mathbf{f}}_l$ and explanatory variable vector $\hat{\mathbf{x}}_l$ by updating the conditioning feature $\hat{\mathbf{f}}[j_m] = \tau_m[]\mathbf{f}[j_m]/\tau_m$. Then the module compares the values between $\mathbf{w}_T^\mathsf{T}\hat{\mathbf{x}} + b_T$ and $\mathbf{w}_F^\mathsf{T}\mathbf{x} + b_F$ to decide whether it should accept the expansion for speeding-up and go through the corresponding branch.

The layer structure expansion and local minima searching process is summarized in Algorithm 3. The algorithm goes through whole tree structure to find out a nearby local minimum that reduces the execution time.

For a whole neural network, each layer goes through the structure expansion and local minima searching process one by one. It is possible that conflicts exist between expanded structures of two neighbouring layers. The module solves these conflicts sequentially by choosing the one having shorter overall execution time.

In addition, we can further analyze the structure expansion process for a particular component on a particular device for a particular application settings. For example, assume that we are compressing the *in_channel* and *out_channel* of a convolutional layer on Nexus 5 with kernel size 3×3, input image size 24×24, and the same padding. We are considering the root condition *in_channel* $\equiv 04$ as shown in Fig. 5.7. According to our execution time model, two linear regression models that fit the two datasets in the left and right child of the root node are:

5 Model Compression for Edge Computing

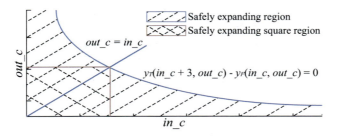

Fig. 5.9 The square region of safely expanding $in\_channel$ for speed up

$$\mathbf{w}_T = [3.41 \times 10^{-8}, 4.03 \times 10^{-6}, 7.11 \times 10^{-25}] b_T = 8.11,$$
$$\mathbf{w}_F = [3.11 \times 10^{-8}, 8.03 \times 10^{-6}, 1.52 \times 10^{-34}] b_F = 12.82. \quad (5.20)$$

Then we can obtain the execution time as a function of $in\_channel$ and $out\_channel$ by substituting the explanatory variable vector $\mathbf{x}$ with definitions illustrated in Table 5.7 as well as the application settings about kernel size, input image size, and padding option.

$$\begin{aligned} y_T(in\_c, out\_c) &= 3.53 \times 10^{-4} \cdot in\_c \cdot out\_c + 8.11 + \\ &\quad 2.32 \times 10^{-2} \cdot in\_c + 2.32 \times 10^{-3} \cdot out\_c, \\ y_F(in\_c, out\_c) &= 3.23 \times 10^{-4} \cdot in\_c \cdot out\_c + 12.82 + \\ &\quad 4.63 \times 10^{-2} \cdot in\_c + 4.63 \times 10^{-3} \cdot out\_c, \end{aligned} \quad (5.21)$$

where we denote $in\_channel$ and $out\_channel$ as in_c and out_c for simplicity.

We are interested in the region where expanding the $in\_channel$ to a nearby multiple of 4 can speed up the execution. This is equivalent to solving

$$y_T(in\_c + 3, out\_c) - y_F(in\_c, out\_c) < 0, \quad (5.22)$$

where its zero contour line is a hyperbola. Therefore, within the region bounded by $in\_channel$ axis, $out\_channel$ axis, and zero contour line, we can safely expand $in\_channel$ to a multiple of 4 to speed up the convolutional layer execution time.

In order to have a more interpretable result, as shown in Fig. 5.9, we can obtain a square region by finding the intersections between the zero contour line and the function $out\_channel = in\_channel$. In this case, within the region $in\_channel \times out\_channel \in [1, 1288] \times [1, 1288]$, we can blindly expand $in\_channel$ to a multiple of 4 to speed up. This region is much larger than the region we are interested in. We can keep analyzing the next condition $out\_channel \equiv 04$ and achieve similar result. Within the region $in\_channel \times out\_channel \in [1, 808] \times [1, 808]$, we can safely expand $in\_channel$ and $out\_channel$ to a nearby multiple of 4 to speed up. In the end, we can obtain a simplified execution time model $\hat{\mathcal{T}}$ as shown in Fig. 5.8.

In summary, the compression steering module compresses the neural network structure for reducing overall execution time with three steps.

1. Compressing neural network with a time-aware objective function (5.19) with execution time model $\mathcal{T}$.
2. Expanding layer structure and searching local minima for further speed up according to Algorithm 3 with execution time model $\mathcal{T}$ or $\hat{\mathcal{T}}$ (if available).
3. Depending on the original compression algorithm, freeze the structure and fine-tune the neural network.

5 The Evaluation of FastDeepIoT

In this section, we evaluate FastDeepIoT through two sets of experiments. The first set evaluates the accuracy of the execution time model generated by our profiling module, while the second set evaluates the performance of our compression steering module. In order to evaluate execution time modeling accuracy, we compare our tree-structured linear regression model to other state-of-the-art regression models on two mobile devices. To evaluate the quality of compression, we present a set of experiments that demonstrate the speed-up of the compressed neural network obtained by the compression steering module with three human-centric interaction and sensing applications.

5.1 Implementation

In this section, we briefly describe the hardware, software, and architecture of FastDeepIoT.

Hardware We test FastDeepIoT on two types of hardware, Nexus 5 phone and Galaxy Nexus phone. Two devices are profiled for each type of hardware. The Nexus 5 phone is equipped with quad-core 2.3 GHz CPU and 2 GB memory. The Galaxy Nexus phone is equipped with dual-core 1.2 GHz CPU and 1 GB memory. We stop the *mpdecision* service and use *userspace* CPU governor for two hardware. We manually set 1.1 GHz for the quad-core CPU on Nexus 5, and 700 MHz for the dual-core CPU on Galaxy Nexus to prevent overheating caused by the constant time profiling. In addition, all profiling and testing neural network models are run solely on CPU. The execution time model building and the compression steering module are implemented on a workstation connected to two phones. All compressing steps are implemented on the workstation.

Software FastDeepIoT utilizes TensorFlow benchmark tool [25], a C++ binary, to profile the execution time of deep learning components. For each neural network, the benchmark tool have one warm up run to initialize the model and then profile all components execution time with 20 runs without internal delay. Mean values are taken as the profiled execution time.

We install Android 5.0.1 on Nexus 5 phone and Android 4.3 on Galaxy Nexus phone. All additional background services are closed during the profiling and testing. All energy consumptions on two devices are measured by an external power meter.

Architecture Given a target device, FastDeepIoT first queries the device and its own database for a pre-generated execution time model with device type and OS version as the key. If the query fails, the profiling module starts its function. FastDeepIoT generates random neural network structures based on the configuration scope in Table 5.6, pushes the Protocol Buffers (.pb file) to the target device, profiles the execution time of components, fetches back and processes the profiling result. Once the profiling process has finished, FastDeepIoT learns tree-structure linear regression execution time models according to Algorithm 2 based on the time profiling dataset. FastDeepIoT pushes the generated execution time models to the target device and its own database for storage.

Then given an original neural network structure and parameters, the compression steering module can automatically generate a compressed structure to speed up inference time for a target device. FastDeepIoT queries the target device and own database for a pre-generated execution time model, and choose a structure compression algorithm, DeepIoT as a default, to reduce the deep learning execution time according to (5.19) and Algorithm 3. The resulting compressed neural network is transferred to the target device used locally.

5.2 Execution Time Model

We implement the following execution time estimation alternatives:

1. **SVR:** support vector regression with radial basis function kernel [26]. This algorithm tries to perform linear separation over a higher dimensional kernel feature space by characterizing the maximal margin.
2. **DT:** classification and regression trees [27]. This is an interpretable model. It groups and predicts execution time by the execution time itself.
3. **RF:** random forest regression [28]. This algorithm trades the interpretability of regression tree for the predictive performance by ensembling multiple trees with random feature selections.
4. **GBRT:** gradient boosted regression trees [29]. This algorithm builds an additive model in a forward stage-wise fashion, which is hard to interpret.

5. **DNN:** multilayer perceptron [30]. Deep neural network is a learning model with high capacity. We build a four-layer fully connected neural network with LeRU as the activation function, except for the output layer. We fine-tune the structure and apply dropout as well as L2 regularization to prevent overfitting. DNN is a black-box model.

We train all the baseline models with the dataset generated by the profiling module in FastDeepIoT (75% for training and 25% for testing). For each deep learning component, such as CNN and LSTM, an individual model is trained. We have trained these models with feature vector $\mathbf{f}$, explanatory variable vector $\mathbf{x}$, and the concatenate of feature and explanatory variable vectors as inputs, where $\mathbf{f}$ and $\mathbf{x}$ are the same as the definitions in Sect. 4.2.2. We find that the model trained with explanatory variable vector $\mathbf{x}$ outperforms other choices consistently in all cases, so we only report the results of models trained with $\mathbf{x}$ for simplicity.

We evaluate these models on convolutional layer, gated recurrent unit, long short term memory, and fully-connected layer with mean absolute percentage error, mean absolute rrror, and coefficient of determination on two hardware. As shown in Table 5.9, FastDeepIoT is consistently among top 2 predictors for all experiments with all three metrics. FastDeepIoT also outperforms the highly capable deep learning model for more than half of the cases, while FastDeepIoT is much more interpretable. There are two reasons for the remarkable performance of FastDeepIoT. On one hand, FastDeepIoT captures the primary characters of deep learning execution time behaviours, which makes an interpretable and accurate model possible. On the other hand, since the profiled dataset is limited (around one thousand samples for training), complex models such as deep neural networks that require large training dataset may not be the best choice here.

5.3 Compression Steering Module

In this section, we evaluate the performance of our compression steering module with three sensing applications. We train the neural networks on traditional benchmark datasets as original models. Then, we compress the original models using FastDeepIoT and the three state-of-the-art baseline algorithms. Finally, we test the accuracy, execution time, and energy consumption of compressed models on mobile devices.

We compare FastDeepIoT with three baseline algorithms:

1. **DeepIoT:** This is a state-of-the-art neural structure compression algorithm [1]. The algorithm designs a compressor neural network with adaptive dropout to explore a succinct structure for the original model.
2. **DeepIoT+localMin:** We enhance DeepIoT with the ability of expanding layer for finding execution-time local minima. This method takes the compressed model of DeepIoT and expands its layers with zero-value elements that can trigger local minima according to Algorithm 3. We use this almost zero-effort

5 Model Compression for Edge Computing

Table 5.9 The Mean Absolute Percentage Error (MAPE), Mean Absolute Error (MAE) in millisecond, and Coefficient of determination (R^2) of execution time models

	FastDeepIoT	SVR	DT	RF	GBRT	DNN
Nexus 5-Convolutional layer						
MAPE	**7.6%**	233.8%	23.8%	19.7%	10.9%	**6.4%**
MAE	**15.2**	227.1	39.2	27.3	20.5	16.4
R^2	**0.991**	−0.229	0.969	0.985	0.988	**0.994**
Nexus 5-Gated recurrent unit						
MAPE	**1.8%**	78.7%	9.4%	6.7%	4.8%	2.0%
MAE	**0.6**	23.6	2.9	1.8	1.5	0.7
R^2	**0.999**	−0.078	0.986	0.995	0.995	**0.999**
Nexus 5-Long short term memory						
MAPE	**2.3%**	73.7%	9.0%	4.7%	4.1%	2.8%
MAE	**0.6**	23.7	3.0	1.4	1.6	0.9
R^2	**0.999**	−0.223	0.977	0.995	0.993	0.998
Nexus 5-Fully-connected layer						
MAPE	**1.9%**	133.5%	22.5%	12.0%	**0.2%**	1.9%
MAE	**0.19**	5.98	1.18	0.38	**0.01**	0.19
R^2	**0.999**	0.065	0.977	0.996	**0.999**	**0.999**
Galaxy Nexus-Convolutional layer						
MAPE	**4.1%**	164.3%	33.1%	23.0%	15.2%	14.5%
MAE	**26.8**	878.7	162.9	123.7	114.6	110.1
R^2	**0.999**	−0.246	0.969	0.980	0.982	0.983
Galaxy Nexus-Gated recurrent unit						
MAPE	**2.9%**	71.5%	10.5%	8.8%	6.0%	4.1%
MAE	**1.1**	27.8	4.8	4.1	3.2	2.2
R^2	**0.997**	−0.065	0.968	0.977	0.984	0.989
Galaxy Nexus-Long short term memory						
MAPE	**2.9%**	66.8%	8.4%	7.8%	6.0%	**2.9%**
MAE	**1.4**	26.2	3.0	3.3	2.7	**1.3**
R^2	**0.997**	−0.196	0.983	0.985	0.987	**0.997**
Galaxy Nexus-Fully-connected layer						
MAPE	**4.0%**	55.0%	12.3%	11.3%	9.5%	4.1%
MAE	**0.3**	6.7	1.2	0.9	1.0	**0.3**
R^2	**0.996**	−0.629	0.944	0.972	0.949	**0.996**

Bold values reflect the evaluation performance of techniques that are presented in the current section

method to show the improvement made on existing compressed models by interpreting deep learning execution time with FastDeepIoT.

3. **DeepIoT+FLOPs:** This method enhances DeepIoT by adding a term that minimizes FLOPs to the original objective function (5.18). Since a large proportion of works use FLOPs as the execution time estimation [3–5], this method shows to what extend FLOPs can be used to compress neural network for reducing execution time.

Table 5.10 VGGNet (hidden units) on CIFAR-10 dataset

Layer	No execution time model				Nexus 5	Galaxy Nexus
	Original	DeepIoT	localMin	FLOPs	FastDeepIoT	FastDeepIoT
conv1-1 (3×3)	64	27	28	19	12	16
conv1-2 (3×3)	64	47	48	17	16	24
conv2-1 (3×3)	128	53	56	33	28	36
conv2-2 (3×3)	128	68	68	50	32	44
conv3-1 (3×3)	256	104	104	89	64	72
conv3-2 (3×3)	256	97	100	79	64	56
conv3-3 (3×3)	256	89	92	77	68	72
conv4-1 (3×3)	512	122	124	115	132	96
conv4-2 (3×3)	512	95	96	112	136	80
conv4-3 (3×3)	512	64	64	112	104	120
conv5-1 (2×2)	512	128	128	143	148	116
conv5-2 (2×2)	512	112	112	132	144	108
conv5-3 (2×2)	512	146	148	182	104	92
fc1	4096	27	27	1097	132	132
fc2	4096	161	161	935	152	123
fc3	1000	10	96	72	157	167
Test accuracy	90.6%	90.6%	90.6%	90.6%	90.6%	90.6%
Execution time t (Nexus 5)	328 ms	31 ms	21 ms	28 ms	**16 ms**	23 ms
Execution time t (Galaxy)	610 ms	72 ms	63 ms	52 ms	36 ms	**34 ms**

Bold values reflect the evaluation performance of techniques that are presented in the current section

5.3.1 Image Recognition on CIFAR-10

This is a vision based task, image recognition based on a low-resolution camera. During this experiment, we use CIFAR-10 as our training and testing dataset. The CIFAR-10 dataset consists of 60,000 32×32 colour images in 10 classes, with 6000 images per class. There are 50,000 training images and 10,000 test images.

During the evaluation, we use VGGNet structure as the original network structure [31]. The detailed structure is shown in Table 5.10, where we also illustrate the best compressed models that keeps the original test accuracy for all algorithms. The compressed model can be even deployed on tiny IoT devices such as Intel Edison.

As shown in Table 5.10, FastDeepIoT achieves the best performance on two hardware with their corresponding execution time models. Compared with the state-of-the-art DeepIoT algorithm, FastDeepIoT can further reduce the model execution time by 48–53%. DeepIoT+localMin outperforms DeepIoT on two hardware, reducing the execution time by 12–32%. This shows that we can decently reduce the neural network execution time by simply expanding the neural network structure

5 Model Compression for Edge Computing

Fig. 5.10 System performance tradeoff for VGGNet on CIFAR-10 dataset

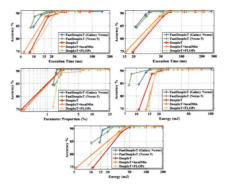

to local execution-time minima. In additional, DeepIoT+FLOPs can speed up the model execution time compared with DeepIoT. However, FastDeepIoT still outperforms DeepIoT+FLOPs by a significant margin. This result highlights that FLOPs is not a proper estimation of time.

Figure 5.10 shows the tradeoff between testing accuracy and execution time for different algorithms. FastDeepIoT consistently outperforms other algorithms by a significant margin. Furthermore, the execution time characters on different hardware can affect the final performance. FastDeepIoT (Nexus 5/Galaxy Nexus) performs better on its corresponding hardware. DeepIoT+localMin achieves a better tradeoff compared with DeepIoT. Therefore, utilizing execution-time local minima is a low-cost strategy to speed up neural network execution. In addition, since FLOPs has different degrees of execution time contribution on different hardware, DeepIoT+FLOPs are not able to achieve a better tradeoff than DeepIoT on all devices.

Figure 5.10 shows the tradeoff between testing accuracy and energy consumption for different algorithms. Although FastDeepIoT is not designed to minimize the energy consumption, FastDeepIoT still achieves the best tradeoff. However, we can see that the characters of energy consumption of deep neural network are different from the execution time. FastDeepIoT with the hardware-specific time models are not always the most energy-saving method on the corresponding hardware. Execution-time local minima cannot consistently help DeepIoT+localMin to outperform DeepIoT. Therefore, further studies on understanding and minimizing deep learning energy consumption are needed.

Figure 5.10 shows the tradeoff between testing accuracy and left proportion of model parameters. Since there is no algorithm targeting at minimizing model parameters, all methods show comparable performances. However, from another perspective, the execution time model learnt by FastDeepIoT empowers existing compression algorithms to reduce more execution time with almost the same amount of parameters.

Table 5.11 VGGNet (hidden units) on ImageNet dataset

Layer	No execution time model				Nexus 5	Galaxy Nexus
	Original	DeepIoT	localMin	FLOPs	FastDeepIoT	FastDeepIoT
conv1-1 (3 × 3)	64	43	44	23	12	16
conv1-2 (3 × 3)	64	47	48	32	12	16
conv2-1 (3 × 3)	128	100	100	65	20	44
conv2-2 (3 × 3)	128	97	100	67	40	40
conv3-1 (3 × 3)	256	164	164	116	88	108
conv3-2 (3 × 3)	256	164	164	135	72	104
conv3-3 (3 × 3)	256	153	156	70	116	108
conv4-1 (3 × 3)	512	235	236	72	268	240
conv4-2 (3 × 3)	512	240	240	181	236	216
conv4-3 (3 × 3)	512	220	240	258	340	200
conv5-1 (3 × 3)	512	255	256	261	376	240
conv5-2 (3 × 3)	512	260	260	303	376	288
conv5-3 (3 × 3)	512	257	260	47	176	216
fc1	4096	436	436	1594	656	920
fc2	4096	1169	1169	824	1150	1189
fc3	1000	297	297	405	287	402
Test top-5 accuracy	88.9%	88.9%	88.9%	88.9%	88.9%	88.9%
Execution time t (Nexus 5)		1682 ms	1605 ms	968.8 ms	**688.8 ms**	725.7 ms
Execution time t (Galaxy)		7773 ms	6991 ms	3930 ms	3211 ms	**2930 ms**

Bold values reflect the evaluation performance of techniques that are presented in the current section

5.3.2 Large-Scale Image Recognition on ImageNet

This is a large-scale vision based task, image recognition based on a high-resolution camera. During this experiment, we use ImageNet as our training and testing dataset. The ImageNet dataset consists of 1.2 million 224 × 224 color images in 1000 classes with 100,000 images for testing.

During the evaluation, we still use VGGNet structure as the original network structure. The detailed structures of best compressed models without accuracy degradation of all algorithms are shown in Table 5.11. Note that the original VGGNet for 224 × 224 colour image input is too large for running on two testing hardware. FastDeepIoT achieves the best performance on the execution time among all methods. Compared with the state-of-the-art DeepIoT method, FastDeepIoT can further reduce the execution time by 59–62%. DeepIoT+localMin still outperforms DeepIoT by reducing around 5–10% of execution time. In addition, FastDeepIoT can further reduce 25–29% of execution time compared with DeepIoT+FLOPs.

Figure 5.11 shows the tradeoff between testing top-5 accuracy and execution time for all algorithms. FastDeepIoT consistently outperforms all other algo-

5 Model Compression for Edge Computing

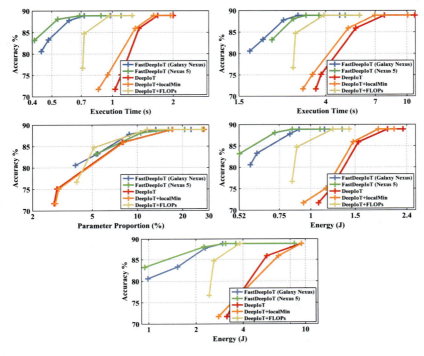

Fig. 5.11 System performance tradeoff for VGGNet on ImageNet dataset

rithms by a significant margin. With the help of execution-time local minima, DeepIoT+localMin can still outperform DeepIoT in all cases. DeepIoT+FLOPs performs better than DeepIoT in this case. As shown in Table 5.8, FLOPs still possess a certain degree of correlation with execution time. However, compared to the execution time model in FastDeepIoT, FLOPs becomes an inferior execution time estimator.

Figure 5.11 illustrates the tradeoff between testing top-5 accuracy and energy consumptions. FastDeepIoT outperforms all algorithms with a large margin. However, FastDeepIoT with the Galaxy Nexus execution time model is not the most energy-saving compression method on the Galaxy Nexus device. Also, DeepIoT+localMin cannot consistently outperforms DeepIoT on energy saving. These two observations witness the discrepancies between the execution time and energy modeling on mobile devices. Figure 5.11 shows the tradeoff between testing accuracy and left proportion of model parameters. Again, all methods show the similar tradeoff, which indicates that FastDeepIoT is a parameter-efficient method on execution time reduction.

5.3.3 Heterogeneous Human Activity Recognition

This is a human-centric context sensing application, recognizing human activities with accelerometer and gyroscope. Especially, we are considering the heterogeneous human activity recognition (HHAR). This task focuses on the generalization ability with human who has not appeared in the training dataset. During this experiment, we use the dataset collected by Allan et al. [23].

During this evaluation, we use DeepSense structure as the original network structure [22]. Table 5.12 illustrates the detailed structure of the original network and final compressed networks generated by four algorithms with no degradation on testing accuracy. As shown in Table 5.12, FastDeepIoT achieves the best performance on two devices with the corresponding execution time models. Compared with DeepIoT, FastDeepIoT can further reduce the model execution time by 22–42%. During the compressing process, we observe that all compressed models tend to approach a model execution time lower bound, which has not been seen in the previous two experiments. In order to obtain the lower bound, we build a DeepSense structure with all hidden units that equal to 1, and then applies Algorithm 3 to find the structure that triggers local minimum. The resulted structure is illustrated in Table 5.12 denoted by t_{min}. If we calculate the deductible model execution time by subtracting t_{min} from the model execution time, compared with DeepIoT, FastDeepIoT can reduce the deductible execution time by 69–78%.

Furthermore, we can attempt to deduce the fundamental cause of the lower bound with our execution time model. As shown in (5.17), the execution time of recurrent layer is partially controlled by the number of *step*, which can be interpreted as an initialization overhead for each *step* in the recurrent layer. We can use an example to illustrate the relationship between the *step* overhead and this lower bound. In our experiment, there are 20 steps in the GRU. The coefficient of *step* on Nexus 5 is 0.666 ms. Therefore, the lower bound is $14.1 \approx 20 \times 0.666$ ms. Thus, only algorithms dealing with reducing recurrent-layer steps can help further reducing the model execution time. Unfortunately, to the best of our knowledge, there is no existing work that solves this problem. However, our empirical observation and execution time model reveal an interesting problem that requires future research.

The tradeoffs between testing accuracy and execution time for different algorithms are illustrated in Fig. 5.12. FastDeepIoT still achieves the best tradeoff for all cases. DeepIoT+localMin still performs better than DeepIoT with the help of our structure expanding and local minima searching algorithm. The performance of DeepIoT+FLOPs is not stable among different devices. The tradeoffs between testing accuracy and energy consumption are illustrated in Fig. 5.12. FastDeepIoT performs better than all other baselines in almost all cases. The tradeoffs between testing accuracy and remaining proportion of model parameters are illustrated in Fig. 5.12. All algorithms show comparable results.

Table 5.12 DeepSense (hidden units) on HHAR dataset

Layer		No execution time model										Nexus 5	Galaxy Nexus
		Original		DeepIoT		localMin		FLOPs		t_{min}		FastDeepIoT	FastDeepIoT
conv1a	conv1b (2×9)	64	64	20	19	20	20	26	25	4	4	16	8	16	16
conv2a	conv2b (1×3)	64	64	20	14	20	16	19	17	4	4	8	12	20	16
conv3a	conv3b (1×3)	64	64	23	23	24	24	22	22	4	4	16	16	16	16
conv4 (2×8)		64		10		12		9		4		12		16
conv5 (1×6)		64		12		12		13		4		16		16
conv6 (1×4)		64		17		18		18		4		12		16
gru1		120		27		27		11		1		15		10
gru2		120		31		31		15		1		17		10
Test accuracy		94.6%		94.7%		94.7%		94.7%		16.7%		94.7%		94.7%
Execution time t (Nexus 5)		26.2 ms		19.5 ms		17.9 ms		18.3 ms		14.1 ms		**15.3 ms**		15.8 ms
$t - t_{min}$ (Nexus 5)		12.1 ms		5.4 ms		3.8 ms		4.2 ms				**1.2 ms**		1.7 ms
Execution time t (Galaxy)		70.9 ms		30.1 ms		27.4 ms		28.2 ms		18.4 ms		22.6 ms		**22.0 ms**
$t - t_{min}$ (Galaxy)		52.5 ms		11.7 ms		9.0 ms		9.8 ms				4.2 ms		**3.6 ms**

Bold values reflect the evaluation performance of techniques that are presented in the current section

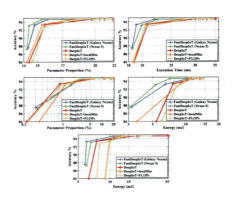

Fig. 5.12 System performance tradeoff for DeepSense on HHAR dataset

References

1. S. Yao, Y. Zhao, A. Zhang, L. Su, T. Abdelzaher, Deepiot: compressing deep neural network structures for sensing systems with a compressor-critic framework, in *Proceedings of the 15th ACM Conference on Embedded Network Sensor Systems*. ACM (2017)
2. Y. Guo, A. Yao, Y. Chen, Dynamic network surgery for efficient DNNs, in *Advances in Neural Information Processing Systems* (2016), pp. 1379–1387
3. X. Zhang, X. Zhou, M. Lin, J. Sun, Shufflenet: an extremely efficient convolutional neural network for mobile devices. CoRR, vol. abs/1707.01083 (2017)
4. A.G. Howard, M. Zhu, B. Chen, D. Kalenichenko, W. Wang, T. Weyand, M. Andreetto, H. Adam, Mobilenets: efficient convolutional neural networks for mobile vision applications. CoRR, vol. abs/1704.04861 (2017)
5. F.N. Iandola, M.W. Moskewicz, K. Ashraf, S. Han, W.J. Dally, K. Keutzer, Squeezenet: Alexnet-level accuracy with 50x fewer parameters and <1mb model size. CoRR, vol. abs/1602.07360 (2016)
6. S. Yao, Y. Zhao, H. Shao, S. Liu, D. Liu, L. Su, T. Abdelzaher, Fastdeepiot: towards understanding and optimizing neural network execution time on mobile and embedded devices, in *Proceedings of the 16th ACM Conference on Embedded Networked Sensor Systems* (2018), pp. 278–291
7. Y. Gal, Z. Ghahramani, Bayesian convolutional neural networks with Bernoulli approximate variational inference. arXiv preprint arXiv:1506.02158 (2015)
8. Y. Gal, Z. Ghahramani, A theoretically grounded application of dropout in recurrent neural networks, in *Advances in Neural Information Processing Systems*, vol. 29, ed. by D. Lee, M. Sugiyama, U. Luxburg, I. Guyon, R. Garnett (2016)
9. N. Srivastava, G.E. Hinton, A. Krizhevsky, I. Sutskever, R. Salakhutdinov, Dropout: a simple way to prevent neural networks from overfitting. J. Mach. Learn. Res. **15**(1), 1929–1958 (2014)
10. W. Zaremba, I. Sutskever, O. Vinyals, Recurrent neural network regularization. arXiv preprint arXiv:1409.2329 (2014)
11. P.W. Glynn, Likelihood ratio gradient estimation for stochastic systems. Commun. ACM **33**(10), 75–84 (1990)
12. J. Peters, S. Schaal, Policy gradient methods for robotics, in *2006 IEEE/RSJ International Conference on Intelligent Robots and Systems*. IEEE (2006), pp. 2219–2225
13. A. Mnih, K. Gregor, Neural variational inference and learning in belief networks, in *Proceedings of the 31st International Conference on Machine Learning (PMLR)*, ed. by E. P. Xing, T. Jebara (Bejing, 2014), pp. 1791–1799
14. S. Gu, S. Levine, I. Sutskever, A. Mnih, Muprop: unbiased backpropagation for stochastic neural networks. arXiv preprint arXiv:1511.05176 (2015)

5 Model Compression for Edge Computing

15. S. Han, H. Mao, W.J. Dally, Deep compression: compressing deep neural network with pruning, trained quantization and Huffman coding. CoRR, abs/1510.00149, vol. 2 (2015)
16. Intel edison compute module. http://www.intel.com/content/dam/support/us/en/documents/edison/sb/edison-module_HG_331189.pdf
17. Loading Debian (ubilinux) on the edison. https://learn.sparkfun.com/tutorials/loading-debian-ubilinux-on-the-edison
18. Theano Development Team, Theano: A Python framework for fast computation of mathematical expressions. arXiv e-prints, vol. abs/1605.02688 (May 2016). Available: http://arxiv.org/abs/1605.02688
19. S. Bhattacharya, N.D. Lane, Sparsification and separation of deep learning layers for constrained resource inference on wearables, in *Proceedings of the 14th ACM Conference on Embedded Network Sensor Systems CD-ROM*. ACM (2016), pp. 176–189
20. V. Panayotov, G. Chen, D. Povey, S. Khudanpur, Librispeech: an ASR corpus based on public domain audio books, in *2015 IEEE International Conference on Acoustics, Speech and Signal Processing (ICASSP)*. IEEE (2015), pp. 5206–5210
21. A. Graves, N. Jaitly, Towards end-to-end speech recognition with recurrent neural networks, in *ICML*, vol. 14 (2014), pp. 1764–1772
22. S. Yao, S. Hu, Y. Zhao, A. Zhang, T. Abdelzaher, Deepsense: a unified deep learning framework for time-series mobile sensing data processing, in *Proceedings of the 26th International Conference on World Wide Web*. International World Wide Web Conferences Steering Committee (2017)
23. A. Stisen, H. Blunck, S. Bhattacharya, T.S. Prentow, M.B. Kjærgaard, A. Dey, T. Sonne, M.M. Jensen, Smart devices are different: assessing and mitigating mobile sensing heterogeneities for activity recognition, in *Proceedings of the 13th ACM Conference on Embedded Networked Sensor Systems*. ACM (2015), pp. 127–140
24. R. Rigamonti, A. Sironi, V. Lepetit, P. Fua, Learning separable filters, in *2013 IEEE Conference on Computer Vision and Pattern Recognition (CVPR)*. IEEE (2013), pp. 2754–2761
25. Tensorflow benchmark tool. https://github.com/tensorflow/tensorflow/tree/r1.4/tensorflow/tools/benchmark
26. H. Drucker, C.J. Burges, L. Kaufman, A.J. Smola, V. Vapnik, Support vector regression machines, in *Advances in Neural Information Processing Systems* (MIT Press, Cambridge, 1997), pp. 155–161
27. L. Breiman, *Classification and Regression Trees* (Routledge, Abingdon, 2017)
28. L. Breiman, Random forests. Mach. Learn. **45**(1), 5–32 (2001)
29. J.H. Friedman, Greedy function approximation: a gradient boosting machine. Ann. Stat. **29**, 1189–1232 (2001)
30. Y. LeCun, Y. Bengio, G. Hinton, Deep learning. Nature **521**(7553), 436 (2015)
31. K. Simonyan, A. Zisserman, Very deep convolutional networks for large-scale image recognition. arXiv preprint arXiv:1409.1556 (2014)

Part II
Distributed Problems

Chapter 6
Communication Efficient Distributed Learning

Navjot Singh, Deepesh Data, and Suhas Diggavi

Abstract Communication bottleneck has been identified as a significant issue in large-scale training of machine learning models over a network. Recently, several approaches have been proposed to mitigate this issue, using gradient compression and infrequent communication based techniques. This chapter summarizes two communication efficient algorithms, *Qsparse-local-SGD* and *SQuARM-SGD*, for *distributed* and *decentralized* settings, respectively. These algorithms utilize *composed* sparsification and quantization operators for aggressive compression, along with local iteration for communication efficiency. We provide theoretical convergence guarantees for these algorithms for smooth non-convex objectives. We also review extensive numerical experiments of our methods for training ResNet architectures and compare them against the state-of-the-art methods in their respective settings. The content of this chapter is based on the papers: Basu et al. (Qsparse-local-SGD: Distributed SGD with quantization, sparsification and local computations. In: NeurIPS, 2019), Basu et al. (IEEE J Sel Areas Inf Theory 1(1):217–226, 2020), Singh et al. (SPARQ-SGD: Event-triggered and compressed communication in decentralized optimization. In: IEEE Control and Decision Conference (CDC), 2020), Singh et al. (IEEE Trans Autom Control, 2022), Singh et al. (IEEE J Sel Areas Inf Theory 2(3):954–969, 2021).

1 Introduction

Recent advancements in computing hardware capabilities and processing has prompted a surge of interest in applying machine learning for edge (wireless) device networks. In such a scenario, data is collected/available locally, and these distributed devices collectively help build a model through communication links with significant (wireless) rate constraints, in contrast to wired datacenter appli-

---

N. Singh · D. Data · S. Diggavi (✉)
Department of Electrical and Computer Engineering, University of California, Los Angeles, CA, USA
e-mail: navjotsingh@ucla.edu; suhas@ee.ucla.edu

© The Author(s), under exclusive license to Springer Nature Switzerland AG 2023
M. Srivatsa et al. (eds.), *Artificial Intelligence for Edge Computing*,
https://doi.org/10.1007/978-3-031-40787-1_6

cations. This presents a fundamental challenge of building learning models on local (heterogeneous) data in a communication-efficient manner as the exchange of information between the devices causes communication to be the bottleneck in training [24, 41]. For example, training the BERT architecture for language models [13], which has about 340 million parameters, via *Stochastic Gradient Decent (SGD)* requires each participating device to exchange around 1.3 GB for one full precision gradient exchange (assuming 32 bit floating point precision).

Consequently, several methods have been developed to tackle this problem and obtain communication efficiency in both *distributed* and *decentralized* model training settings. In a distributed setting, there typically exists a central server which coordinates the information exchange among the training devices (also called *clients*). Decentralized training does not assume the presence of such a central facilitator; instead, clients are connected with each-other through a (sparse) graph and iteratively exchange their updates with the neighboring clients to train collaboratively[1] [36]. The methods to achieve communication efficient training in both these settings can be broadly divided into two categories: *compression* based techniques, and *local iterations* for skipping communication rounds. In particular, for distributed training, participating clients *compress* information/gradients before communicating—either with *sparsification* [2, 4, 37, 51, 52], *quantization* [3, 7, 25, 53, 61]. Local iterations based schemes comprise of skipping communication rounds while performing a certain number of *local SGD* steps [11, 50, 64]. Similarly, compressed communication, with either quantization or sparsification schemes, has been studied recently for decentralized training as well [29, 32, 54, 55], and local iteration schemes (without compression) in decentralized setting have been studied in [31, 58]. There are a few important aspects to the works mentioned above: (i) They rely on strong set of assumptions for their theoretical analyses: all of them assume a uniform bound on variance of stochastic gradients and also on the gradient dissimilarity across the clients, while [29, 32, 55] assume a bound on the second moment of stochastic gradients. (ii) None of these works incorporates momentum in their theoretical analyses, which has been very successful in achieving good generalization error in training large-scale machine learning models. (iii) They use *either* quantization or sparsification for compressing the exchanged information.

In this chapter, we provide a unified framework for communication efficient training in distributed and decentralized scenarios. The content presented is based on the papers [5, 6, 46, 48, 49] and seeks to address the issues presented in the preceding paragraph. We first develop the idea of a composed compression operation, which combine sparsification and quantization to yield a relatively more aggressive compression than its individual components. We derive the compression coefficients for the resulting composed operator when used with popular sparsification [4, 51] and quantization [3, 7] notions in literature. We also formalize the notion of local iterations and introduce *event-triggering* mechanisms that can be applied in

---

[1] This can also be motivated through learning over local wireless mesh (or ad hoc) networks.

6 Communication Efficient Distributed Learning

combination with compressed communication. Further discussion on triggering based communication can be found in [46, 49].

A *synchronous* algorithm, titled *Qsparse-local-SGD* (from [5, 6]), based on SGD for distributed training utilizing these composed operations and local iterations is then developed, which uses an error-feedback mechanism that locally accumulates the non-communicated components of the gradient vectors at clients and use them in later communication rounds along with the true gradient update in that round. We then provide convergence guarantees for the resulting algorithm on general (smooth) objectives utilizing typically made assumptions about the objective functions in literature. A *non-synchronous* version for the algorithm, along with its convergence guarantees can be found in [5, 6]. Similarly, for the decentralized setting we present the algorithm *SQuARM-SGD* (from [48]) which utilizes the notions of compression and local iteration while implicitly following an error-feedback mechanism. A version of this algorithm with event-based triggering can be found in [46, 49]. SQuARM-SGD also incorporates Nesterov momentum updates, and we analyze its theoretical convergence guarantees. Importantly, the convergence results for the decentralized setting in this chapter are presented with a relatively weaker set of assumptions than what is typically considered in decentralized optimization literature. The use of relaxed assumptions can be extended similarly to the distributed training in [5, 6].

1.1 Chapter Organization

In Sect. 2, we formally state the optimization problem for *distributed* and *decentralized* training, and present the notation we use throughout the chapter. Section 3 discusses techniques for communication efficient learning with compression and local iterations, where we also discuss composed compression operators. Section 4 describes the *Qsparse-local-SGD* algorithm for efficient training in distributed setting and provides its theoretical convergence rates. Section 5 introduces *SQuARM-SGD* algorithm for communication efficient training in decentralized settings with momentum and provide its convergence results under a weak assumption set. Section 6 provides experimental results for both the algorithms.

2 Problem Setup and Notation

We now formalize the setting for distributed and decentralized training we work with and establish the notation we follow throughout the chapter. We assume that there are n number of *clients*, also interchangeably called *worker nodes*. Associated with each client $i \in [n]$ is a local dataset denoted by $\mathcal{D}_i$ and an objective function $f_i : \mathbb{R}^d \to \mathbb{R}$. The datasets and objective functions are allowed to be different for each client and we assume that for $i \in [n]$, the objective function f_i has the form $f_i(\mathbf{x}) = \mathbf{E}_{\xi_i \sim \mathcal{D}_i}[F_i(\mathbf{x}, \xi_i)]$ where $\xi_i \sim \mathcal{D}_i$ denotes a random sample from $\mathcal{D}_i$, $\mathbf{x}$ denotes the parameter vector, and $F_i(\mathbf{x}, \xi_i)$ denotes the risk associated with sample ξ_i with respect to (w.r.t.) the parameter vector $\mathbf{x}$. The clients are allowed to collaborate constrained by the underlying topology (discussed in the following paragraph) and seek to optimize the following empirical risk objective, where $f : \mathbb{R}^d \to \mathbb{R}$ is called the *global* objective function:

$$\min_{\mathbf{x} \in \mathbb{R}^d} \left(f(\mathbf{x}) := \frac{1}{n} \sum_{i=1}^{n} f_i(\mathbf{x}) \right), \tag{6.1}$$

The goal of each client is thus to reach the optimal parameter $\mathbf{x}^* \in \arg\min_{\mathbf{x} \in \mathbb{R}^d} f(x)$ while exchanging compressed information about its local objective with the other clients. In the following, we discuss the two underlying communication topologies (distributed and decentralized) we present our convergence results on and establish additional notation related to the connectivity structure.

Distributed Topology In distributed training, one typically assumes the presence of a central coordinator, also called a *parameter server* (PS), which facilitates the exchange of information among the clients. In our proposed distributed training algorithm given in Sect. 4, each client is allowed to communicate (compressed) updates to the PS, which in turn aggregates this information and broadcasts it back to all the clients.

Decentralized Topology For the general case of decentralized training, we assume that the underlying connectivity of the clients can be represented by a graph structure $\mathcal{G} := (\mathcal{V}, \mathcal{E})$, with the clients forming the nodes of the graph $\mathcal{V} := [1, 2, \dots, n]$. We denote the set of neighboring clients for a node i as $\mathcal{N}_i := \{j : (i, j) \in \mathcal{E}\}$. We denote the weighted connectivity matrix of $\mathcal{G}$ by $W \in \mathbb{R}^{n \times n}$, with w_{ij} for every $i, j \in [n]$ being its (i, j)th entry, which denotes the weight on the link between client i and j. W is assumed to be symmetric and doubly stochastic, which implies that all its eigenvalues $\lambda_i(W)$, $i = 1, 2, \dots, n$, lie in $[-1, 1]$. Without loss of generality, assume that $|\lambda_1(W)| > |\lambda_2(W)| \geq \dots \geq |\lambda_n(W)|$. Since W is doubly stochastic, we have $\lambda_1(W) = 1$, and since $\mathcal{G}$ is connected, we have $\lambda_2(W) < \lambda_1(W)$. We denote the spectral gap of W as $\delta := 1 - |\lambda_2(W)|$. Since $|\lambda_2(W)| \in [0, 1)$ we have that $\delta \in (0, 1]$. Simple matrices W with $\delta > 0$ are known to exist for every connected graph, [29]. Note that one may consider the distributed scenario presented above as a special case of decentralized training scenario with the underlying graph structure among the nodes as the fully connected graph K_n.

3 Techniques for Communication-Efficient Training

Traditionally, distributed stochastic gradient descent affords to send full precision (32 or 64 bit) unbiased gradient updates across workers to peers or to a central

6 Communication Efficient Distributed Learning

server that helps with aggregation. Moreover, these updates are communicated in every iteration. We can save on communication in two ways: First, by compressing the update vectors, and second, by not communicating the update in every iteration. We will explain the compression part in Sect. 3.1 and infrequent communication part in Sects. 3.2 and 3.3.

3.1 Compression Operation

Communication bottlenecks that arise in bandwidth limited networks limit the applicability of full-precision distributed SGD at a large scale when the parameter size is massive or the data is widely distributed on a very large number of worker nodes. In such settings, one could think of updates which not only result in convergence, but also require less bandwidth thus making the training process faster. In the following sections we discuss several useful operators from literature and enhance their use by proposing a novel class of composed operators.

We first consider two different techniques used in the literature for mitigating the communication bottleneck in distributed optimization, namely, quantization and sparsification.

3.1.1 Quantization

In quantization, we reduce precision of the gradient vector by mapping each of its components by a deterministic [7, 25] or randomized [3, 53, 61, 65] map to a finite number of quantization levels.

Definition 1 (Randomized Quantizer) We say that $Q_s : \mathbb{R}^d \rightarrow \mathbb{R}^d$ is a randomized quantizer with s quantization levels, if the following holds for every $\mathbf{x} \in \mathbb{R}^d$: (i) $\mathbb{E}_Q[Q_s(\mathbf{x})] = \mathbf{x}$; (ii) $\mathbb{E}_Q[\|Q_s(\mathbf{x})\|^2] \leq (1 + \beta_{d,s})\|\mathbf{x}\|^2$, where $\beta_{d,s} > 0$ could be a function of d and s. Here expectation is taken over the randomness of Q_s.

Examples of randomized quantizers include

1. *QSGD* [3, 61], which independently quantizes components of $\mathbf{x} \in \mathbb{R}^d$ into s levels, with $\beta_{d,s} = \min(\frac{d}{s^2}, \frac{\sqrt{d}}{s})$.
2. *Stochastic s-level Quantization* [53, 65], which independently quantizes every component of $\mathbf{x} \in \mathbb{R}^d$ into s levels between $\text{argmax}_i x_i$ and $\text{argmin}_i x_i$, with $\beta_{d,s} = \frac{d}{2s^2}$.
3. *Stochastic Rotated Quantization* [53], which is a stochastic quantization, preprocessed by a random rotation, with $\beta_{d,s} = \frac{2 \log_2(2d)}{s^2}$.

Instead of quantizing randomly into s levels, we can take a deterministic approach and round off each component of the vector to the nearest level. In particular, we can just take the sign, which has shown promise in [7, 25].

Definition 2 (Deterministic Sign Quantizer) A deterministic quantizer $Sign$: $\mathbb{R}^d \to \{+1, -1\}^d$ is defined as follows: for every vector $\mathbf{x} \in \mathbb{R}^d$, the i'th component of $Sign(\mathbf{x})$, for $i \in [d]$, is defined as $\mathbb{1}\{x_i \geq 0\} - \mathbb{1}\{x_i < 0\}$.

Such methods drew interest since RPROP [43], which only used the temporal behavior of the sign of the gradient. This is an example where the biased 1-bit quantizer as in Definition 2 is used. This further inspired optimizers, such as RMSPROP [56], ADAM [28], which incorporate appropriate adaptive scaling with momentum acceleration and have demonstrated empirical superiority in non-convex applications.

3.1.2 Sparsification

In sparsification, we sparsify the gradient vector before using it to update the parameter vector. We consider two important examples of sparsification operators: Top_k and Rand_k [29, 51]. For any $\mathbf{x} \in \mathbb{R}^d$, $\text{Top}_k(\mathbf{x})$ is equal to a d-length vector, which has at most k non-zero components whose indices correspond to the indices of the largest k components (in absolute value) of $\mathbf{x}$. Similarly, $\text{Rand}_k(\mathbf{x})$ is a d-length (random) vector, which is obtained by selecting k components of $\mathbf{x}$ uniformly at random. Both of these satisfy a so-called "compression" property as defined below, with $\omega = k/d$.

Definition 3 (Compression, [51]) A (possibly randomized) function $C : \mathbb{R}^d \to \mathbb{R}^d$ is called a *compression* operator, if there exists a positive constant $\omega < 1$, such that the following holds for every $\mathbf{x} \in \mathbb{R}^d$:

$$\mathbb{E}_C[\|\mathbf{x} - C(\mathbf{x})\|_2^2] \leq (1 - \omega)\|\mathbf{x}\|_2^2, \tag{6.2}$$

where expectation is taken over the randomness of C. We assume $C(\mathbf{0}) = \mathbf{0}$.

Note that stochastic quantizers, as defined in Definition 1, also satisfy this regularity condition in Definition 3 for $\beta_{d,s} \leq 1$. Now we give a simple but important corollary, which allows us to apply different contraction operators to different coordinates of a vector. As an application, in the case of training neural networks, we can apply different operators to different layers.

Corollary 1 (Piecewise Compression, [5]) *Let $C_i : \mathbb{R}^{d_i} \to \mathbb{R}^{d_i}$ for $i \in [L]$ denote possibly different contraction operators with contraction coefficients ω_i. Let $\mathbf{x} = [\mathbf{x}_1 \mathbf{x}_2 \dots \mathbf{x}_L]$, where $\mathbf{x}_i \in \mathbb{R}^{d_i}$ for all $i \in [L]$. Then $C(\mathbf{x}) := [C_1(\mathbf{x}_1) C_2(\mathbf{x}_2) \dots C_L(\mathbf{x}_L)]$ is a contraction operator with the contraction coefficient being equal to $\omega_{min} = \min\limits_{i \in [L]} \omega_i$.*

3.1.3 Composition of Quantization and Sparsification

Now we show that we can compose deterministic/randomized quantizers with sparsifiers and the resulting operator is a compression operator. First we compose a general stochastic quantizer with an explicit sparsifier such as $\text{Top}_k(\mathbf{x})$ and $\text{Rand}_k(\mathbf{x})$.

Lemma 1 (Composing Sparsification with Stochastic Quantization, [5]) *Let $Comp_k \in \{\text{Top}_k, \text{Rand}_k\}$. Let $Q_s : \mathbb{R}^d \to \mathbb{R}^d$ be a stochastic quantizer with parameter s that satisfies Definition 1. Let $Q_s Comp_k : \mathbb{R}^d \to \mathbb{R}^d$ be defined as $Q_s Comp_k(\mathbf{x}) := Q_s(Comp_k(\mathbf{x}))$ for every $\mathbf{x} \in \mathbb{R}^d$. Then $\frac{Q_s Comp_k(\mathbf{x})}{1+\beta_{k,s}}$ is a compression operator with the compression coefficient being equal to $\omega = \frac{k}{d(1+\beta_{k,s})}$, i.e., for every $\mathbf{x} \in \mathbb{R}^d$*

$$\mathbb{E}_{C,Q}\left[\left\|\mathbf{x} - \frac{Q_s Comp_k(\mathbf{x})}{1+\beta_{k,s}}\right\|_2^2\right] \leq \left[1 - \frac{k}{d(1+\beta_{k,s})}\right]\|\mathbf{x}\|_2^2,$$

where expectation is taken over the randomness of the contraction operator $Comp_k$ as well as of the quantizer Q_s.

Note that $\frac{Q_s Comp_k(\mathbf{x})}{1+\beta_{k,s}}$ is always a compression operator for all values of $\beta_{k,s} > 0$. The examples of randomized quantizers stated after Definition 1 are all applicable here (by replacing d by k).

We can also compose a *deterministic* 1-bit quantizer $Sign$ with $Comp_k$. For that we need some notations first. For $Comp_k \in \{\text{Top}_k, \text{Rand}_k\}$ and given vector $\mathbf{x} \in \mathbb{R}^d$, let $S_{Comp_k(\mathbf{x})} \in \binom{[d]}{k}$ denote the set of k indices chosen for defining $Comp_k(\mathbf{x})$. For example, if $Comp_k = \text{Top}_k$, then $S_{Comp_k(\mathbf{x})}$ denote the set of k indices corresponding to the largest k components of $\mathbf{x}$; if $Comp_k = \text{Rand}_k$, then $S_{Comp_k(\mathbf{x})}$ denote a set of random set of k indices in $[d]$. The composition of $Sign$ with $Comp_k \in \{\text{Top}_k, \text{Rand}_k\}$ is denoted by $SignComp_k : \mathbb{R}^d \to \mathbb{R}^d$, and for $i \in [d]$, the i'th component of $SignComp_k(\mathbf{x})$ is defined as

$$(SignComp_k(\mathbf{x}))_i := \begin{cases} \mathbb{1}\{x_i \geq 0\} - \mathbb{1}\{x_i < 0\} & \text{if } i \in S_{Comp_k(\mathbf{x})}, \\ 0 & \text{otherwise.} \end{cases}$$

In the following lemma we show that $SignComp_k$ is a compression operator.

Lemma 2 (Composing Sparsification and Sign Quantization, [5]) *For $Comp_k \in \{\text{Top}_k, \text{Rand}_k\}$, the operator*

$$\frac{\|Comp_k(\mathbf{x})\|_m \, SignComp_k(\mathbf{x})}{k}$$

for any $m \in \mathbb{Z}_+$ is a compression operator with the compression coefficient ω_m being equal to

$$
\omega_m = \begin{cases} \max \left\{ \frac{1}{d}, \frac{k}{d} \left(\frac{\|Comp_k(\mathbf{x})\|_1}{\sqrt{d}\|Comp_k(\mathbf{x})\|_2} \right)^2 \right\} & \text{if } m = 1, \\ \frac{k^{\frac{2}{m}-1}}{d} & \text{if } m \geq 2. \end{cases}
$$

Remark 1 Observe that for $m = 1$, depending on the value of k, either of the terms inside the max can be bigger than the other term. For example, if $k = 1$, then $\|Comp_k(\mathbf{x})\|_1 = \|Comp_k(\mathbf{x})\|_2$, which implies that the second term inside the max is equal to $1/d^2$, which is much smaller than the first term. On the other hand, if $k = d$ and the vector $\mathbf{x}$ is dense, then the second term may be much bigger than the first term.

3.2 Local Iterations

An effective scheme for reducing the amount of communication is to communicate less frequently. Specifically, we can relax the requirement for the worker nodes to communicate at each iteration of the training process and instead allow them to communicate with each other after some prescribed number of iterations. During these skipped communication iterations, the workers take gradient steps based on their locally available data to make progress on optimizing their local objectives. For the proposed algorithms in this chapter and the corresponding convergence results, we will assume that we are in the synchronous setting, that is, each worker has the same prescribed number of iterations which it skips.

For a given number of iterations T, let H be the number of iterations the worker node skips before the next communication. We define $\mathcal{I}_T := \{H, 2H, \ldots, T\}$ as the set of synchronization indices at which the workers are allowed to communicate. Note that $H = 1$ is equivalent to the case when workers communicate in each iteration as in standard vanilla distributed training.

3.3 Triggering Based Updates

Another related idea to reduce the amount of communication among workers is to communicate only in the event there is a significant difference in the exchanged parameter since the last communication round. In this scenario, the communication criterion for each worker is thus governed by a locally computable trigger condition. Such a triggering condition based on thresholding the parameter norm difference is analyzed in [46, 49] for the *decentralized* setting. We remark that such a technique can be similarly extended to the *distributed* setting in [5, 6].

6 Communication Efficient Distributed Learning 207

4 Distributed Training—Qsparse-Local-SGD

Let $\mathcal{I}_T \subseteq [T] := \{1, \ldots, T\}$ with $T \in \mathcal{I}_T$ denote a set the indices for which the worker nodes synchronize with the central server. Every worker node $i \in [n]$ maintains a local parameter vector $\mathbf{x}_i^{(t)}$ which is updated in each iteration t. If $t \in \mathcal{I}_T$, every worker $i \in [n]$ sends the compressed and error-compensated update $\mathbf{q}_i^{(t)}$ computed on the net progress made since the last synchronization to the central server, and updates its local memory $\mathbf{m}_i^{(t)}$. Upon receiving $\mathbf{q}_i^{(t)}$, $i = 1, 2, \ldots, n$, the central server aggregates them, updates the global parameter vector, and sends the new model $\bar{\mathbf{x}}^{(t+1)}$ to all the worker nodes. Upon receiving this update, the nodes set their local parameter vector $\mathbf{x}_i^{(t+1)}$ to be equal to the global parameter vector $\bar{\mathbf{x}}^{(t+1)}$. Our resulting algorithm is summarized in Algorithm 1.

Algorithm 1 Qsparse-local-SGD

1: Initialize parameters $\bar{\mathbf{x}}^{(0)} = \mathbf{x}_i^{(0)} = \mathbf{m}_i^{(0)} = \mathbf{0}$ for each node $i \in [n]$, SGD step sizes $\{\eta_t\}_{t \geq 0}$, compression operator C having parameter ω, set of synchronization indices $\mathcal{I}_T$.

2: **for** $t = 0$ to $T - 1$ **do**

3:     **On Workers:**

4:     **for** $i = 1$ to n **do**

5:         Sample $\xi_i^{(t)}$; define stochastic gradient $\mathbf{g}_i^{(t)} := \nabla F_i(\mathbf{x}_i^{(t)}, \xi_i^{(t)})$

6:         $\mathbf{y}_i^{(t+1)} = \mathbf{x}_i^{(t)} - \eta_t \mathbf{g}_i^{(t)}$

7:         **if** $t + 1 \notin \mathcal{I}_T$ **then**

8:             $\bar{\mathbf{x}}^{(t+1)} = \bar{\mathbf{x}}^{(t)}, \mathbf{m}_i^{(t+1)} = \mathbf{m}_i^{(t)}$ and $\mathbf{x}_i^{(t+1)} = \mathbf{y}_i^{(t+1)}$

9:         **else**

10:             $\mathbf{q}_i^{(t)} = C\left(\mathbf{m}_i^{(t)} + \bar{\mathbf{x}}^{(t)} - \mathbf{y}_i^{(t+1)}\right)$, send $\mathbf{q}_i^{(t)}$ to the central server

11:             $\mathbf{m}_i^{(t+1)} = \mathbf{m}_i^{(t)} + \bar{\mathbf{x}}^{(t)} - \mathbf{y}_i^{(t+1)} - \mathbf{q}_i^{(t)}$

12:             Receive $\bar{\mathbf{x}}^{(t+1)}$ from the central server and set $\mathbf{x}_i^{(t+1)} = \bar{\mathbf{x}}^{(t+1)}$

13:         **end if**

14:     **end for**

15:     **At Central Server:**

16:     **if** $(t + 1) \notin \mathcal{I}_T$ **then**

17:         $\bar{\mathbf{x}}^{(t+1)} = \bar{\mathbf{x}}^{(t)}$

18:     **else**

19:         Receive $\mathbf{q}_i^{(t)}$ from all workers $i \in [n]$ and compute $\bar{\mathbf{x}}^{(t+1)} = \bar{\mathbf{x}}^{(t)} - \frac{1}{n}\sum_{i=1}^{n} \mathbf{q}_i^{(t)}$

20:         Broadcast $\bar{\mathbf{x}}^{(t+1)}$ to all workers

21:     **end if**

22: **end for**

4.1 Error Compensation

Sparsified gradient methods, where workers send the largest k coordinates of the updates based on their magnitudes have been investigated in the literature and serves as a communication efficient strategy for distributed training of learning models.

However, the convergence rates are subpar to distributed vanilla SGD. Together with some form of error compensation, these methods have been empirically observed to converge as fast as vanilla SGD in [2, 4, 25, 37, 51, 52]. In [4, 51], sparsified SGD with such feedback schemes has been carefully analyzed. Reference [4] prove the convergence of distributed Top_k SGD with error feedback under significantly strong assumptions. The net error in the system is accumulated by each worker locally on a per iteration basis and this is used as feedback for generating the future updates. [51] did the analysis for the *centralized* Top_k SGD for strongly convex objectives. Our work in [5, 6] significantly extends the results in these papers by considering both synchronous and asynchronous *distributed* setting, without strong assumptions on the sparsification as in [4].

In Algorithm 1, the error introduced in every iteration is accumulated into the memory of each worker, which is compensated for in the future rounds of communication. This feedback is the key to recovering the convergence rates matching vanilla SGD. The operators employed provide a controlled way of using both the current update as well as the compression errors from the previous rounds of communication. Under the assumption of the uniform boundedness of the gradient we analyze the controlled evolution of memory through the optimization process; the results are summarized in Lemma 3 below.

In the following lemma, we show that if we run Algorithm 1 with a fixed learning rate $\eta_t = \eta$, $\forall t$, then the local memory at each worker is bounded.

Lemma 3 (Bounded Memory, [5]) *Let* $gap(\mathcal{I}_T) \le H$. *Then the following holds for every worker* $r \in [R]$ *and for every* $t \in \mathbb{Z}^+$:

$$\mathbb{E}\|m_t^{(r)}\|_2^2 \le 4 \frac{\eta^2(1-\omega^2)}{\omega^2} H^2 G^2. \tag{6.3}$$

Note that, for fixed ω, H, the memory is upper bounded by a constant $O(\eta^2)$. Observe that since the memory accumulates the past errors due to compression and local computation, in order to asymptotically reduce the memory to zero, the learning rate would have to be reduced once in a while throughout the training process.

4.2 Theoretical Results

We first present the key set of assumptions our results rely upon.

A.1 (*L*-**Smoothness**): Each local function f_i for $i \in [n]$ is *L*-smooth, i.e., $\forall \mathbf{x}, \mathbf{y} \in \mathbb{R}^d$, we have $f_i(\mathbf{y}) \le f_i(\mathbf{x}) + \langle \nabla f_i(\mathbf{x}), \mathbf{y} - \mathbf{x} \rangle + \frac{L}{2}\|\mathbf{y} - \mathbf{x}\|^2$.

A.2 (**Bounded Variance**): For every $i \in [n]$, we have $\mathbb{E}_{\xi_i}\|\nabla F_i(\mathbf{x}, \xi_i) - \nabla f_i(\mathbf{x})\|^2 \le \sigma_i^2$, for some finite σ_i, where $\nabla F_i(\mathbf{x}, \xi_i)$ is the unbiased gradient at worker i

6 Communication Efficient Distributed Learning

such that $\mathbb{E}_{\xi_i}[\nabla F_i(\mathbf{x}, \xi_i)] = \nabla f_i(\mathbf{x})$. We define the average variance across all workers as $\bar{\sigma}^2 := \frac{1}{n} \sum_{i=1}^{n} \sigma_i^2$.

A.3 (Bounded second moment):[2] For every $i \in [n]$, we have $\mathbb{E}_{\xi_i} \|\nabla F_i(\mathbf{x}, \xi_i)\|^2 \leq G^2$, for some finite G.

Under the set of assumptions presented above, the following convergence result holds for Algorithm 1 when training with smooth non-convex objectives.

Theorem 1 (Convergence Rate of Qsparse-Local-SGD for Smooth Non-convex Objectives, [5]) *Let f_i be L-smooth for every $i \in [n]$. Let $C : \mathbb{R}^d \to \mathbb{R}^d$ be a compression operator with compression coefficient equal to $\omega \in (0, 1]$. Let $\{\mathbf{x}_i^{(t)}\}_{t=0}^{T-1}$ be generated according to Algorithm 1 with C, for step sizes $\eta = \frac{\hat{C}}{\sqrt{T}}$ (where $\hat{C}$ is a constant such that $\frac{\hat{C}}{\sqrt{T}} \leq \frac{1}{2L}$) and $gap(\mathcal{I}_T) \leq H$. Then we have*

$$\mathbb{E}\|\nabla f(\mathbf{z}_T)\|_2^2 \leq \left(\frac{\mathbb{E}[f(\bar{\mathbf{x}}^{(0)})] - f^*}{\hat{C}} + \hat{C}L \left(\frac{\sum_{i=1}^{n} \sigma_i^2}{n^2} \right) \right) \frac{4}{\sqrt{T}} + 8 \left(4 \frac{(1-\omega^2)}{\omega^2} + 1 \right) \frac{\hat{C}^2 L^2 G^2 H^2}{T}.$$

Here $\mathbf{z}_T$ is a random variable which samples a previous parameter from $\{\mathbf{x}_i^{(t)}\}$ for $i \in [n]$ and $t \in [T]$ with probability $1/nT$.

Corollary 2 *Let $\mathbb{E}[f(\bar{\mathbf{x}}^{(0)})] - f^* \leq J^2$, where $J < \infty$ is a constant,[3] $\sigma_{max} = \max_{r \in [R]} \sigma_r$, and $\hat{C}^2 = \frac{n(\mathbb{E}[f(\bar{\mathbf{x}}^{(0)})] - f^*)}{\sigma_{max}^2 L}$, we have*

$$\mathbb{E}\|\nabla f(\mathbf{z}_T)\|_2^2 \leq O \left(\frac{J \sigma_{max}}{\sqrt{nT}} \right) + O \left(\frac{J^2 n G^2 H^2}{\sigma_{max}^2 \omega^2 T} \right).$$

Remark 2 The convergence rate in Corollary 2 demonstrates a dominant rate of $O(1/\sqrt{T})$, which matches that of vanilla distributed SGD. Thus, quantizing and sparsifying the gradient, even after local iterations asymptotically yields an almost "free" efficiency gain as the factors ω, H pertaining to communication efficiency only appear in the higher order term. As Theorem 1 provides non-asymptotic guarantees where we are required to decide the horizon T before running the algorithm, to ensure that the dominant rate is not affected by these factors, we need to fix $T \geq T_0 := C_0 \times \left(\frac{J^2 n^3 G^4 H^4}{\sigma_{max}^6 \omega^4} \right)$ for sufficiently large constant C_0. It can be seen that increasing the amount of local iterations (H) or compressing the gradients more aggressively (making ω small) will lead to a larger value of T_0, as expected.

Remark 3 The convergence rate of $O(1/\sqrt{T})$ presented in Theorem 1 relies on assuming a bounded second moment of stochastic gradients (Assumption **A.3**). Recently, [26] provided analysis of distributed SGD using a bounded gradient

---

[2] Bounded second moment has been a standard assumption in stochastic optimization with *compressed* communication [4, 51]. There are also works that relax this assumption; see [26, 30, 48].

[3] Even classical SGD requires knowing an upper bound on $\|\bar{\mathbf{x}}^{(0)} - \bar{\mathbf{x}}^*\|$ in order to choose the learning rate. Smoothness of f translates this to the difference of the function values.

dissimilarity assumption, which is weaker than the bounded second moment assumption. Similarly, [19] provided analysis for distributed optimization with compression via gradient tracking without the second moment assumption, although only for *unbiased* compression operators with further additional assumptions for the heterogeneous settings. In the next section, we provide an algorithm for compressed *decentralized* training which allows for *biased* compression operators, and present its theoretical result without using the bounded second moment assumption. We remark that this relaxation can similarly be extended to the result of Theorem 1 of [5, 6] presented above.

5 Decentralized Training—SQuARM-SGD

In this section, we describe our algorithm for decentralized training with compression and local iterations to optimize the objective in (6.1). We provide the gradient updates with Nesterov acceleration, which is also analyzed theoretically in the corresponding convergence result for the algorithm presented in Sect. 5.1. The momentum based updates can be similarly extended and theoretically analyzed to the specific case of distributed training in the previous section.

For a given connected graph G with connectivity matrix $\mathbf{W}$, we first initialize a consensus step-size γ (see Theorem 2 for definition), momentum factor β, learning rate η, and momentum vector $\mathbf{v}_i$ for each node i initialized to $\mathbf{0}$. We initialize the copies of all the nodes $\hat{\mathbf{x}}_i = \mathbf{0}$ and allow each node to communicate in the first round. At each time step t, each worker $i \in [n]$ samples a stochastic gradient $\nabla F_i(\mathbf{x}_i^{(t)}, \xi_i)$ and takes a local SGD step on parameter $\mathbf{x}_i^{(t)}$ using Nesterov's momentum to form an intermediate parameter $\mathbf{y}_i^{(t+1)}$ (lines 3–5). If the next iteration corresponds to a synchronization index, i.e., $(t+1) \in I_T$, the worker communicates the compressed change in its copy to all its neighbors $\mathcal{N}_i$ (lines 8–9). After receiving the compressed updates of copies from all its neighbors, the node i updates the locally available copies and its own copy (line 10). With these updated copies, the worker nodes finally take a consensus (line 12) with appropriate weighting decided by entries of $\mathbf{W}$. In the case when $(t+1) \notin I_T$, the nodes maintain their copies and move on to next iteration (line 14); thus no communication takes place.

5.1 Theoretical Results

We now provide our main results for decentralized training with Algorithm 2 for non-convex smooth objectives. As mentioned earlier during the algorithm development, our results are derived while taking into account the Nesterov momentum coefficient β in the analysis. Further, we provide the following result with a relatively weaker set of assumptions than the ones used in Sect. 4.2 for

6 Communication Efficient Distributed Learning

Algorithm 2 SQuARM-SGD: Sparsified and quantized action regulated momentum SGD

1: Initial parameter values $\mathbf{x}_i^{(0)} \in \mathbb{R}^d$ for each node $i \in [n]$, consensus step size γ, SGD step size η, momentum parameter β, compression operator C having parameter ω, communication graph $G = ([n], E)$ and mixing matrix W, set of synchronization indices $\mathcal{I}_T$, initialize $\hat{\mathbf{x}}_i^{(0)} := \mathbf{0}$ for all $i \in [n]$
2: **for** $t = 0$ **to** $T - 1$ in parallel for all workers $i \in [n]$ **do**
3:     Sample $\xi_i^{(t)}$; define stochastic gradient $\mathbf{g}_i^{(t)} := \nabla F_i(\mathbf{x}_i^{(t)}, \xi_i^{(t)})$
4:     $\mathbf{v}_i^{(t)} = \beta \mathbf{v}_i^{(t-1)} + \mathbf{g}_i^{(t)}$
5:     $\mathbf{y}_i^{(t+1)} := \mathbf{x}_i^{(t)} - \eta(\beta \mathbf{v}_i^{(t)} + \mathbf{g}_i^{(t)})$
6:     **if** $(t + 1) \in \mathcal{I}_T$ **then**
7:         Set $\mathbf{q}_i^{(t)} := C(\mathbf{y}_i^{(t+1)} - \hat{\mathbf{x}}_i^{(t)})$
8:         **for** neighbors $j \in \mathcal{N}_i \cup i$ **do**
9:             Send $\mathbf{q}_i^{(t)}$ and receive $\mathbf{q}_j^{(t)}$
10:            $\hat{\mathbf{x}}_j^{(t+1)} := \mathbf{q}_j^{(t)} + \hat{\mathbf{x}}_j^{(t)}$
11:        **end for**
12:        $\mathbf{x}_i^{(t+1)} = \mathbf{y}_i^{(t+1)} + \gamma \sum_{j \in \mathcal{N}_i} w_{ij}(\hat{\mathbf{x}}_j^{(t+1)} - \hat{\mathbf{x}}_i^{(t+1)})$
13:    **else**
14:        $\hat{\mathbf{x}}_i^{(t+1)} = \hat{\mathbf{x}}_i^{(t)}$ and $\mathbf{x}_i^{(t+1)} = \mathbf{y}_i^{(t+1)}$
15:    **end if**
16: **end for**

---

distributed training. Along with assumption **A.1** on the smoothness of objectives given in Sect. 4.2, we assume the following:

A.4 (Bounded Variance): We assume that there exists finite constants $\sigma, M \geq 0$, such that for all $\mathbf{x} \in \mathbb{R}^d$ we have:

$$\frac{1}{n} \sum_{i=1}^n \mathbb{E}_{\xi_i} \|\nabla F_i(\mathbf{x}_i, \xi_i) - \nabla f_i(\mathbf{x}_i)\|_2^2 \leq \sigma^2 + \frac{M^2}{n} \sum_{i=1}^n \|\nabla f_i(\mathbf{x}_i)\|_2^2, \qquad (6.4)$$

where $\nabla F_i(\mathbf{x}, \xi_i)$, $i \in [n]$, denotes an unbiased stochastic gradient, that is, $\mathbb{E}_{\xi_i}[\nabla F_i(\mathbf{x}, \xi_i)] = \nabla f_i(\mathbf{x})$.

A.5 (Bounded Gradient Dissimilarity): We assume that there exists finite constants $G \geq 0$ and $B \geq 1$, such that for all $\mathbf{x} \in \mathbb{R}^d$ we have:

$$\frac{1}{n} \sum_{i=1}^n \|\nabla f_i(\mathbf{x})\|_2^2 \leq G^2 + B^2 \|\nabla f(\mathbf{x})\|_2^2. \qquad (6.5)$$

Remark 4 (Comparison with Existing Assumptions) Note that Assumption **A.2** presented in Sect. 4.2 considers a uniform variance bound for all worker nodes and is stronger than Assumption **A.4** presented above. It can be seen that **A.2** implies **A.4** where we set $\sigma^2 = \frac{1}{n} \sum_{i=1}^n \sigma_i^2$ and $M = 0$. Similarly, Assumption **A.5** is weaker than assuming a uniform bound on the gradient dissimilarity [63], i.e., $\frac{1}{n} \sum_{i=1}^{n} \|\nabla f_i(\mathbf{x}) - \nabla f(\mathbf{x})\|_2^2 \leq \kappa^2$, which implies Assumption **A.5** with $G = \kappa$ and $B = 1$ – this follows from the identity $\frac{1}{n} \sum_{i=1}^{n} \|\nabla f_i(\mathbf{x}) - \nabla f(\mathbf{x})\|_2^2 = \frac{1}{n} \sum_{i=1}^{n} \|\nabla f_i(\mathbf{x})\|_2^2 - \|\nabla f(\mathbf{x})\|_2^2$.

Both assumptions **A.4** and **A.5** are weaker than assumption **A.3** which assumes uniformly bounded second moment for gradients and has been standard in the stochastic optimization with compressed gradients [5, 32, 51, 66].

We remark above relaxed assumptions can similarly be adopted for the case of distributed training presented in Sect. 4 to extend the theoretical results of that setting; see Remark 3. In the following, we present our results for decentralized training using Algorithm 2 with the relaxed assumptions. As before, we perform the training with a fixed step size η, which is dependent on the number of iterations T the algorithm is run for.

Theorem 2 (Convergence Rate of SQuARM-SGD for Smooth Non-convex Objectives, [48]) *Let C be a compression operator with parameter $\omega \in (0, 1]$ and* $gap(\mathcal{I}_T) = H$. *Consider running SQuARM-SGD for T iterations with consensus step-size* $\gamma = \frac{2\delta\omega^3}{48^2\omega^2 + \delta^2 + 128\lambda^2 + 24\omega^2\lambda^2}$, *(where $\lambda = max_i\{1 - \lambda_i(\mathbf{W})\}$), momentum coefficient $\beta \in [0, 1)$, and constant learning rate $\eta = (1-\beta)\sqrt{\frac{n}{T}}$. Let the algorithm generate $\{\mathbf{x}_i^{(t)}\}_{t=0}^{T-1}$ for $i \in [n]$. Running Algorithm 2 for $T \geq U_0$ for some universal constant U_0, the averaged iterates $\overline{\mathbf{x}}^{(t)} := \frac{1}{n} \sum_{i=0}^{n} \mathbf{x}_i^{(t)}$ satisfy:*

$$\frac{1}{T} \sum_{t=0}^{T-1} \mathbb{E}\left\|\nabla f(\overline{\mathbf{x}}^{(t)})\right\|_2^2 \leq \frac{Z_1(f(\overline{\mathbf{x}}^{(0)}) - f^*)}{\sqrt{nT}} + \frac{Z_2 L}{\sqrt{T}} \left(\frac{\sigma^2 + 2(M^2 + n)G^2}{\sqrt{n}}\right)$$
$$+ \frac{\beta^2 n L^2 H^2((M^2 + 1)G^2 + \sigma^2)Z_3}{T\delta\gamma} + \frac{L^2 n^2 H^2((M^2 + 1)G^2 + \sigma^2)Z_4}{\delta\gamma T^2}$$

where Z_1, Z_2, Z_3, Z_4 are universal constants.

The rate expression above can be simplified to yield the following corollary:

Corollary 3 *Let $J^2 < \infty$ be a constant such that $\mathbb{E}[f(\overline{\mathbf{x}}^{(0)})] - f^* \leq J^2$. Hiding constants (including L) in the O notation and substituting the value of γ, we can simplify the rate expression in Theorem 2 to the following:*

$$\frac{\sum_{t=0}^{T-1} \mathbb{E}\|\nabla f(\overline{\mathbf{x}}^{(t)})\|_2^2}{T} = O\left(\frac{J^2 + \sigma^2 + (M^2 + n)G^2}{\sqrt{nT}}\right)$$
$$+ O\left(\frac{\beta^2 n H^2((M^2 + 1)G + \sigma^2)}{T\delta^2\omega^3}\right) + O\left(\frac{n^2 H^2((M^2 + 1)G + \sigma^2)}{\delta^2\omega^3 T^2}\right)$$

6 Communication Efficient Distributed Learning

Remark 5 Observe that ω, H, δ do not affect the dominating term $O\left(\frac{J^2+\sigma^2+(M^2+n)G^2}{\sqrt{nT}}\right)$ as they all appear in the higher order terms. Since Theorem 2 provides non-asymptotic guarantee, we need to decide the horizon T before running the algorithm. To ensure that the dominant term does not get affected by these different factors, we require $T \geq T_1 := C_1 \times \max\left\{\frac{n^3\beta^4H^4[(M^2+1)G+\sigma^2]^2}{[J^2+\sigma^2+(M^2+n)G^2]^2\delta^4\omega^6}, \frac{nH^{4/3}[(M^2+1)G+\sigma^2]^{2/3}}{[J^2+\sigma^2+(M^2+n)G^2]^{2/3}\omega^2\delta^{4/3}}\right\}$ for sufficiently large constant C_1. This implies that for large enough T, we get the benefits of all these techniques in saving the communication bits, essentially for "free", without affecting the convergence rate by too much. Further, we can observe the effect of ω, H, c_0, δ on T_0 as following: (i) if we compress the communication more, i.e., smaller ω, then T_0 increases, as expected; (ii) if we take more number of local iterations H, T_0 would again increase, as expected, because increasing H means communicating less frequently; (iii) if the spectral gap $\delta \in (0, 1]$ is closer to 1, which implies that the graph is well-connected, then the threshold T_0 decreases, which is also expected, as good connectivity means faster spreading of information, resulting in faster consensus.[4]

6 Experimental Results

We now present numerical simulations to demonstrate the communication efficiency of Algorithm 1 (Qsprse-Local-SGD) for distributed training scenarios and Algorithm 2 (SQuARM-SGD/SPARQ-SGD)[5] for decentralized training scenarios. The presented results are from the papers [5, 6, 46, 48, 49], and the interested reader may refer to these papers for additional results. These results show that Algorithm 1 and Algorithm 2 achieve similar performance as vanilla SGD training in their respective scenarios, as also seen in the theoretical results presented in Sects. 4.2 and 5.1 while yielding significant improvements towards communication efficiency arising from the use of compression and local iteration techniques. We remark that the compression operator defined in (6.2) allows for *lossy* compression of parameters or gradients in Algorithms 1 and 2 for the learning task. Such schemes are thus different from the *lossless* compression techniques for reconstruction available in modern GPU communication libraries, e.g. NVCOMP.[6]

---

[4] If we are to design the underlying communication graph, one possible choice is to consider the *expander graphs*, [10], that will simultaneously give low communication and faster convergence, as they have constant degree and large spectral gap, [23].

[5] We remark that from an implementation perspective SPARQ-SGD [46, 49] and SQuARM-SGD [48] are the same, with both algorithms empirically employing momentum based updates to speedup training (which is also the case for Qsparse-Local-SGd presented in the previous subsection). However, the key difference lies in the theoretical results where SQuARM-SGD also includes momentum in the convergence results and further works with a weaker set of assumptions than SPARQ-SGD.

[6] https://developer.nvidia.com/nvcomp.

6.1 Distributed Training

6.1.1 Setup

We train ResNet-50 [21] (which has $d = 25,610,216$ parameters) on the ImageNet dataset [12], using 8 NVIDIA Tesla V100 GPUs. We use a learning rate schedule consisting of 5 epochs of linear warmup, followed by a piecewise decay of 0.1 at epochs 30, 60 and 80, with a batch size of 256 per GPU. The optimization is done via SGD with Nesterov momentum coefficient of 0.9, applied on the local iterations of the workers. The compression scheme is built into the Horovod framework [44]. We use the composed operators $SignTop_k$ and $QTop_k$ operators as defined in Sect. 3 where the quantization operator Q is from [3]. In Top_k, we only update $k_t = \min(d_t, 1000)$ elements per step for each tensor t, where d_t is the number of elements in the tensor. For ResNet-50 architecture, this amounts to updating a total of $k = 99,400$ elements per step.

6.1.2 Results

From Fig. 6.1a, we observe that quantization and sparsification, both individually and combined, when error compensation is enabled through accumulating errors, has almost no penalty in terms of convergence rate, with respect to vanilla SGD. We observe $QTopK$-SGD employing a 4 bit quantizer and the $TopK$ sparsifier, demonstrating superior performance over other schemes, both in terms of loss as well as test accuracy. In particular we observe $QTopK$ dominating over the sign based composed operator, and this is because of the implicit sparsity induced by the Q operator from [3] which results in fewer than k coordinates being transmitted.

Observe that the incorporation of local iterations in Fig. 6.2a, c has very little impact on the convergence rates, as compared to vanilla SGD with the corresponding number of local iterations. This in fact provides an added advantage in terms of savings in communicated bits for achieving target loss as seen in Fig. 6.2b, d by a factor of 6 to 8 times, as compared to the $Qsparse$ operator.

Figure 6.3b, c, d show the training loss, top-1 and top-5 convergence rates[7] respectively, with respect to the total number of bits of communication used. We observe that $Qsparse$-$local$-SGD combines the bit savings of either the deterministic sign based operator or the stochastic quantizer (QSGD), and aggressive sparsifier along with infrequent communication, thereby outperforming the cases where these techniques are individually used. In particular, the required number of bits to achieve the same loss or top-1 accuracy in the case of $Qsparse$-$local$-SGD is around 1/128 in comparison with TopK-SGD and over 1000× less than vanilla SGD. This also verifies that error compensation through memory can be used to mitigate not only

---

[7] Here top-i refers to the accuracy of the top i predictions by the model from the list of possible classes, see [35].

6 Communication Efficient Distributed Learning   215

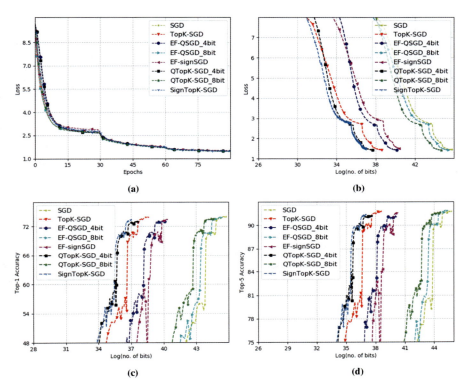

Fig. 6.1 Figure a–d demonstrate the gains in performance achieved by our Qsparse operators in the non-convex setting. (**a**) Comparison of training loss against epochs. (**b**) Comparison of training loss with $\log_2$ of communication budget. (**c**) top-1 accuracy [35] for schemes in (a). (**d**) top-5 accuracy [35] for schemes in (a)

the missing components from updates in previous synchronization rounds, but also explicit quantization error.

6.2 Decentralized Training

6.2.1 Setup

We provide numerical simulations on the CIFAR-10 [34] dataset and train a ResNet-20 [60] model with $n = 8$ nodes connected in a ring topology. Learning rate is initialized to 0.1, following a schedule consisting of a warmup period of 5 epochs followed by piecewise decay of 5 at epoch 200 and 300 and we stop training at epoch 400. The SGD algorithm is implemented with momentum with a factor of 0.9 and minibatch size of 256. SPARQ-SGD (Algorithm 2) consists of $H = 5$

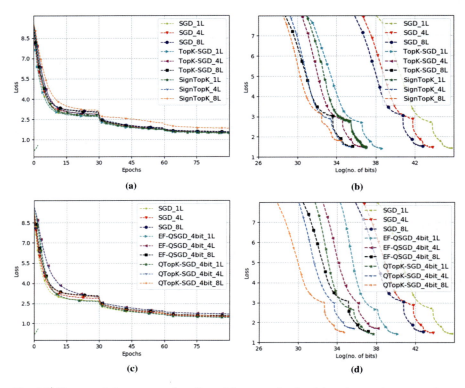

Fig. 6.2 Figure a–d demonstrate the effect of incorporating local iterations and compare these effects across vanilla SGD, the sparsifier *TopK*, as well as its composition with the *Sign* operator. Similar comparisons are also made between vanilla SGD, the quantizer QSGD with error accumulation, as well as its composition with the *TopK* sparsifier. (**a**) Comparison of training loss against epochs. (**b**) Comparison of training loss with $\log_2$ of communication budget. (**c**) Comparison of training loss against epochs. (**d**) Comparison of training loss with $\log_2$ of communication budget

local iterations and then communicating with the composed SignTopK operator, where we take top 1% elements of each tensor and only transmit the sign and norm of the result. We compare performance of SPARQ-SGD against CHOCO-SGD [29, 32], which is another decentralized training algorithm utilizing compressed updates with $Sign$, $TopK$ compression (taking top 1% of elements of the tensor), and decentralized vanilla SGD [36].

6.2.2 Results

We plot the global loss function evaluated at the average parameter vector across nodes in Fig. 6.4a, where we observe SPARQ-SGD converging at a similar rate as CHOCO-SGD and vanilla decentralized SGD. Figure 6.4b shows the performance

6 Communication Efficient Distributed Learning   217

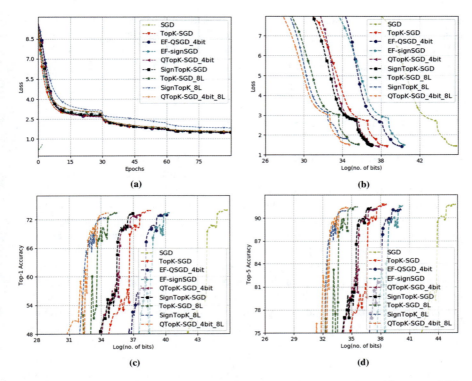

Fig. 6.3 Figure a–d demonstrate the performance of our scheme in comparison with EF-SIGNSGD [25], TopK-SGD [4, 51] and local SGD [50, 64] in the non-convex setting. (**a**) Comparison of training loss against epochs. (**b**) Comparison of training loss with $\log_2$ of communication budget. (**c**) top-1 accuracy [35] for schemes in (a). (**d**) top-5 accuracy [35] for schemes in (a)

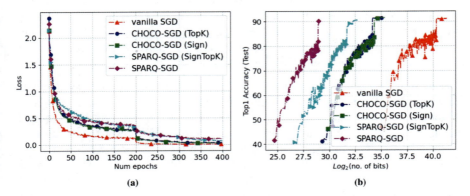

Fig. 6.4 Figure (**a**) and (**b**) are for non-convex objective showing loss vs epochs and Top-1 accuracy vs total no. of bits communicated, respectively

for a given bit-budget, where we show the Top-1 test accuracy as a function of the total number of bits communicated. For target Top-1 test-accuracy of around 90%, SPARQ-SGD requires about $40\times$ less bits than CHOCO-SGD with $Sign$ or $TopK$ compression, and around $3K\times$ less bits than vanilla decentralized SGD.

7 Other Related Works and Discussion

Other Related Works We first discuss other related works in literature which were not mentioned earlier. Methods in [57, 59] also make use of the induced sparsity in the quantized gradients, although without explicitly sparsifying the gradients. [62] analyzed error compensation for QSGD, without sparsification while focusing on quadratic functions. Further discussion on event-triggered communication, from a perspective of control theory, can be found in [14, 16, 22, 38, 45] and in the optimization literature [9, 15, 27]. These papers focus on continuous-time, deterministic optimization algorithms for convex problems; in contrast, [46, 49] are event-driven stochastic gradient descent algorithms for both convex and general (non-convex) smooth objectives, e.g., neural network training for large-scale deep learning. [8] proposed an adaptive scheme to skip gradient computations in a *distributed* setting for *deterministic* gradients; moreover, their focus is on saving communication rounds, without compressed communication. Recently, [42] proposed EF21, a new error compensation framework for distributed heterogeneous settings and does not use any strong assumptions on the gradients. Incorporating local iterations and triggered communication in this new framework would be an interesting direction. Distributed training for heterogeneous client training was recently studied in [26] which uses control variates to account for client drift, and provides theoretical convergence under relaxed assumptions on client objectives. Compressed learning with pairwise constrained objectives has been studied in [47]. There are also many works that combine compression and learning with privacy, see [17, 18]. Extending these ideas to work with compression and local iterations schemes presented in chapter contents above is an interesting direction.

Discussion We now discuss some important extension and possible avenues for research related to the content presented in the chapter. The algorithms presented in the chapter take a *stateful* approach to learning, where each client is required to maintain and utilize a *state*, which carries information about previous updates to guide the training process. Such states can be also be important in heterogeneous client training to combat the client drift problem; see [26]. In large-scale applications, it is possible that some clients may drop out of training process and can only participate for a few rounds. Thus it is of interest to see how staleness or unavailability of client states can affect the training process, and in effect, the generalization capability of the trained model. Design and analysis of algorithms which can reduce the dependence on client states is an interesting future direction. There has been significant interest in developing *lower bounds* for computation

6 Communication Efficient Distributed Learning

and communication complexity for distributed optimization problems. [1] were among the first to develop bounds for computation complexity for stochastic convex optimization in an oracle model. Recently, [39] derived information theoretic lower bounds for gradient compression with quantization. Reference [33] consider communication complexity bounds for number of bits exchanged between devices in a distributed setting. Recently, [20] present an alternate approach to learning in communication constrained environments by compressing the client data itself. Taking advantage of data dependency during the quantization process, they are able to achieve lower rates than the ones suggested in gradient compression works [39, 40]. Merging the data quantization framework with the communication efficiency techniques presented in this chapter, and deriving corresponding lower bounds is a promising direction to understand the limitations of efficient large-scale training. As discussed in contents of the chapter, error compensation is a key mechanism to ensure convergence with (biased) compression schemes. The error compensation is realized by maintaining a *memory* for the training clients. Understanding the convergence of memory based methods from a perspective of developing corresponding lower bounds on computation and communication complexity is also of significant interest.

Acknowledgments This work was partially supported by NSF grants #2007714, #2139304, #1955632, and by Army Research Laboratory under Cooperative Agreement W911NF-17-2-0196. Views and conclusions contained in this document are those of the authors and should not be interpreted as representing the official policies, either expressed or implied, of Army Research Laboratory or the U.S. Government. The U.S. Government is authorized to reproduce and distribute reprints for Government purposes notwithstanding any copyright notation here on.

References

1. A. Agarwal, M.J. Wainwright, P. Bartlett, P. Ravikumar, Information-theoretic lower bounds on the oracle complexity of convex optimization, in *NIPS*, vol. 22 (2009)
2. A.F. Aji, K. Heafield, Sparse communication for distributed gradient descent, in *EMNLP* (2017), pp. 440–445
3. D. Alistarh, D. Grubic, J. Li, R. Tomioka, M. Vojnovic, QSGD: communication-efficient SGD via gradient quantization and encoding, in *NIPS* (2017), pp. 1709–1720
4. D. Alistarh, T. Hoefler, M. Johansson, N. Konstantinov, S. Khirirat, C. Renggli, The convergence of sparsified gradient methods, in *NeurIPS* (2018), pp. 5973–5983
5. D. Basu, D. Data, C. Karakus, S. Diggavi, Qsparse-local-SGD: distributed SGD with quantization, sparsification and local computations, in *NeurIPS* (2019), pp. 14668–14679
6. D. Basu, D. Data, C. Karakus, S.N. Diggavi, Qsparse-local-sgd: distributed sgd with quantization, sparsification, and local computations. IEEE J. Sel. Areas Inf. Theory **1**(1), 217–226 (2020)
7. J. Bernstein, Y.-X. Wang, K. Azizzadenesheli, A. Anandkumar, signSGD: compressed optimisation for non-convex problems, in *ICML* (2018), pp. 560–569
8. T. Chen, G. Giannakis, T. Sun, W. Yin, Lag: lazily aggregated gradient for communication-efficient distributed learning, in *NeurIPS* (2018), pp. 5050–5060

9. W. Chen, W. Ren, Event-triggered zero-gradient-sum distributed consensus optimization over directed networks. Automatica **65**, 90–97 (2016)
10. Y.T. Chow, W. Shi, T. Wu, W. Yin, Expander graph and communication-efficient decentralized optimization, in *ACSSC* (2016), pp. 1715–1720
11. G.F. Coppola, Iterative parameter mixing for distributed large-margin training of structured predictors for natural language processing. PhD thesis, University of Edinburgh (2015)
12. J. Deng, W. Dong, R. Socher, L.-J. Li, K. Li, L. Fei-Fei, ImageNet: a large-scale hierarchical image database, in *2009 IEEE Conference on Computer Vision and Pattern Recognition* (IEEE, 2009), pp. 248–255
13. J. Devlin, M.-W. Chang, K. Lee, K. Toutanova, BERT: Pre-training of deep bidirectional transformers for language understanding, in *Proceedings of the 2019 Conference of the North American Chapter of the Association for Computational Linguistics: Human Language Technologies* (Association for Computational Linguistics, 2019), pp. 4171–4186
14. D.V. Dimarogonas, E. Frazzoli, K.H. Johansson, Distributed event-triggered control for multi-agent systems. IEEE TAC **57**(5), 1291–1297 (2012)
15. W. Du, X. Yi, J. George, K.H. Johansson, T. Yang, Distributed optimization with dynamic event-triggered mechanisms, in *IEEE CDC* (2018), pp. 969–974
16. A. Girard, Dynamic triggering mechanisms for event-triggered control. IEEE TAC **60**(7), 1992–1997 (2015)
17. A. Girgis, D. Data, S. Diggavi, P. Kairouz, A.T. Suresh, Shuffled model of differential privacy in federated learning, in *International Conference on Artificial Intelligence and Statistics, PMLR* (2021), pp. 2521–2529
18. A.M. Girgis, D. Data, S. Diggavi, P. Kairouz, A.T. Suresh, Shuffled model of federated learning: privacy, accuracy and communication trade-offs. IEEE J. Sel. Areas Inf. Theory **2**(1), 464–478 (2021)
19. F. Haddadpour, M.M. Kamani, A. Mokhtari, M. Mahdavi, Federated learning with compression: Unified analysis and sharp guarantees, in *International Conference on Artificial Intelligence and Statistics, PMLR* (2021), pp. 2350–2358
20. O.A. Hanna, Y.H. Ezzeldin, C. Fragouli, S. Diggavi, Quantization of distributed data for learning. IEEE J. Sel. Areas Inf. Theory **2**(3), 987–1001 (2021)
21. K. He, X. Zhang, S. Ren, J. Sun, Deep residual learning for image recognition, in *2016 IEEE Conference on Computer Vision and Pattern Recognition, CVPR 2016, Las Vegas, NV, June 27–30, 2016*, pp. 770–778
22. W.P.M.H. Heemels, K.H. Johansson, P. Tabuada, An introduction to event-triggered and self-triggered control, in *IEEE CDC* (2012), pp. 3270–3285
23. S. Hoory, N. Linial, A. Wigderson, Expander graphs and their applications. Bull. Am. Math. Soc. **43**(4), 439–561 (2006)
24. P. Kairouz, H.B. McMahan, B. Avent, A. Bellet, M. Bennis, A.N. Bhagoji, K. Bonawitz, Z. Charles, G. Cormode, R. Cummings, et al., Advances and open problems in federated learning. Found. Trends Mach. Learn. **14**(1–2), 1–210 (2021)
25. S.P. Karimireddy, Q. Rebjock, S. Stich, M. Jaggi, Error feedback fixes signSGD and other gradient compression schemes, in *International Conference on Machine Learning, PMLR* (2019), pp. 3252–3261
26. S.P. Karimireddy, S. Kale, M. Mohri, S. Reddi, S. Stich, A.T. Suresh, Scaffold: stochastic controlled averaging for federated learning, in *International Conference on Machine Learning, PMLR* (2020), pp. 5132–5143
27. S.S. Kia, J. Cortés, S. Martínez, Distributed convex optimization via continuous-time coordination algorithms with discrete-time communication. Automatica **55**, 254–264 (2015)
28. D.P. Kingma, J. Ba, ADAM: a method for stochastic optimization, in *3rd International Conference on Learning Representations, ICLR 2015, San Diego, CA, May 7–9, 2015, Conference Track Proceedings*
29. A. Koloskova, S.U. Stich, M. Jaggi, Decentralized stochastic optimization and gossip algorithms with compressed communication, in *ICML* (2019)

6 Communication Efficient Distributed Learning

30. A. Koloskova, N. Loizou, S. Boreiri, M. Jaggi, S. Stich, A unified theory of decentralized sgd with changing topology and local updates, in *International Conference on Machine Learning, PMLR* (2020), pp. 5381–5393
31. A. Koloskova, N. Loizou, S. Boreiri, M. Jaggi, S.U. Stich, A unified theory of decentralized SGD with changing topology and local updates, in *International Conference on Machine Learning (ICML)*, vol. 119. Proceedings of Machine Learning Research, PMLR (2020), pp. 5381–5393
32. A. Koloskova, T. Lin, S.U. Stich, M. Jaggi, Decentralized deep learning with arbitrary communication compression, in *ICLR* (2020)
33. J.H. Korhonen, D. Alistarh, Towards tight communication lower bounds for distributed optimisation. Adv. Neural Inf. Proces. Syst. **34** (2021)
34. A. Krizhevsky, G. Hinton, et al., Learning multiple layers of features from tiny images. Technical Report (2009). https://www.cs.toronto.edu/~kriz/learning-features-2009-TR.pdf
35. M. Lapin, M. Hein, B. Schiele, Top-k multiclass SVM, in *NIPS* (2015), pp. 325–333
36. X. Lian, C. Zhang, H. Zhang, C.-J. Hsieh, W. Zhang, J. Liu, Can decentralized algorithms outperform centralized algorithms? A case study for decentralized parallel stochastic gradient descent, in *NIPS* (2017), pp. 5330–5340
37. Y. Lin, S. Han, H. Mao, Y. Wang, W.J. Dally, Deep gradient compression: reducing the communication bandwidth for distributed training, in *ICLR* (2018)
38. Y. Liu, C. Nowzari, Z. Tian, Q. Ling, Asynchronous periodic event-triggered coordination of multi-agent systems, in *IEEE CDC* (2017), pp. 6696–6701
39. P. Mayekar, H. Tyagi, Limits on gradient compression for stochastic optimization, in *2020 IEEE International Symposium on Information Theory (ISIT), IEEE* (2020), pp. 2658–2663
40. P. Mayekar, H. Tyagi, Ratq: a universal fixed-length quantizer for stochastic optimization, in *International Conference on Artificial Intelligence and Statistics, PMLR* (2020), pp. 1399–1409
41. B. McMahan, E. Moore, D. Ramage, S. Hampson, B.A. y Arcas, Communication-efficient learning of deep networks from decentralized data, in *Artificial Intelligence and Statistics* (2017), pp. 1273–1282
42. P. Richtárik, I. Sokolov, I. Fatkhullin, Ef21: a new, simpler, theoretically better, and practically faster error feedback, in *NeurIPS*, vol. 34 (2021)
43. M. Riedmiller, H. Braun, A direct adaptive method for faster backpropagation learning: the rprop algorithm, in *IEEE International Conference on Neural Networks*, vol. 1, March 1993, pp. 586–591
44. A. Sergeev, M.D. Balso, Horovod: fast and easy distributed deep learning in tensorflow. CoRR, abs/1802.05799 (2018)
45. G.S. Seyboth, D.V. Dimarogonas, K.H. Johansson, Event-based broadcasting for multi-agent average consensus. Automatica **49**(1), 245–252 (2013)
46. N. Singh, D. Data, J. George, S. Diggavi, SPARQ-SGD: event-triggered and compressed communication in decentralized optimization, in *IEEE Control and Decision Conference (CDC)* (2020)
47. N. Singh, X. Cao, S. Diggavi, T. Basar, Decentralized multi-task stochastic optimization with compressed communications. arXiv preprint arXiv:2112.12373 (2021)
48. N. Singh, D. Data, J. George, S. Diggavi, SQuARM-SGD: communication-efficient momentum SGD for decentralized optimization. IEEE J. Sel. Areas Inf. Theory **2**(3), 954–969 (2021)
49. N. Singh, D. Data, J. George, S. Diggavi, SPARQ-SGD: Event-triggered and compressed communication in decentralized optimization. IEEE Trans. Autom. Control **68**(2), 721–736 (2022)
50. S.U. Stich, Local SGD converges fast and communicates little, in *ICLR* (2019)
51. S.U. Stich, J.-B. Cordonnier, M. Jaggi, Sparsified SGD with memory, in *NeurIPS* (2018), pp. 4447–4458
52. N. Strom, Scalable distributed DNN training using commodity GPU cloud computing, in *INTERSPEECH 2015, 16th Annual Conference of the International Speech Communication Association, Dresden, September 6–10, 2015*, pp. 1488–1492

53. A.T. Suresh, F.X. Yu, S. Kumar, H.B. McMahan, Distributed mean estimation with limited communication, in *ICML* (2017), pp. 3329–3337
54. H. Tang, S. Gan, C. Zhang, T. Zhang, J. Liu, Communication compression for decentralized training, in *Advances in Neural Information Processing Systems 31: Annual Conference on Neural Information Processing Systems 2018, NeurIPS 2018*, 3–8 December 2018, Montréal, pp. 7663–7673
55. H. Tang, C. Yu, X. Lian, T. Zhang, J. Liu, Doublesqueeze: parallel stochastic gradient descent with double-pass error-compensated compression, in *International Conference on Machine Learning, ICML* (2019), pp. 6155–6165
56. T. Tieleman, G. Hinton, Lecture 6.5-rmsprop, coursera: Neural networks for machine learning. University of Toronto, Technical Report 6 (2012)
57. H. Wang, S. Sievert, S. Liu, Z.B. Charles, D.S. Papailiopoulos, S. Wright, ATOMO: communication-efficient learning via atomic sparsification, in *NeurIPS* (2018), pp. 9872–9883
58. J. Wang, V. Tantia, N. Ballas, M. Rabbat, SlowMo: improving communication-efficient distributed SGD with slow momentum, in *International Conference on Learning Representations, ICLR* (2020)
59. J. Wangni, J. Wang, J. Liu, T. Zhang, Gradient sparsification for communication-efficient distributed optimization, in *NIPS* (2018), pp. 1306–1316
60. W. Wen, C. Wu, Y. Wang, Y. Chen, H. Li, Learning structured sparsity in deep neural networks, in *NIPS* (2016), pp. 2074–2082
61. W. Wen, C. Xu, F. Yan, C. Wu, Y. Wang, Y. Chen, H. Li, Terngrad: ternary gradients to reduce communication in distributed deep learning, in *NIPS* (2017), pp. 1508–1518
62. J. Wu, W. Huang, J. Huang, T. Zhang, Error compensated quantized SGD and its applications to large-scale distributed optimization, in *ICML* (2018), pp. 5321–5329
63. H. Yu, R. Jin, S. Yang, On the linear speedup analysis of communication efficient momentum SGD for distributed non-convex optimization, in *International Conference on Machine Learning, PMLR* (2019), pp. 7184–7193
64. H. Yu, S. Yang, S. Zhu, Parallel restarted SGD with faster convergence and less communication: demystifying why model averaging works for deep learning, in *AAAI* (2019), pp. 5693–5700
65. Y. Zhang, J.C. Duchi, M.I. Jordan, M.J. Wainwright, Information-theoretic lower bounds for distributed statistical estimation with communication constraints, in *Advances in Neural Information Processing Systems*, vol. 26 (2013), pp. 2328–2336
66. S. Zheng, Z. Huang, J. Kwok, Communication-efficient distributed blockwise momentum SGD with error-feedback, in *Advances in Neural Information Processing Systems*, vol. 32 (2019)

Chapter 7
Coreset-Based Data Reduction for Machine Learning at the Edge

Hanlin Lu, Ting He, and Shiqiang Wang

Abstract This chapter covers the issue of reducing the communication cost for machine learning at the edge from the perspective of data compression. Unlike traditional data compression schemes that aim at supporting the reconstruction of the original data, here the compression only needs to support the learning of the models that need to be learned from the original data, in order to support AI applications in a bandwidth-limited edge network. This lowered goal opens the door to a variety of application-specific lossy compression schemes designed to support machine learning. The focus in this chapter is on a subset of such schemes that can construct a weighted dataset much smaller than the original dataset that can function as a replacement of the original dataset in learning tasks, known as *coreset*. It reviews the history of coresets and their limitations, and then details two recently proposed improvements on (1) robust coreset construction and (2) integration of coreset construction and quantization.

1 Introduction

In the recent decade, we have observed a dramatic growth in distributed data generation, which was powered by the rapid development and deployment of data-capturing devices, including smart phones, wearables, smart cameras, and other Internet of Things (IoT) devices. This results in a large amount of data distributed across edge devices, which opens up opportunities for new AI-based applications,

---

H. Lu (✉)
Bytedance Inc., Mountain View, CA, USA
e-mail: hanlin.lu@bytedance.com

T. He (✉)
Pennsylvania State University, University Park, PA, USA
e-mail: tinghe@psu.edu

S. Wang
IBM T. J. Watson Research Center, Yorktown Heights, NY, USA
e-mail: wangshiq@us.ibm.com

© The Author(s), under exclusive license to Springer Nature Switzerland AG 2023
M. Srivatsa et al. (eds.), *Artificial Intelligence for Edge Computing*,
https://doi.org/10.1007/978-3-031-40787-1_7

such as intelligent traffic management, smart health, and smart environmental monitoring. Meanwhile, unleashing the value of such distributed big data via AI faces a number of challenges, such as limited computation capacity and battery at edge devices, limited network bandwidth, heterogeneous data distributions, and privacy concerns. The first two challenges are particularly important to address in applications with relatively weak data-capturing devices (e.g., IoT and smart health applications). For such applications, the data-capturing devices typically do not have the resources to run complicated computations, ruling out approaches that push the computation to devices (e.g., federated/decentralized learning); meanwhile, they often do not have the bandwidth to report all the collected data to edge servers to run the computation there. This creates a challenging research question: *how can we offload machine learning (ML) computation from data-capturing devices (a.k.a. data sources) to nearby edge servers without significantly degrading the quality of learning or congesting the edge network?*

In this chapter, we will examine a line of techniques, originally designed for a different purpose of speeding up centralized learning, that are proved promising in addressing the above question. These techniques construct a variety of small weighted datasets in the same sample space as the original dataset, called *coresets*, that are designed to preserve the relevant information in the original dataset with respect to target models. They can be thought of as generalizations of subsampling (indeed subsampling is a specific way of constructing coresets), as points in a coreset do not have to come from the original dataset. Originally used as a preprocessing step to reduce the data size in order to speed up the execution of high-complexity algorithms, coreset construction is designed to be fast, with an output size that is much smaller than the input size. This makes it an attractive approach for communication-efficient offloading of ML computation: instead of running full-fledged ML tasks locally or reporting all the collected data, data sources only need to construct and report small coresets of their local data, leaving the rest of the computation to edge servers. Moreover, if the constructed coreset is the so-called ϵ-coreset for the objective function of interest (i.e., the cost function of the target model), we can also guarantee the quality of coreset-based learning in the sense that the model learned from an ϵ-coreset will achieve an objective value that differs by at most a factor of $1 + O(\epsilon)$ from the objective value achieved by the model learned from the original dataset.

However, to apply this idea in practice, we face a fundamental challenge: in order to achieve substantial data reduction while preserving useful information, existing coresets are all tailor-made to support specific models by approximating the corresponding cost functions (see Sect. 2.3), meaning that different coresets are required to learn different models, which multiplies the communication cost and limits the broad application of coresets. This motivates our search for a "robust coreset" that can provide a good approximation for a broad range of cost functions.

Following the same research direction, we note that coreset construction algorithms can only reduce the number of data points, but not the number of bits required to represent each data point. The latter is the goal of *quantization*, where various techniques, from simple rounding-based quantizers to sophisticated vector quantizers, have been proposed to transform the data points from arbitrary values in the sample space to a set of discrete values that can be encoded by a smaller number of bits [1]. We will show the first framework to integrate quantization into the coreset construction process to improve the tradeoff between communication cost and approximation error compared to using coreset construction or quantization alone.

Roadmap The rest of the chapter will be organized as follows: Sect. 2 gives the background of coreset in comparison with other approaches for learning over distributed data, Sect. 3 details the algorithms and performances for robust coreset construction, Sect. 4 further combines coreset construction with quantization, and Sect. 5 concludes the chapter with a summary and a remark.

2 Background and Preliminaries

2.1 General Approaches for Learning over Distributed Data

Driven by the rapid growth in distributed data generation and the proliferation of learning-based applications, there is a high demand for techniques that can allow ML to utilize the information in data distributed across different nodes, without imposing excessive burdens on any individual node or the communication network. To achieve this goal, existing solutions can be roughly categorized into three approaches as follows.

Sharing Output The first approach is to let each node learn its own model, but collect and aggregate the predictions of the local models to produce a global prediction at inference time. This approach was popular in early works on distributed classification. By treating the outputs of local models as the inputs of the global model, [2] proposed a hierarchical classification method, which was further improved in [3] to achieve high prediction accuracy and efficient scaling-up for a variety of distributed datasets. In [4], the researchers examined multiple heuristic methods to aggregate the outputs of local classifiers, including the commonly-used majority vote method. While avoiding the problem of distributed model training, this approach cannot utilize all the data effectively, as no model is constructed based on global data.

Sharing Model Aiming at constructing a model that incorporates information in all the data, this approach lets nodes exchange information throughout the training process. Each node is still in charge of learning its own model (e.g., via gradient descent), but it exchanges model parameters or other related quantities with other nodes to incorporate the information in their data, either through a centralized parameter server (federated learning) [5, 6] or through peer-to-peer connections (decentralized learning) [7]. This process repeats iteratively until a certain convergence criterion is satisfied.

Sharing Summary Also aiming at constructing a model based on information in all the data, the third approach lets nodes construct and report summaries of their data to a server (or a cluster of servers), such that the intended learning tasks can be performed on the data summaries. Compared to the second approach, this approach shifts the majority of computation to the server, and is thus more suitable in scenarios where the data-capturing devices are weak in computation power, as is the case in many IoT applications. In addition, sharing data summaries instead of model parameters can potentially save communication cost, as the data summaries only need to be transferred once, and one summary can potentially be used to compute multiple models, amortizing the cost. There have been a variety of data summarization techniques such as coresets, sketches, and projections [8–10]. Among these, coresets, which reside in the same sample space as the original dataset, have received particular attention, as they can be viewed as "smaller versions" of the original dataset and hence used as natural replacements. For example, coresets can be used to identify "representative data points" for prioritized data labeling and engineer features within the original sample space, and a model trained on a coreset can be used to process a new sample without further transformation. For these reasons, we will focus on coreset-based data summarization.

2.2   Cost Function and Coreset

To formally introduce coreset, we need to first formulate the ML problem of interest. Given a d-dimensional dataset $P \subseteq \mathbb{R}^d$, a space of model parameters X, and a *cost function* $\text{cost}(P, x)$ that describes how badly a model $x \in X$ fits the dataset P, the goal of ML is to find the model $x^* \in X$ with the best fit, defined as

$$x^* := \arg \min_{x \in X} \text{cost}(P, x). \tag{7.1}$$

In the case of supervised learning, the cost function is commonly known as the loss function, but we will use the term "cost function" to incorporate both supervised and unsupervised learning.

The overall cost function $\text{cost}(P, x)$ is an aggregation of the *per-point cost function* $\text{cost}(p, x)$ that describes the fitting error of model $x \in X$ for a data point $p \in P$. For the sake of generality, we assume that P is a weighted dataset, with $w_p > 0$ denoting the weight for $p \in P$; $w_p \equiv 1$ denotes the case of an unweighted dataset. We use $w_{\min} := \min_{p \in P} w_p$ to denote the minimum weight in P. Typically, the overall cost function is in one of the following two forms: (1) *sum cost*, i.e., $\text{cost}(P, x) = \sum_{p \in P} w_p \text{cost}(p, x)$, (2) *maximum cost*, i.e., $\text{cost}(P, x) = \max_{p \in P} \text{cost}(p, x)$. Most ML problems use sum cost. For example, in deep learning with cross entropy loss, the overall cost, which represents the negative log-likelihood of the entire set of samples, equals the summation of the negative log-likelihoods of the individual samples. Most unsupervised learning

7 Coreset-Based Data Reduction for Machine Learning at the Edge

problems, e.g., k-means and PCA, also adopt sum cost. Meanwhile, there are ML problems adopting maximum cost, which are typically about finding extreme values or outliers. For example, the *minimum enclosing ball (MEB)* problem [11], which aims to find a center that minimizes the maximum distance between each data point and the center, is an example for maximum cost with the overall cost function defined as $\mathrm{cost}(P, x) = \max_{p \in P} \mathrm{dist}(p, x)$, the per-point cost function as $\mathrm{cost}(p, x) = \mathrm{dist}(p, x)$, and the solution space as $X = \mathbb{R}^d$. Here $\mathrm{dist}(p, q)$ denotes the Euclidean distance between points p and q.

Given a cost function $\mathrm{cost}(P, x)$, a coreset S is a weighted set of data points that can be used to approximate the original dataset P in solving (7.1), formalized below.

Definition 1 ([12]) Give a cost function $\mathrm{cost}(P, x)$ with $P \subseteq \mathbb{R}^d$ and $x \in X$, a dataset $S \subseteq \mathbb{R}^d$ with weights u_q ($q \in S$) is an ϵ-coreset for P with respect to (w.r.t.) $\mathrm{cost}(P, x)$ if it satisfies that

$$(1 - \epsilon)\mathrm{cost}(P, x) \leq \mathrm{cost}(S, x) \leq (1 + \epsilon)\mathrm{cost}(P, x), \quad \forall x \in X, \tag{7.2}$$

where $\mathrm{cost}(S, x)$ is defined in the same way as $\mathrm{cost}(P, x)$, i.e., $\mathrm{cost}(S, x) = \sum_{q \in S} u_q \mathrm{cost}(q, x)$ for sum cost, and $\mathrm{cost}(S, x) = \max_{q \in S} \mathrm{cost}(q, x)$ for maximum cost.

2.3 Overview of Coreset Construction Algorithms

By Definition 1, the quality of a coreset clearly depends on the cost function under consideration. Accordingly, most of the existing coreset construction algorithms are tailored-made to specific cost functions that represent specific ML problems. Based on the design principle, we can categorize existing coreset construction algorithms into three common approaches, briefly summarized below; we refer to [8, 9] for more detailed surveys.

(1) Gradient Descent Algorithms These algorithms were first derived for solving MEB in [11, 13], where a point farthest from the current center is selected and added into the current coreset, and the procedure is repeated until all the data points could be covered by expanding the enclosing ball of the current coreset by $1 + \epsilon$ times. Similar algorithms have been developed for probabilistic MEB [14], *support vector machine (SVM)* [15, 16], dimensionality reduction [17], etc.

(2) Random Sampling Algorithms These algorithms randomly sample data points and assign weights to form a coreset according to some predefined probabilities, and differ from each other in how the sampling probabilities are designed. *Sensitivity sampling* [18] basically defines the sampling probability of each data point proportionally to its per-point cost. Based on this idea, a unified framework with applications including k-means/median, *principle component analysis (PCA)* [12], dictionary learning [19], and dependency networks [20], was developed in [12].

Although different cost functions can be plugged into this framework to support different ML problems, the generated coresets are still in one-one correspondence with the cost functions.

(3) Geometric Decomposition Algorithms These algorithms divide the input space into several partitions, and then select representative points in each partition, which are used to form a coreset. Possible applications of this approach include weighted facility problems [21], Euclidean graph problems [22], k-means/median [23, 24], etc.

There are a few coreset construction algorithms that do not fall squarely into any of the above three approaches, e.g., SVD-based algorithms in [25, 26], but most existing coreset construction algorithms can be covered.

Although often designed for the centralized setting, the above coreset construction algorithms can be easily extended to the distributed setting using a *merge-and-reduce approach* proposed in [27]. This property makes coresets particularly suitable for supporting learning over distributed data.

3 Robust Coreset Construction

As mentioned in Sect. 2.3, existing coresets are each tailor-made to support a specific ML problem. Meanwhile, collecting a separate coreset for each new problem we want to solve will be clearly wasteful in network bandwidth, as these coresets are all generated from the same original dataset. This dilemma triggers the following research question: *Is there a coreset that can simultaneously support a broad set of ML problems?* In other words, can we construct a coreset that provides a robust approximation for many different cost functions? An affirmative answer to this question will provide us a data summary that can simultaneously support multiple learning tasks, thus amortizing the communication cost in collecting the summary. This is the main question we addressed in [28].

Below, we will first present our results in a centralized setting in Sect. 3.1, where the original data reside at a single node (that wishes to offload ML computation to a server by sending a summary of its data). We will then extend the solution to a distributed setting in Sect. 3.2, where the original data are split across multiple nodes (that collectively report a summary to a server to enable ML over the union of their data).

3.1 Centralized Construction of Robust Coreset

We start with a humble goal of comparing the robustness of existing representative coresets when intentionally misused for problems they are not designed for. Perhaps surprisingly, one coreset stands out with superior robustness. We then analyze the root cause of this phenomenon, which leads to a robust coreset construction algorithm and the conditions under which it will give guaranteed performance.

3.1.1 Motivating Experiment

Before diving into theoretical analysis, we will start with a motivating experiment, which compares the robustness of existing coresets. To this end, we select four representative coreset construction algorithms based on the categorization in Sect. 2.3: 'farthest point' algorithm from [13] as a representative of gradient descent algorithms, 'nonuniform sampling' algorithm from [12] as a representative of random sampling algorithms, 'decomposition' algorithm from [23] as a representative of geometric decomposition algorithms, and 'uniform sampling' algorithm, which serves as our baseline for comparison. Since the selected algorithms [12, 13, 23] were originally designed for either MEB or k-means, we test their performance in supporting both MEB and k-means.

In this experiment, our input dataset was synthesized with 5000 points uniformly distributed in $[0, 100]^3$. We evaluate the performance of a coreset S by the *normalized cost* of the model x_S learned on S, defined as $\text{cost}(P, x_S)/\text{cost}(P, x^*)$, where x^* is the optimal model calculated from the original dataset as in (7.1). Ideally, we want the normalized cost to be close to 1, meaning that the model learned from the coreset is nearly as good as the model learned from the original dataset. Due to the randomness in constructing the coresets and solving the ML problems, we repeat each experiment for 100 Monte Carlo runs and plot the *cumulative distribution function (CDF)* of the normalized costs in Fig. 7.1.

The results validate the expectation that a coreset tailored-made for one ML problem can perform poorly for another problem, e.g., the 'farthest point' algorithm designed for MEB performs poorly for k-means. What is less expected is that the 'decomposition' algorithm from [23] shows robust performance for both MEB and

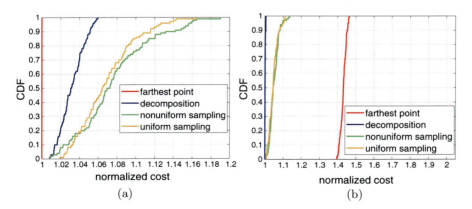

Fig. 7.1 Comparison of coreset construction algorithms (coreset size: 50)

230 H. Lu et al.

k-means. This triggers a research question of whether this phenomenon indicates a fundamental law that applies to more general datasets and ML problems. The rest of this section will answer this question based on rigorous analysis.

3.1.2 The k-Clustering Problem

The idea of the 'decomposition' algorithm from [23] is to partition the sample space by k-clustering. To understand why this algorithm is so robust, we formally define the k-clustering problem as follows.

Suppose that we have a weighted dataset $P \subseteq \mathbb{R}^d$ with weight w_p ($\forall p \in P$), and a set $Q = \{q_1, \ldots, q_k\} \subseteq \mathbb{R}^d$ of $k \geq 1$ centers. Then the k-clustering cost for P with centers Q is defined as:

$$c(P, Q) = \sum_{p \in P} w_p (\min_{q \in Q} \text{dist}(p, q))^z, \tag{7.3}$$

and the k-clustering problem is to find the set of centers that minimizes (7.3). Different values of z lead to different clustering problems. If $z = 1$, this is the k-median problem; if $z = 2$, this is the k-means problem. In the following discussion, we will use $c(P, \cdot)$ to denote the k-clustering cost, and $\text{cost}(P, \cdot)$ to denote the cost function of a generic ML problem.

For ease of presentation, we will use $\mu(P)$ to denote the optimal 1-clustering center of P. It is known that for $z = 2$, $\mu(P)$ is the sample mean:

$$\mu(P) = \frac{1}{\sum_{p \in P} w_p} \sum_{p \in P} w_p p. \tag{7.4}$$

and for $z = 1$, $\mu(P)$ is the sample median.

While it is NP-hard to solve k-means and k-median problems [29, 30], there exist heuristic methods to efficiently calculate approximate solutions. We will use $\text{opt}(P, k)$ and $\text{approx}(P, k)$ to denote the optimal and approximate costs for k-clustering problems, respectively.

Each k-clustering solution could be understood as a partition $\{P_1, \ldots, P_k\}$ induced by k centers $Q = \{q_i\}_{i=1}^k$, where data points whose closest center in Q is q_i form P_i (ties broken arbitrarily). In the rest of Sect. 3.1, we will use $\{P_i\}_{i \in [k]}$ ($[k] := \{1, \ldots, k\}$) to denote the partition under the optimal k-clustering, and $\{\widetilde{P}_i\}_{i \in [k]}$ to denote the partition under a suboptimal k-clustering.

3.1.3 Coreset Construction by Optimal k-Clustering

We now explain the underlying reasons for the robust performance of the 'decomposition' algorithm from [23] observed in Sect. 3.1.1. Roughly speaking, this

7 Coreset-Based Data Reduction for Machine Learning at the Edge

algorithm partitions the space by k-clustering and constructs a coreset point to represent each partition. As the cluster center is by definition a representation of its cluster, the question is how well the cluster centers will work as a coreset. Below, we will first answer this question for the optimal centers and then extend our answer to suboptimal centers computable by practical clustering algorithms (see Sect. 3.1.4).

Sketch of Analysis In short, if we find that doubling the number of centers will not cause much reduction in the clustering cost, then the current centers will form a coreset that closely approximates the original dataset w.r.t. any smooth cost function, based on the following reasons:

1. the small overall cost reduction implies that in any cluster, adding another center will not decrease the clustering cost by much compared with having a single center (Lemma 1);
2. if little cost reduction is obtained by adding another center in a cluster, then all the data points in this cluster must be very close to the current center, hence well approximated by the center (Lemma 2);
3. if each data point is well approximated by its center, then the set of centers, each weighted by the total weight of points in its cluster, gives a coreset that can provide a good approximation for any smooth cost function (Lemmas 3).

Formal Analysis Formally, the complete analysis is stated as followings. We refer to [28] for the proofs.

Lemma 1 *If $\forall \epsilon' > 0$, $opt(P, k) - opt(P, 2k) \leq \epsilon'$, then $\forall i \in [k]$, we have $opt(P_i, 1) - opt(P_i, 2) \leq \epsilon'$, where $\{P_i\}_{i=1}^k$ is the partition of P generated by the optimal k-clustering solution.*

Lemma 2 *If in one cluster P_i, $opt(P_i, 1) - opt(P_i, 2) \leq \epsilon'$, then $\forall p \in P_i$, $dist(p, \mu(P_i)) \leq (\frac{\epsilon'}{w_{\min}})^{\frac{1}{z}}$.*

Lemma 3 *For any ML problem with a summation (or maximum) cost function: $cost(P, x) = \sum_{p \in P} w_p cost(p, x)$ (or $cost(P, x) = \max_{p \in P} cost(p, x)$), if there exists a partition $\{P_i\}_{i=1}^k$ of P such that $\forall x \in X$, $i \in [k]$, and $p \in P_i$,*

$$(1 - \epsilon)cost(p, x) \leq cost(\mu(P_i), x) \leq (1 + \epsilon)cost(p, x), \tag{7.5}$$

then $S = \{\mu(P_i)\}_{i=1}^k$ with weight $u_{\mu(P_i)} = \sum_{p \in P_i} w_p$ is an ϵ-coreset for P w.r.t. $cost(P, x)$.

With the above lemmas, we are now ready to prove our main theorem:

Theorem 1 *If $opt(P, k) - opt(P, 2k) \leq w_{\min}(\frac{\epsilon}{\rho})^z$, then the optimal k-clustering centers of P, each weighted by the total weight of points in its cluster, form an ϵ-coreset for P w.r.t. both the sum cost and the maximum cost, as long as the per-point cost function satisfies (i) $cost(p, x) \geq 1$, and (ii) $cost(p, x)$ is ρ-Lipschitz-continuous in p, $\forall x \in X$.*

Proof Let us use ϵ' denote $w_{\min}(\frac{\epsilon}{\rho})^z$. According to Lemma 1, if $\mathrm{opt}(P, k) - \mathrm{opt}(P, 2k) \leq \epsilon'$, then $\forall$ cluster P_i generated by the optimal k-clustering, $\mathrm{opt}(P_i, 1) - \mathrm{opt}(P_i, 2) \leq \epsilon'$. This in turn implies $\forall p \in P_i$, $\mathrm{dist}(p, \mu(P_i)) \leq (\frac{\epsilon'}{w_{\min}})^{\frac{1}{z}}$. Moreover, we assume that $\forall x \in X$, $\mathrm{cost}(p, x)$ is ρ-Lipschitz-continuous in p, and thus

$$| \mathrm{cost}(p, x) - \mathrm{cost}(\mu(P_i), x) | \leq \rho(\frac{\epsilon'}{w_{\min}})^{\frac{1}{z}}, \quad \forall x \in X, p \in P_i. \tag{7.6}$$

Now by plugging in $\epsilon' = w_{\min}(\frac{\epsilon}{\rho})^z$ and assuming $\mathrm{cost}(p, x) \geq 1$, we have

$$\frac{| \mathrm{cost}(p, x) - \mathrm{cost}(\mu(P_i), x) |}{\mathrm{cost}(p, x)} \leq \rho(\frac{\epsilon'}{w_{\min}})^{\frac{1}{z}} = \epsilon. \tag{7.7}$$

As (7.7) satisfies the condition in Lemma 3, the lemma implies that the weighted k-clustering centers form an ϵ-coreset w.r.t. the sum/maximum cost. □

A direct corollary from Theorem 1 is the quality of a k-point coreset formed by k-clustering centers, stated as follows:

Corollary 1 *Given a coreset size $k \in \mathbb{Z}^+$ (positive integers), for any cost function satisfying the conditions in Theorem 1, the optimal k-clustering centers, each weighted by the total weight of its cluster, give an ϵ-coreset for P w.r.t. this cost function, where*

$$\epsilon = \rho \left(\frac{\mathit{opt}(P, k) - \mathit{opt}(P, 2k)}{w_{\min}} \right)^{\frac{1}{z}}. \tag{7.8}$$

Discussion From the analysis of Theorem 1, we can see that if every data point $p \in P$ is within a distance of ϵ/ρ to its nearest center, i.e. $\mathrm{dist}(p, \mu(P_i)) \leq \epsilon/\rho$ for all $i \in [k]$ and $p \in P_i$, then the theorem holds. This also implies that the centers form an ϵ-coreset with

$$\epsilon = \rho \left(\max_{i \in [k]} \max_{p \in P_i} \mathrm{dist}(p, \mu(P_i)) \right). \tag{7.9}$$

Although (7.9) can give a tighter bound on the approximation error than (7.8), (7.8) has the advantage that it can be used to derive a bound that is easily computable without solving the k-clustering problem, which can be beneficial for large datasets; we defer the details on this point to Sect. 4.

In Theorem 1, for any ML problem with non-negative per-point cost, Condition (i) is easily satisfied by adding "+1" to the cost function, while maintaining the same optimal solution. We note that this condition is just an artificial requirement to satisfy the relative-error-based definition of ϵ-coreset in Definition 1. Without this condition, similar analysis will show that the approximation error is bounded as $|\text{cost}(P, x) - \text{cost}(S, x)| \leq \widetilde{\epsilon}$ ($\forall x \in X$), where $\widetilde{\epsilon} = \epsilon \sum_{p \in P} w_p$ for the sum cost, and $\widetilde{\epsilon} = \epsilon$ for the maximum cost. Meanwhile, Condition (ii) limits the ML problems that can be supported. Nevertheless, a number of common ML problems satisfy Condition (ii). Take MEB as an example, whose per-point cost function is $\text{cost}(p, x) = \text{dist}(p, x)$. By the triangle inequality,

$$|\text{dist}(p, x) - \text{dist}(p', x)| \leq \text{dist}(p, p'), \quad \forall p, p', x \in \mathbb{R}^d. \tag{7.10}$$

Therefore, MEB has a 1-Lipschitz-continuous cost function, i.e. $\rho = 1$.

3.1.4 Coreset Construction by Suboptimal k-Clustering

However, solving k-clustering problems to optimality is NP-hard [29, 30], which means that in practice we have to resort to suboptimal clustering algorithms. Nevertheless, we will show that under a few intuitive assumptions, the centers computed by such an algorithm still form a robust coreset with guaranteed performance for a broad set of ML problems.

We now list the assumptions about the clustering algorithm under consideration, where $\{\widetilde{P}_i\}_{i=1}^k$ denotes the partition of P generated by the clustering algorithm, and $\text{approx}(P, k)$ denotes the k-clustering cost achieved by the algorithm for dataset P.

Assumption 4 (Local Optimality) In each cluster $\widetilde{P}_i$ generated by the algorithm, the center selected by the algorithm is the optimal 1-clustering center, i.e. $\mu(\widetilde{P}_i)$.

Assumption 5 (Self-consistency) The $2k$-clustering cost of the algorithm is no worse than the cost from a two-step procedure: k-clustering followed by 2-clustering in each cluster (all by the same algorithm), i.e.,

$$\text{approx}(P, 2k) \leq \sum_{i=1}^k \text{approx}(\widetilde{P}_i, 2). \tag{7.11}$$

Assumption 6 (Greedy Dominance) The 2-clustering cost of the algorithm is no worse than the cost from a two-step procedure: 1-clustering followed by the addition of the point farthest from the center, i.e.,

$$\text{approx}(P, 2) \leq c(P, \{\mu(P), p^*\}), \tag{7.12}$$

where $c(P, Q)$ is defined in (7.3), and p^* is the farthest point defined as $p^* := \arg\max_{p \in P} w_p \text{dist}(p, \mu(P))^z$.

Remark These assumptions are relatively easy to satisfy for any good clustering algorithms, and more discussions will come in Sect. 3.1.5.

With the help of these assumptions, we will show that similar results as in Sect. 3.1.3 hold for coresets formed by suboptimal clustering centers. We refer to [28] for the detailed proofs.

Lemma 4 *For any $\epsilon' > 0$, if $approx(P, k) - approx(P, 2k) \leq \epsilon'$, then $approx(\widetilde{P}_i, 1) - approx(\widetilde{P}_i, 2) \leq \epsilon'$ for any $i \in [k]$.*

Lemma 5 *If $approx(\widetilde{P}_i, 1) - approx(\widetilde{P}_i, 2) \leq \epsilon'$, then $dist(p, \mu(\widetilde{P}_i)) \leq (\frac{\epsilon'}{w_{min}})^{\frac{1}{z}}$, $\forall p \in \widetilde{P}_i$.*

Theorem 2 *Given a suboptimal clustering algorithm satisfying Assumptions 4–6, if $approx(P, k) - approx(P, 2k) \leq w_{min}(\frac{\epsilon}{\rho})^z$, then the k centers computed by the algorithm, each weighted by the total weight of its cluster, form an ϵ-coreset for P w.r.t. both the sum cost and the maximum cost, as long as the per-point cost function satisfies Conditions (i–ii) in Theorem 1.*

Proof The theorem is proved by similar arguments as in the proof of Theorem 1, replacing Lemma 1 by Lemma 4, and Lemma 2 by Lemma 5. □

In practice, the coreset size is often limited by the maximum tolerable communication cost, in which case a corollary similar to Corollary 1 follows.

Corollary 2 *Given a coreset size $k \in \mathbb{Z}^+$, for any cost function satisfying the conditions in Theorem 1 and any k-clustering algorithm satisfying Assumptions 1– 3, the k-clustering centers given by the algorithm, each weighted by the total weight of its cluster, form an ϵ-coreset for P w.r.t. the cost function, where*

$$\epsilon = \rho \left(\frac{approx(P, k) - approx(P, 2k)}{w_{min}} \right)^{\frac{1}{z}}. \tag{7.13}$$

3.1.5 Coreset Construction Algorithm

Based on Theorem 2, we propose our *Robust Coreset Construction (RCC)* (Algorithm 1) in the centralized setting. RCC constructs a coreset by invoking a given clustering algorithm as subroutine in lines 2 and 4, and provides performance guarantee in the sense of ϵ-coreset for any ML problem whose cost function is ρ-Lipschitz-continuous. Here, we assume that the subroutine 'k-clustering(P, k)' returns the set of centers and the corresponding clusters for k-clustering of P. If the coreset size is given, then our algorithm will start from line 4.

Discussion We note that when $z = 2$, i.e., the subroutine is a k-means algorithm, it is easy to satisfy Assumptions 4–6 required by Theorem 2. The standard Lloyd's algorithm runs in an iterative way by assigning each data point to its nearest clustering center and calculating new centers as the mean of each cluster, which already satisfies Assumption 4. To satisfy the other two assumptions, we will initialize the clustering algorithm in the following way. For an odd number of

Algorithm 1: Robust coreset construction (P, ϵ, ρ)

input : A weighted set P with minimum weight $w_{\min}$, approximation error $\epsilon > 0$, Lipschitz constant ρ

output: An ϵ-coreset S for P w.r.t. a cost function satisfying Theorem 2

1 foreach $k = 1, \ldots, |P|$ **do**

2   **if** $approx(P, k) - approx(P, 2k) \leq w_{\min}(\frac{\epsilon}{\rho})^z$ **then**

3   break;

4 $(\{\mu(\widetilde{P}_i)\}_{i=1}^k, \{\widetilde{P}_i\}_{i=1}^k) \leftarrow k\text{-clustering}(P, k)$;

5 $S \leftarrow \{\mu(\widetilde{P}_i)\}_{i=1}^k$, where $\mu(\widetilde{P}_i)$ has weight $\sum_{p \in \widetilde{P}_i} w_p$;

6 return S;

clusters, the initial centers are chosen randomly. For an even number (say $2k$) of clusters,

1. if $k = 1$, we use the mean $\mu(P)$ and the point p^* with the highest clustering cost as the initial centers, which guarantees (7.12) and thus satisfies Assumption 6;
2. if $k > 1$, we compute k-clustering of P and 2-clustering of each of the k clusters, all by the given subroutine. We then use all the computed $2k$ clustering centers as the initial centers, which guarantees (7.11) and thus satisfies Assumption 5.

The same holds for k-median, except that the update step computes the new centers as the geometric median of each cluster (e.g., by Cohen et al. [31]). Generally, Lloyd's algorithm with the above initialization will satisfy Assumptions 4–6 for an arbitrary $z > 0$.

3.2 Distributed Construction of Robust Coreset

We now extend our solution to a distributed setting with n nodes, where each node v_i $(i \in [n])$ has a subset P_i of the entire dataset P. From Sect. 3.1, we already know that k-clustering centers give a robust coreset with general applicability. However, naively applying clustering at each node and returning the local centers as the global coreset can lead to poor performance, because different nodes may select nearly the same local centers and thus report duplicate data, wasting communication bandwidth. Below, we will show a principled approach to address this issue.

3.2.1 Algorithm for Distributed Robust Coreset Construction

CDCC Algorithm We start by examining a state-of-the-art algorithm for distributed clustering. The algorithm, called *Communication-aware Distributed Coreset Construction (CDCC)* [32], works as follows:

1. each node reports its local k-clustering cost $c(P_i, B_i)$ to the server (B_i: set of local centers), and the server allocates a sample size t_i to each node, where $t_i :=$ $t \cdot c(P_i, B_i)/(\sum_{j=1}^{n} c(P_j, B_j))$ and t is the total sample size (a design parameter).
2. each node v_i reports a local coreset D_i, which includes the local centers B_i and t_i randomly sampled data points, generated by i.i.d. sampling from P_i with sampling probabilities proportional to the per-point clustering costs (evaluated w.r.t. centers B_i).
3. the server collects a global coreset $D = \bigcup_{i=1}^{n} D_i$ and computes a set of k-clustering centers based on D.

Adaptation for Coreset Construction CDCC was designed for distributed k-clustering, but we want to construct a robust coreset to support a broader set of ML problems. Thus, we need to adapt CDCC to better serve our purpose.

First, we skip step (3), and directly use $D = \bigcup_{i=1}^{n} D_i$ as the final coreset. This is because the coreset of a coreset cannot be better than the original coreset [27].

Next, we notice that CDCC fixes the number of local centers k in the algorithm design. In our context, there is no predetermined value of k and no reason to require the same number of centers from each node. Therefore, we treat the number of local centers k_j for node v_j as a design parameter and allow it to differ from node to node. In Theorem 3 below, we will show that given a coreset size N, $(k_j)_{j=1}^{n}$ influences the coreset approximation error through $\dfrac{1}{\sqrt{N-\sum_{j=1}^{n} k_j}} \sum_{j=1}^{n} \text{approx}(P_j, k_j)$, which provides an objective function for optimizing $(k_j)_{j=1}^{n}$. Once the numbers of local centers $k_1, \ldots, k_n$ are fixed, we can plug them into steps (1–2) of CDCC to construct the coreset.

The above reasoning leads to a coreset construction algorithm in the distributed setting, named *Distributed Robust Coreset Construction (DRCC)* (Algorithm 2). At a high level, DRCC works in the following way: (1) each node reports its local k-clustering cost for $k = 1, \ldots, K$, where K is a design parameter specifying the possible values for the number of local centers per node (lines 2–3), (2) the server uses the reported costs to configure the number of local centers k_j and the number of random samples t_j for each node v_j (lines 5–7), and (3) each node independently constructs a local coreset using a combination of local samples and local centers (lines 9–11) and reports the coreset to the server (line 12). DRCC extends CDCC because: (1) DRCC can take a weighted input dataset; (2) DRCC optimizes the number of local centers for each node, while allowing different numbers of local centers for different nodes.

3.2.2 Performance Analysis for Distributed Robust Coreset Construction

Communication Overhead For efficient computation, K should be a small integer. Therefore, measured by the total number of scalars reported by the nodes *besides the coreset itself*, the communication overhead of DRCC is $O(Kn)$, which is

Algorithm 2: Distributed robust coreset construction $((P_j)_{j=1}^n, N, K)$

---

input : A distributed dataset $(P_j)_{j=1}^n$, global coreset size N, maximum number of local centers K

output: A coreset $D = \bigcup_{j=1}^n (S_j \cup B_j^{k_j})$ for $P = \bigcup_{j=1}^n P_j$

1 each v_j $(j \in [n])$:
2   compute local approximate k-clustering centers B_j^k on P_j for $k = 1, \ldots, K$;
3   report $(c(P_j, B_j^k))_{k=1}^K$ to the server;

4 the server:
5   find $(k_j)_{j=1}^n$ that minimizes $\frac{1}{\sqrt{N - \sum_{j=1}^n k_j}} \sum_{j=1}^n c(P_j, B_j^{k_j})$ s.t. $k_j \in [K]$ and $\sum_{j=1}^n k_j \leq N$;
6   randomly allocate $t = N - \sum_{j=1}^n k_j$ points i.i.d. among $v_1, \ldots, v_n$, where each point belongs to v_j with probability $\frac{c(P_j, B_j^{k_j})}{\sum_{j=1}^n c(P_j, B_j^{k_j})}$;
7   communicate $(k_j, t_j, \frac{C}{t})$ to each v_j $(j \in [n])$, where t_j is the number of points allocated to v_j and $C = \sum_{l=1}^n c(P_l, B_l^{k_l})$;

8 each v_j $(j \in [n])$:
9   sample a set S_j of t_j points i.i.d. from P_j, where each sample equals $p \in P_j$ with probability $\frac{m_p}{c(P_j, B_j^{k_j})}$ for $m_p = c(\{p\}, B_j^{k_j})$;
10   set the weight of each $q \in S_j$ to $u_q = \frac{C w_q}{t m_q}$;
11   set the weight of each $b \in B_j^{k_j}$ to $u_b = \sum_{p \in P_b} w_p - \sum_{q \in P_b \cap S_j} u_q$, where P_b is the set of points in P_j whose closest center in $B_j^{k_j}$ is b;
12   report each point $q \in S_j \cup B_j^{k_j}$ and its weight u_q to the server;

---

much smaller than the $O((n-1)Nd)$ overhead of the merge-and-reduce approach in [27]. Specifically, the merge-and-reduce approach works by applying a centralized coreset construction algorithm (e.g., Algorithm 1) repeatedly to combine local coresets into a global coreset. Given n local coresets computed by the n nodes, each containing N points in $\mathbb{R}^d$, the centralized algorithm needs to be applied $n-1$ times, each time requiring a local coreset to be transmitted to the node holding another local coreset. This results in a total communication overhead of $(n-1)Nd$.

Quality of Coreset Define $f_x(p) := w_p(\text{cost}(p, x) - \text{cost}(b_p, x) + \rho \text{dist}(p, b_p))$, where for a point $p \in P_j$, b_p is the center in $B_j^{k_j}$ closest to $p \in P_j$. Denote the *dimension of the function space* $F := \{f_x(p) : x \in X\}$ as $\dim(F, P)$ [32]. We have the following bound on the absolute approximation error.

Theorem 3 *For any per-point cost function* $\text{cost}(p, x)$ *that is* ρ-*Lipschitz-continuous in* p *for any* $x \in X$, $\exists t = O(\frac{1}{\epsilon^2}(\dim(F, P) + \log \frac{1}{\delta}))$ *such that with probability at least* $1 - \delta$, *the coreset* D *constructed by DRCC based on local* k-*median clustering with* k_j *local centers from node* v_j $(j \in [n])$ *and* t *random* *samples satisfies that for all $x \in X$,*

$$\left| \sum_{p \in P} w_p cost(p, x) - \sum_{q \in D} u_q cost(q, x) \right| \leq 2\epsilon\rho \sum_{j=1}^{n} c(P_j, B_j^{k_j}). \tag{7.14}$$

Replacing the parameter ϵ by a function of the total coreset size N and the numbers of local centers $(k_j)_{j=1}^{n}$, the error bound on the right-hand side of (7.14) is

$$O\left(\frac{\rho\sqrt{\dim(F, P) + \log\frac{1}{\delta}}}{\sqrt{N - \sum_{j=1}^{n} k_j}} \cdot \sum_{j=1}^{n} c(P_j, B_j^{k_j}) \right). \tag{7.15}$$

This bound captures the tradeoff in configuring the numbers of local centers. While the number of local centers k_j increases, $1/\sqrt{N - \sum_{j=1}^{n} k_j}$ increases and $\sum_{j=1}^{n} c(P_j, B_j^{k_j})$ decreases, which represents a tradeoff. Thus, we select $(k_j)_{j=1}^{n}$ to minimize this error bound in line 5 of Algorithm 2. As a design parameter, increasing K will enlarge the solution space, which may lead to a better coreset, but this will introduce more computation and communication overhead. On the opposite, decreasing K will make DRCC more efficient at the risk of constructing a potentially worse coreset.

Remark We note that the bound in Theorem 3 is on the absolute error, instead of the relative error as defined for ϵ-coreset. Nevertheless, under some regularity conditions on the cost function, (7.14) implies that DRCC constructs an ϵ-coreset with probability at least $1 - \delta$ if the total coreset size satisfies $N = O(\frac{\rho^2}{\epsilon^2}(\dim(F, P) + \log\frac{1}{\delta}) + \sum_{j=1}^{n} k_j)$. We refer to [28] for a more detailed discussion. We also note that such bound is only known for k-median-based DRCC; theoretical guarantees for DRCC based on other k-clustering variants remain open.

3.3 Performance Evaluation for Robust Coreset Construction

We now compare different coreset construction algorithms on multiple datasets and ML problems in both centralized and distributed settings. We will only present representative results below for conciseness, and refer to [28] for the results from more experiments.

3.3.1 Experiment Setup

Datasets (1) MNIST data [34], which consists of 60,000 images of handwritten digits for training plus 10,000 images for testing, each trimmed to 20×20 pixels,

7 Coreset-Based Data Reduction for Machine Learning at the Edge

Table 7.1 Parameters of datasets

Dataset	Size ($|P|$)	Dimension (d)	#Distinct labels (L)
MNIST	70,000	401	10
HAR	10,299	562	6

and (2) Human Activity Recognition (HAR) using Smartphones data [35], which contains 10,299 samples (70% for training and 30% for testing) of smartphone sensor readings during 6 different activities, each sample containing 561 readings.

As coreset construction views each sample as a point in a high-dimensional Euclidean space, we need some preprocessing to align the dimensions and map nonnumerical values to numerical values. In this experiment, we normalize each numerical dimension to $[0, 1]$, and map labels to numbers such that the distance between two points with different labels is no smaller than the distance between two points with the same label. Given a d-dimensional dataset (including labels) with L types of labels, we map type-l label to $(l - 1)\tau$ ($l \in [L]$) for $\tau = \lceil \sqrt{d - 1} \rceil$. Each data point has a unit weight. See Table 7.1 for a summary of the datasets.

To generate distributed datasets, we use three data distribution schemes: (1) *uniform*, where the points are uniformly distributed across n nodes, (2) *specialized*, where each node is associated with one label and contains all the data points with this label, and (3) *hybrid*, where the first n_0 nodes are "specialized" as in (2), and the remaining data are randomly partitioned among the remaining nodes.

Machine Learning Problems We evaluate three unsupervised learning problems, i.e. MEB, k-means, and PCA, and a supervised learning problem by a Neural Network (NN). In our NN experiment, we define a three-layer network with one hidden layer that contains 100 neurons. Table 7.2 gives their cost functions, where for a data point $p \in \mathbb{R}^d$, $p_{1:d-1} \in \mathbb{R}^{d-1}$ denotes the numerical portion and $p_d \in \mathbb{R}$ denotes the label (mapped to a numerical value). The meaning of the model parameter x is problem-specific, as explained in the footnote. We also provide (upper bounds of) the Lipschitz constant ρ except for NN, since it is NP-hard to evaluate ρ for even a two-layer network [33]. Here l is the number of principle components computed by PCA, and Δ is the diameter of the sample space. In our experiments, $\Delta = \sqrt{(d - 1)(L^2 - 2L + 2)}$, which is 181.1 for MNIST and 120.8 for HAR. While NN does not have a meaningful ρ, we still include it to test the robustness of the constructed coresets.

Coreset Construction Algorithms In the centralized setting, we evaluate RCC based on k-median clustering ('RCC-kmedian') and RCC based on k-means clustering ('RCC-kmeans'), together with several benchmarks. These include the algorithm in [13] ('farthest point'), the framework in [12] instantiated for k-means ('nonuniform sampling'), and uniform sampling ('uniform sampling'). We note that the algorithm in [23] ('decomposition' in Fig. 7.1) is essentially RCC-kmeans, hence omitted.

Table 7.2 Machine learning cost functions

Problem	Overall cost function[a]	ρ
MEB	$\max_{p \in P} \text{dist}(x, p)$	1
k-means	$\sum_{p \in P} w_p \cdot \min_{q_i \in x} \text{dist}(q_i, p)^2$	2Δ
PCA	$\sum_{p \in P} w_p \cdot \text{dist}(p, xp)^2$	$2\Delta(l+1)$
Neural Net	$\sum_{p \in P} (-p_d) \cdot \log(o_p)$, where o_p is the output for input $p_{1:d-1}$	NP-hard [33]

[a] The model x denotes the center of enclosing ball for MEB and the set of centers for k-means. For PCA, $x = WW^T$, where W is a $d \times l$ matrix consisting of the first l ($l < d$) principle components as columns

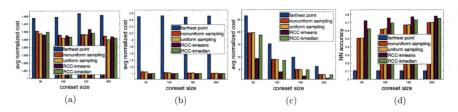

Fig. 7.2 Evaluation on MNIST with varying coreset size (label: 'labels'). (**a**) MEB. (**b**) k-means ($k = 2$). (**c**) PCA (200 components). (**d**) NN

In the distributed setting, we take the best-performing algorithm in the centralized setting ('RCC-kmeans') and evaluate its distributed extensions – including CDCC [32] and DRCC.

Performance Metrics For the unsupervised learning problems (MEB, k-means, and PCA), we evaluate the performance by the normalized cost as explained in Sect. 3.1.1. For the supervised learning problem (NN), we evaluate the performance by the accuracy in predicting the labels of testing data. MNIST and HAR datasets are already divided into training set and testing set.

3.3.2 Experiment Results

Results in Centralized Setting Figures 7.2 and 7.3 show the performances achieved at a variety of coreset sizes, averaged over 100 Monte Carlo runs. Better performance is indicated by lower cost for unsupervised learning or higher accuracy for supervised learning. Note that even the largest coresets generated in these experiments are much smaller (by 97.2–99.7%) than the original (training) dataset, implying dramatic reduction in the communication cost by reporting a coreset instead of the raw data.

We see that the clustering-based coreset construction algorithms ('RCC-kmeans' and 'RCC-kmedian') perform either the best or comparably to the best across all the evaluated datasets and ML problems, demonstrating good robustness. In comparison, the other algorithms are less robust, e.g., the farthest point algorithm

7 Coreset-Based Data Reduction for Machine Learning at the Edge

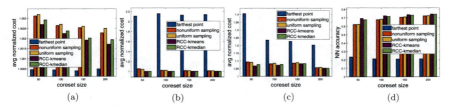

Fig. 7.3 Evaluation on HAR with varying coreset size (label: 'labels'). (**a**) MEB. (**b**) k-means ($k = 2$). (**c**) PCA (7 components). (**d**) NN

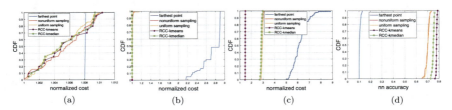

Fig. 7.4 Detailed evaluation on MNIST dataset (label: 'labels', coreset size: 200). (**a**) MEB. (**b**) k-means ($k = 2$). (**c**) PCA (200 components). (**d**) NN

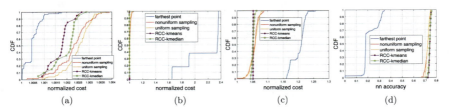

Fig. 7.5 Detailed evaluation on HAR dataset (label: 'labels', coreset size: 200). (**a**) MEB. (**b**) k-means ($k = 2$). (**c**) PCA (7 components). (**d**) NN

in [13] that was designed for MEB performs very poorly for other ML problems, and the sampling-based algorithms ('nonuniform sampling' [12] and 'uniform sampling') that work well for k-means perform relatively poorly for other problems. Moreover, we see that the advantage of the RCC algorithms is more significant at small coreset sizes.

Besides the average performance, we also evaluated the CDFs of the results, shown in Figs. 7.4 and 7.5. The results indicate that in addition to robust average performance, the clustering-based algorithms ('RCC-kmeans' and 'RCC-kmedian') also have significantly less performance variation than the benchmarks, especially the sampling-based algorithms ('nonuniform sampling' and 'uniform sampling'). This means that the quality of the coresets constructed by the proposed algorithms is more reliable.

Between the proposed algorithms, 'RCC-kmeans' often outperforms 'RCC-kmedian', e.g., Fig. 7.2(c and d). Moreover, we note that 'RCC-kmeans' can be

Table 7.3 Average running time (s)

Algorithm	MNIST	HAR
Farthest point	12.35	3.44
Nonuniform sampling	1.61	0.29
Uniform sampling	0.0021	0.0013
RCC-kmeans	42.05	1.75
RCC-kmedian	127.18	11.64

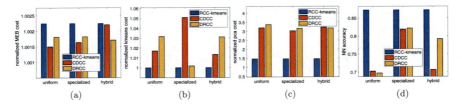

Fig. 7.6 Evaluation on MNIST in distributed setting (label: 'labels', coreset size: 200, $K = 10$). (**a**) MEB. (**b**) k-means ($k = 2$). (**c**) PCA (200 components). (**d**) NN

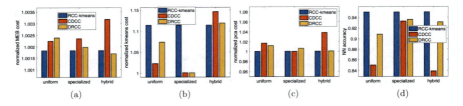

Fig. 7.7 Evaluation on HAR in distributed setting (label: 'activity', coreset size: 200, $K = 10$). (**a**) MEB. (**b**) k-means ($k = 2$). (**c**) PCA (7 components). (**d**) NN

an order of magnitude faster than 'RCC-kmedian', as shown in Table 7.3. Given that our primary goal in designing a robust coreset is to reduce the communication cost while supporting diverse ML problems, this result shows that such robustness may come with certain penalty in running time.

Results in Distributed Setting We use MNIST and HAR to generate distributed datasets at $n = L$ nodes according to the three aforementioned data distribution schemes ('uniform', 'specialized', 'hybrid'), where node j ($j \in [n]$) is associated with label '$j - 1$', and $n_0 = \lfloor n/2 \rfloor$. We parameterize CDCC with $k = 2$ according to the evaluated k-means problem. For DRCC (Algorithm 2), we solve line 5 by a greedy heuristic. As a reference, we also show the performance of RCC based on k-means clustering ('RCC-kmeans'), which is the best-performing algorithm in the centralized setting. We generate 5 distributed datasets using each scheme and repeat each coreset construction algorithm for 5 times on each dataset.

Figures 7.6 and 7.7 show the average results under fixed coreset size N and parameter K (defined in Algorithm 2). Recall that this coreset size means a 97.2–99.7% reduction compared to the size of the full training dataset. Overall, DRCC achieves better performance than CDCC across the experiments, even though both

Table 7.4 Average running time (s)

Algorithm	MNIST	HAR
CDCC	10.09	1.08
DRCC	26.35	2.09

are based on clustering. This is because CDCC blindly sets an equal number of local centers at each node, which leads to poor approximation when the local datasets are highly heterogeneous (e.g., under the 'specialized' or 'hybrid' data distribution scheme), while DRCC is able to customize its configuration to the dataset at hand by automatically tuning the numbers of local centers according to the local clustering costs, and thus achieves a more robust performance. We note that as CDCC was designed to support k-means, its performance for k-means can be slightly better than DRCC (e.g., under some data distributions in Figs. 7.6 and 7.7b).

Meanwhile, the robustness of DRCC comes at certain costs. First, DRCC incurs a slightly higher communication overhead by requiring each node to report K scalars instead of one scalar as in CDCC, in addition to the coreset itself. However, this difference is negligible compared to the communication cost of reporting the coreset (e.g., 200×401 scalars for MNIST and 200×562 scalars for HAR). Moreover, DRCC has a higher running time than CDCC as shown in Table 7.4, due to the need to compute multiple instances of k-clustering ($k = 1, \ldots, K$) on local datasets. This is the real cost for achieving the better robustness, similar to the observation in the centralized setting.

4 Joint Coreset Construction and Quantization

Although coreset-based data reduction as considered in Sect. 3 has exhibited great promise in reducing the communication cost in learning over remote/distributed data, it has a fundamental limitation: coresets only reduce the cardinality of the dataset, leaving the dimensionality and the precision unchanged. While various dimensionality reduction techniques (e.g., feature selection/extraction) can be applied to reduce the data dimensionality, these techniques will change the sample space (to a lower-dimensional space), hence introducing some negative side effects such as harming the interpretability of the reduced features and the models learned on such features. In contrast, reducing the data precision while preserving the sample space, as targeted by *quantization techniques*, can avoid such negative side effects.

In contrast to coreset construction that reduces the number of points in the dataset, quantization reduces the data size in an orthogonal way by reducing the number of bits needed to represent each point. Thus, combining both techniques can potentially further reduce the communication cost when learning over remote/distributed data. Nevertheless, achieving this benefit without significantly degrading the quality of the learned models requires a careful configuration of both techniques, which was not formally tackled until our work in [36]. Intuitively, under a given communication budget specifying the total number of bits to collect (by the edge server), there is a tradeoff between collecting more data points at a lower precision and collecting fewer data points at a higher precision. *Jointly* configuring the quantizer and the coreset construction algorithm to achieve the best tradeoff can potentially achieve a smaller approximation error (and hence a better learned model) than using quantization or coreset construction alone. Our goal in [36] is to realize this potential by developing efficient algorithms to compute the optimal configuration parameters explicitly.

4.1 Background on Coreset and Quantization

As reviewed in Sect. 2.3, coresets have been widely applied in shape fitting and clustering problems. However, most existing coreset construction algorithms are model-specific. In Sect. 3, we have presented a family of *robust coreset construction (RCC)* algorithms that can simultaneously support a variety of ML problems with provable error bounds. Therefore, in this section, we will use RCC as our choice of the coreset construction algorithm.

Quantization techniques [1] aim to quantize the data points to a set of discrete values so that each quantized value can be encoded by a smaller number of bits. Recently, quantization has been leveraged to reduce the size of ML models without seriously degrading the model accuracy [37–39]. Existing quantizers can be classified into *scalar quantizers* and *vector quantizers*, where scaler quantizers apply quantization operations to each attribute of a data point, and vector quantizers [40] apply quantization to each data point as a whole. In this section, we will focus on a simple rounding-based scalar quantizer due to its simplicity and broad applicability. However, we note that our analysis can be easily extended to any given quantizer.

What was missing is a general framework to optimally integrate coreset construction and quantization. To this end, we will present an optimization-based framework to jointly configure a given coreset construction algorithm and a given quantizer to achieve the optimal tradeoff between the ML error and the communication cost.

4.2 Preliminaries

For self-contained presentation, we briefly review several definitions and algorithms that will be used in Sect. 4. Frequently used notations in this section are listed in Table 7.5.

7 Coreset-Based Data Reduction for Machine Learning at the Edge

Table 7.5 Main notations

Variable	Definition
CS	The operation of coreset construction
QT	The operation of quantization
ϵ, ϵ_i	Overall/local ML error
B, B_i	Global/local communication budget
$\mathcal{Y}, \mathcal{Y}_i$	Total/local original dataset
S	Coreset
n, k	Cardinalities of $\mathcal{Y}$ and S
y_i, y_{ij}	One data point and one attribute of the data point
b_0, b	#bits for representing each attribute in $\mathcal{Y}$ and S
Δ	Maximum quantization error
$\mathbf{x}, \mathcal{X}$	One solution and solution space for the ML task
$\text{cost}(\mathcal{Y}, \mathbf{x})$	Cost function of the ML task
ρ	Lipschitz constant for the ML cost function
$\text{opt}(k)$	Optimal k-means clustering cost for $\mathcal{Y}$
$\text{opt}_\infty(k)$	Optimal k-center clustering cost for $\mathcal{Y}$
m_e	#exponent bits in the floating point representation of an attribute
N	Number of nodes in distributed setting

4.2.1 Data Representation

Let $\mathcal{Y}$ denote the original dataset with cardinality $n := |\mathcal{Y}|$, dimension d, and precision b_0. Each data point $\mathbf{y}_i \in \mathcal{Y}$ is a column vector in d-dimensional space, and each attribute y_{ij} is represented as a floating point number with a sign bit, an m_e-bit exponent, and a $(b_0 - 1 - m_e)$-bit significand. Let $\mathbf{Y} := [\mathbf{y}_1, \ldots, \mathbf{y}_i, \ldots, \mathbf{y}_n]$ denote the matrix with column vectors $\mathbf{y}_i$. For simplicity of analysis, we assume that y_{ij}'s have been normalized to $[-1, 1]$ with zero mean (i.e., $\frac{1}{n}\sum_i y_{ij} = 0$ for each j). Let $\mu(\mathcal{Y}) := \frac{1}{n}\sum_{\mathbf{y}_i \in \mathcal{Y}} \mathbf{y}_i$ denote the sample mean of $\mathcal{Y}$.

4.2.2 Coreset Construction

A generic ML task can be considered as a cost minimization problem. Using $\mathcal{X}$ to denote the set of possible models, and $\text{cost}(\mathcal{Y}, \mathbf{x})$ to denote the mismatch between the dataset $\mathcal{Y}$ and a candidate model $\mathbf{x}$, the problem seeks to find the model that minimizes $\text{cost}(\mathcal{Y}, \mathbf{x})$. The cost function $\text{cost}(\mathcal{Y}, \mathbf{x})$ is usually in the form of a summation $\text{cost}(\mathcal{Y}, \mathbf{x}) = \sum_{\mathbf{y} \in \mathcal{Y}} \text{cost}(\mathbf{y}, \mathbf{x})$ or a maximization $\text{cost}(\mathcal{Y}, \mathbf{x}) = \max_{\mathbf{y} \in \mathcal{Y}} \text{cost}(\mathbf{y}, \mathbf{x})$, where $\text{cost}(\mathbf{y}, \mathbf{x})$ is the per-point cost that is model-specific. For example, minimum enclosing ball (MEB) [41] minimizes a maximum cost and k-means minimizes a sum cost.

A coreset S is a weighted (and often smaller) dataset that approximates $\mathcal{Y}$ in terms of costs.

Definition 2 (ϵ_{CS}-Coreset [12]) A set $S \subseteq \mathbb{R}^d$ with weights u_q ($\forall q \in S$) is an ϵ_{CS}-coreset for $\mathcal{Y}$ w.r.t. $\text{cost}(\mathcal{Y}, x)$ ($x \in \mathcal{X}$) if $\forall x \in \mathcal{X}$,

$$(1 - \epsilon_{CS})\text{cost}(\mathcal{Y}, x) \leq \text{cost}(S, x) \leq (1 + \epsilon_{CS})\text{cost}(\mathcal{Y}, x), \tag{7.16}$$

where $\text{cost}(S, \mathbf{x})$ is defined as $\text{cost}(S, \mathbf{x}) = \sum_{q \in S} u_q \text{cost}(\mathbf{q}, \mathbf{x})$ if $\text{cost}(\mathcal{Y}, \mathbf{x})$ is a sum cost, and $\text{cost}(S, \mathbf{x}) = \max_{q \in S} \text{cost}(\mathbf{q}, \mathbf{x})$ if $\text{cost}(\mathcal{Y}, \mathbf{x})$ is a maximum cost.

Definition 2 also provides a performance measure for coresets: $\epsilon_{CS,S} := \sup_{x \in \mathcal{X}} |\text{cost}(\mathcal{Y}, \mathbf{x}) - \text{cost}(S, \mathbf{x})|/\text{cost}(\mathcal{Y}, \mathbf{x})$ measures the maximum relative error in approximating the ML cost function by coreset S, called the *ML error* of S. The smaller $\epsilon_{CS,S}$, the better S is in supporting the ML task.

We have proved RCC achieves guaranteed performances for a broad class of ML tasks with Lipschitz-continuous cost functions. In the sequel, we denote the optimal k-means clustering cost for $\mathcal{Y}$ by $\text{opt}(k)$. It is known that the optimal 1-means center of $\mathcal{Y}$ is the sample mean $\mu(\mathcal{Y})$.

4.2.3 Quantization

Quantization reduces the number of bits required to encode each data point by transforming it to the nearest point in a set of discrete points, the selection of which largely defines the quantizer. Our solution will utilize the maximum quantization error, defined as $\Delta := \max_{y \in \mathcal{Y}} \text{dist}(\mathbf{y}, \mathbf{y}')$, where $\mathbf{y}'$ denotes the quantized version of data point $\mathbf{y}$ and $\text{dist}(\mathbf{y}, \mathbf{y}')$ is their Euclidean distance. Given a quantizer, Δ depends on the number of bits used to represent each quantized value. Below we analyze Δ for a simple but practical *rounding-based quantizer* as a concrete example, but our framework also allows other quantizers.

Let y_{ij} denote the j-th attribute of the i-th data point. The b_0-bit binary floating point representation of y_{ij} is given by $(-1)^{\text{sign}(y_{ij})} \times 2^{e_{ij}} \times (a_{ij}(0) + a_{ij}(1) \times 2^{-1} + \ldots + a_{ij}(b_0 - 1 - m_e) \times 2^{-(b_0-1-m_e)})$ [42]. Here, $\text{sign}(y_{ij})$ is the sign of y_{ij} (0: nonnegative, 1: negative), e_{ij} is an m_e-bit exponent, and $a_{ij}(\cdot) \in \{0, 1\}$ are the significant digits, where $a_{ij}(0) \equiv 1$ and does not need to be stored explicitly.

Consider a scalar quantizer that rounds each y_{ij} to s significant digits. The quantized value equals $y'_{ij} = (-1)^{\text{sign}(y_{ij})} \times 2^{e_{ij}} \times (a_{ij}(0) + a_{ij}(1) \times 2^{-1} + \ldots + a_{ij}(s) \times 2^{-s} + a'_{ij}(s) \times 2^{-s})$, where $a'_{ij}(s) \in \{0, 1\}$ is the result of rounding the remaining digits (0: round down, 1: round up). As $|y_{ij} - y'_{ij}| \leq 2^{e_{ij}-s}$ and $|y_{ij}| \geq 2^{e_{ij}}$, we have $|y_{ij} - y'_{ij}|/|y_{ij}| \leq 2^{-s}$. Hence, for $\mathcal{Y}$ in $\mathbb{R}^d$ where each attribute y_{ij} is normalized to $[-1, 1]$, the maximum quantization error of this quantizer is bounded by

$$\Delta \leq 2^{-s} \cdot \max_{y_i \in \mathcal{Y}} \|\mathbf{y}_i\|. \tag{7.17}$$

7 Coreset-Based Data Reduction for Machine Learning at the Edge

4.3 Budgeted Optimization of Coreset Construction and Quantization

We first analyze the ML error bounds based on the data summary computed by a combination of coreset construction and quantization, and then formulate an optimization problem to minimize the ML error under a given budget of communication cost.

4.3.1 Workflow Design

The first question in the integration of quantization (QT) into coreset construction (CS) is to determine the order of these two operations. Intuitively, QT is needed after CS since the CS algorithm can result in arbitrary values that cannot be represented using b bits as specified for the quantizer. Therefore, we consider a pipeline where CS is followed by QT.

4.3.2 Error Bound Analysis

The error bound for CS + QT is stated as follows.

Theorem 4 *After applying a Δ-maximum-error quantizer to an ϵ_{CS}-coreset S of the original dataset $\mathcal{Y}$, the quantized coreset S' is an $(\epsilon_{CS} + \rho\Delta + \epsilon_{CS} \cdot \rho\Delta)$-coreset for $\mathcal{Y}$ w.r.t. any cost function satisfying:*

1. *$cost(\mathbf{y}, \mathbf{x}) \geq 1$*
2. *$cost(\mathbf{y}, \mathbf{x})$ is ρ-Lipschitz-continuous in $\mathbf{y} \in \mathcal{Y}, \forall \mathbf{x} \in \mathcal{X}$.*

Theorem 4 is directly implied by the following Lemma 6, which gives the ML error after one single quantization.

Lemma 6 *Given a set of points $\mathcal{Y} \subseteq \mathbb{R}^d$, let $\mathcal{Y}'$ be the corresponding set of quantized points with a maximum quantization error of Δ. Then, $\mathcal{Y}'$ is an $\rho\Delta$-coreset of $\mathcal{Y}$ w.r.t. any cost function satisfying the conditions in Theorem 4.*

Proof of Lemma 6 For each $\mathbf{y} \in \mathcal{Y}$, we know $\text{dist}(\mathbf{y}, \mathbf{y}') \leq \Delta$. By the ρ-Lipschitz-continuity of $cost(\cdot, \mathbf{x})$, we have

$$|cost(\mathbf{y}, \mathbf{x}) - cost(\mathbf{y}', \mathbf{x})| \leq \rho\Delta. \tag{7.18}$$

Moreover, since $cost(\mathbf{y}, \mathbf{x}) \geq 1$, we have

$$\frac{|cost(\mathbf{y}, \mathbf{x}) - cost(\mathbf{y}', \mathbf{x})|}{cost(\mathbf{y}, \mathbf{x})} \leq \rho\Delta, \tag{7.19}$$

and thus

$$(1 - \rho\Delta)\text{cost}(\mathbf{y}, \mathbf{x}) \leq \text{cost}(\mathbf{y}', \mathbf{x}) \leq (1 + \rho\Delta)\text{cost}(\mathbf{y}, \mathbf{x}). \tag{7.20}$$

If $\text{cost}(\mathcal{Y}, \mathbf{x}) = \sum_{\mathbf{y} \in \mathcal{Y}} \text{cost}(\mathbf{y}, \mathbf{x})$, then treating $\mathcal{Y}'$ as a coreset with unit weights, its cost is $\text{cost}(\mathcal{Y}', \mathbf{x}) = \sum_{\mathbf{y}' \in \mathcal{Y}'} \text{cost}(\mathbf{y}', \mathbf{x})$. Summing (7.20) over all $\mathbf{y} \in \mathcal{Y}$ (or $\mathbf{y}' \in \mathcal{Y}'$), we have

$$(1 - \rho\Delta)\text{cost}(\mathcal{Y}, \mathbf{x}) \leq \text{cost}(\mathcal{Y}', \mathbf{x}) \leq (1 + \rho\Delta)\text{cost}(\mathcal{Y}, \mathbf{x}). \tag{7.21}$$

Similarly for max cost function, if $\text{cost}(\mathcal{Y}, \mathbf{x}) = \max_{\mathbf{y} \in \mathcal{Y}} \text{cost}(\mathbf{y}, \mathbf{x})$, then the cost of $\mathcal{Y}'$ is $\text{cost}(\mathcal{Y}', \mathbf{x}) = \max_{\mathbf{y}' \in \mathcal{Y}'} \text{cost}(\mathbf{y}', \mathbf{x})$. Suppose that the maximum is achieved at $\mathbf{y}_1$ for $\text{cost}(\mathcal{Y}, \mathbf{x})$, and $\mathbf{y}_2'$ for $\text{cost}(\mathcal{Y}', \mathbf{x})$. Based on (7.20), we have

$$(1 - \rho\Delta)\text{cost}(\mathbf{y}_1, \mathbf{x}) \leq \text{cost}(\mathbf{y}_1', \mathbf{x}) \leq \text{cost}(\mathbf{y}_2', \mathbf{x}) \tag{7.22a}$$

$$\leq (1 + \rho\Delta)\text{cost}(\mathbf{y}_2, \mathbf{x}) \leq (1 + \rho\Delta)\text{cost}(\mathbf{y}_1, \mathbf{x}) \tag{7.22b}$$

which again leads to (7.21) as $\text{cost}(\mathcal{Y}, \mathbf{x}) = \text{cost}(\mathbf{y}_1, \mathbf{x})$ and $\text{cost}(\mathcal{Y}', \mathbf{x}) = \text{cost}(\mathbf{y}_2', \mathbf{x})$. $\qquad\square$

Proof of Theorem 4 By Lemma 6 and Definition 2, we have that $\forall \mathbf{x} \in \mathcal{X}$,

$$(1 - \epsilon_{\text{CS}})(1 - \rho\Delta)\text{cost}(\mathcal{Y}, \mathbf{x}) \leq (1 - \rho\Delta)\text{cost}(\mathcal{S}, \mathbf{x})$$
$$\leq \text{cost}(\mathcal{S}', \mathbf{x}) \leq (1 + \rho\Delta)\text{cost}(\mathcal{S}, \mathbf{x})$$
$$\leq (1 + \epsilon_{\text{CS}})(1 + \rho\Delta)\text{cost}(\mathcal{Y}, \mathbf{x}), \tag{7.23}$$

which yields the result. $\qquad\square$

4.3.3 Configuration Optimization

(1) Abstract Formulation Our objective is to minimize the ML error under bounded communication costs, through the joint configuration of coreset construction and quantization. Given a n-point dataset in $\mathbb{R}^d$ and a communication budget of B, we aim to find a quantized coreset $\mathcal{S}$ with k points and a precision of b bits per attribute, that can be represented by no more than B bits. Our goal is to *Minimize the Error under a given Communication Budget (MECB)*, formulated as

$$\min_{b,k} \quad \epsilon_{\text{CS}}(k) + \rho\Delta(b) + \epsilon_{\text{CS}}(k) \cdot \rho\Delta(b) \tag{7.24a}$$

$$\text{s.t.} \quad b \cdot k \cdot d \leq B, \tag{7.24b}$$

$$b, k \in \mathbb{Z}^+, \tag{7.24c}$$

where $\epsilon_{\text{CS}}(k)$ represents the ML error of a k-point coreset constructed by the given coreset construction algorithm, and $\Delta(b)$ is the maximum quantization error of b-bit

7 Coreset-Based Data Reduction for Machine Learning at the Edge

quantization by the given quantizer. We want to find the optimal values of k and b to minimize the error bound (7.24a) according to Theorem 4, under the given budget B. Note that our focus is on finding the optimal configuration of known CS/QT algorithms instead of developing new algorithms.

(2) Concrete Formulation We now concretely formulate and solve an instance of MECB for two practical CS/QT algorithms. Suppose that the CS operation is by the k-means based *robust coreset construction (RCC)* algorithm in [43], which is proven to yield a $\rho\sqrt{f(k)}$-coreset for all ML tasks with ρ-Lipschitz-continuous cost functions, where $f(k) := \text{opt}(k) - \text{opt}(2k)$ is the difference between the k-means and the $2k$-means costs. Moreover, suppose that the QT operation is by the rounding-based quantizer defined in Sect. 4.2.3, which has a maximum quantization error of $\Delta(b) := 2^{-(b-1-m_e)} \max_{\mathbf{y}_i \in \mathcal{Y}} \|\mathbf{y}_i\|$ to generate a b-bit quantization with $s = b - 1 - m_e$ significant digits according to (7.17). Then, by Theorem 4, the MECB problem in this case becomes:

$$\min_{b,k} \quad \rho\sqrt{f(k)} + \rho\Delta(b) + \rho^2\Delta(b)\sqrt{f(k)} \tag{7.25a}$$

$$\text{s.t.} \quad b \cdot k \cdot d \leq B, \tag{7.25b}$$

$$b, k \in \mathbb{Z}^+. \tag{7.25c}$$

(3) Straightforward Solution In (7.25), only b (or k) is the free decision variable. Thus, a straightforward way to solve (7.25) is to evaluate the objective function (7.25a) for each possible value of b (or k) and then select b^* (or k^*) that minimizes the objective value. We refer to this solution as the *EMpirical approach (EM)* later in the paper. However, this approach is computationally expensive for large datasets, as evaluating $f(k)$ requires solving k-means problems for large values of k. In fact, computing the k-means and the $2k$-means costs $\text{opt}(k)$ and $\text{opt}(2k)$ for a candidate value of k is already NP-hard[1] [29]. To address this challenge, we will develop efficient heuristic algorithms for approximately solving (7.25) in the following section by identifying proxies of the objective function that can be evaluated efficiently.

4.4 Efficient Algorithms for MECB

Below, we propose two algorithms to efficiently solve the concrete MECB problem formulated in (7.25).

---

[1] One can compute an approximation using existing k-means heuristics, e.g., [44], which is what we have done in evaluating EM. Nevertheless, this algorithm is still inefficient as shown in Sect. 4.6.

250   H. Lu et al.

4.4.1 Eigenvalue Decomposition Based Algorithm for MECB (EVD-MECB)

(1) Re-formulating the Optimization Problem This algorithm is motivated by the following bound derived in [45].

Theorem 5 (Bound for k-Means Costs [45]) *The optimal k-means cost for $\mathcal{Y}$ is bounded by*

$$opt(k) \geq n\overline{\mathbf{y}^2} - \sum_{i=1}^{k-1} \lambda_i, \tag{7.26}$$

where $n\overline{\mathbf{y}^2} := \sum_{i=1}^{n} \mathbf{y}_i^T \mathbf{y}_i$ is the total variance and λ_i is the i-th principal eigenvalue of the covariance matrix $\mathbf{YY}^T$.

Algorithm 3: EVD-MECB

---

    input : A dataset $\mathcal{Y}$, Lipschitz constant ρ for the targeted ML task, communication budget B.

    output: Optimal (k^*, b^*) to configure a quantized ϵ-coreset $\mathcal{S}'$ for $\mathcal{Y}$ within budget B.

1 Calculate eigenvalues $\{\lambda_i\}_{i=1}^{d}$ for $\mathbf{YY}^T$;

2 $\Lambda_j \leftarrow \sum_{i=1}^{j} \lambda_i, \forall 1 \leq j \leq d$;

3 **foreach** $b = [1 + m_e, 2 + m_e, \ldots, b_0]$ **do**

4 $k \leftarrow \lfloor B/d/b \rfloor$;

5 $f(k) \leftarrow \Lambda_{2k-1} - \Lambda_{k-1}$;

6 $\Delta(b) \leftarrow 2^{-(b-1-m_e)} \max_{\mathbf{y}_i \in \mathcal{Y}} \|\mathbf{y}_i\|$;

7 $\epsilon(k, b) \leftarrow \rho \cdot \sqrt{f(k)} + \rho \cdot \Delta(b) + \rho^2 \cdot \Delta(b) \cdot \sqrt{f(k)}$;

8 $(k^*, b^*) \leftarrow \arg\min \epsilon(k, b)$;

9 **return** (k^*, b^*);

We use the bound in (7.26) as an approximation of k-means cost that is much faster to evaluate than the exact k-means cost. Replacing $opt(k)$ by this bound, we obtain an approximation of (7.25), where $f(k)$ is approximated by

$$f(k) \approx n\overline{\mathbf{y}^2} - \sum_{i=1}^{k-1} \lambda_i - (n\overline{\mathbf{y}^2} - \sum_{i=1}^{2k-1} \lambda_i) = \sum_{i=k}^{2k-1} \lambda_i. \tag{7.27}$$

(2) EVD-MECB Algorithm The righthand side of (7.27) is easier to calculate than the exact value of $f(k)$, as we can compute the eigenvalue decomposition once [46], and use the results to evaluate $\sum_{i=k}^{2k-1} \lambda_i$ for all possible values of k. As each number in $\mathcal{Y}$ has $b_0 - 1 - m_e$ significant digits, the number of feasible values for b (and hence k) is $b_0 - 1 - m_e$. By enumerating all the feasible values, we can easily

7 Coreset-Based Data Reduction for Machine Learning at the Edge

find the optimal solution (k^*, b^*) to this approximation of (7.25). We summarize the algorithm in Algorithm 3.

Algorithm 4: k-center cost computation

input : A dataset $\mathcal{Y}$, the maximum number of centers K.
output: The costs $(g(k))_{k=1}^K$ for greedy k-center clustering for $k = 1, \ldots, K$.

1 $\mathcal{G} \leftarrow \emptyset$;
2 **foreach** $y \in \mathcal{Y}$ **do**
3     $d(y) \leftarrow \infty$;
4 **while** $|\mathcal{G}| < K$ **do**
5   find $\mathbf{y} \leftarrow \arg\max_{\mathbf{q} \in \mathcal{Y}} d(\mathbf{q})$;
6     $\mathcal{G} \leftarrow \mathcal{G} \cup \{\mathbf{y}\}$;
7   **foreach** $j = 1, \ldots, |\mathcal{Y}|$ **do**
8        $d(\mathbf{y}_j) \leftarrow \min(d(\mathbf{y}_j), \text{dist}(\mathbf{y}_j, \mathbf{y}))$;
9     $g(|\mathcal{G}|) \leftarrow \max_{\mathbf{y}_j \in \mathcal{Y}} d(\mathbf{y}_j)$;
10 **return** $(g(k))_{k=1}^K$;

4.4.2 Max-Distance Based Algorithm for MECB (MD-MECB)

(1) Re-formulating the Optimization Problem Alternatively, we can bound the ML error based on the maximum distance between each data point and its corresponding point in the coreset. Let $f_2(k) := \max_{i=1,\ldots,k} \max_{\mathbf{y} \in \mathcal{Y}_i} \text{dist}(\mathbf{y}, \mu(\mathcal{Y}_i))$ be the maximum distance between any data point and its nearest k-means center, where $\{\mu(\mathcal{Y}_i)\}_{i=1}^k$ are the k-means clusters. Then, a similar proof as that of Lemma 6 implies the following.

Lemma 7 *The centers of the optimal k-means clustering of $\mathcal{Y}$, each weighted by the number of points in its cluster, provide a $\rho f_2(k)$-coreset for $\mathcal{Y}$ w.r.t. any cost function satisfying the conditions in Theorem 4.*

This lemma provides an alternative error bound for the RCC algorithm in [43], which constructs the coreset as in Lemma 7. Using $\epsilon_{CS} = \rho f_2(k)$, if we apply the rounding-based quantization after RCC, we can apply Theorem 4 to obtain an alternative error bound for the resulting quantized coreset, which is $\rho f_2(k) + \rho \Delta(b) + \rho^2 f_2(k) \Delta(b)$. We note that minimizing $f_2(k)$ is exactly the objective of *k-center clustering* [47–49]. Hence, we would like to use the optimal k-center cost, denoted by $\text{opt}_\infty(k)$, as a heuristic to calculate $f_2(k)$. By using this alternative error bound and approximating $f_2(k) \approx \text{opt}_\infty(k)$, we can reformulate the MECB problem as follows:

$$\min_{b,k} \quad \rho \cdot \text{opt}_\infty(k) + \rho \Delta(b) + \rho^2 \cdot \text{opt}_\infty(k) \Delta(b) \tag{7.28a}$$

$$\text{s.t.} \quad b \cdot k \cdot d \le B, \tag{7.28b}$$

$$b, k \in \mathbb{Z}^+, \tag{7.28c}$$

where $\Delta(b)$ is defined as in (7.25).

(2) MD-MECB Algorithm The re-formulation (7.28) allows us to leverage algorithms for k-center clustering to efficiently evaluate $\text{opt}_\infty(k)$. Although k-center clustering is a NP-hard problem [50], a number of efficient heuristics have been proposed. In particular, it has been proved [50] that the best possible approximation for k-center clustering is 2-approximation, achieved by a simple greedy algorithm [51] that keeps adding the point farthest from the existing centers to the set of centers until k centers are selected. The beauty of this algorithm is that we can modify it to compute the k-center clustering costs for all possible values of k in one pass, as shown in Algorithm 4. Specifically, after adding each center (lines 5–6) and updating the distance from each point to the nearest center (line 8), we record the clustering cost (line 9). As the greedy algorithm achieves 2-approximation [50], the returned costs satisfy $\text{opt}_\infty(k) \le g(k) \le 2\text{opt}_\infty(k)$, where $g(\cdot)$ is defined in line 9. Based on this algorithm, the MD-MECB algorithm, shown in Algorithm 5, solves an approximation of (7.28) with $\text{opt}_\infty(k)$ approximated by $g(k)$.

Algorithm 5: MD-MECB

input : A dataset $\mathcal{Y}$, Lipschitz constant ρ for the targeted ML task; communication budget B.

output: Optimal (k^*, b^*) to configure a quantized ϵ-coreset S' for Y within budget B.

1 Run Algorithm 4 with input $\mathcal{Y}$ and $K = \min\{\lfloor \frac{B}{d \cdot (1+m_e)} \rfloor, n\}$;

2 **foreach** $b = [1 + m_e, 2 + m_e, \ldots, b_0]$ **do**

3 $\quad k \leftarrow \lfloor B/(d \cdot b) \rfloor$;

4 $\quad \text{opt}_\infty(k) \leftarrow g(k)$ by the output of Algorithm 4;

5 $\quad \Delta(b) \leftarrow 2^{-(b-1-m_e)} \max_{\mathbf{y}_i \in \mathcal{Y}} \|\mathbf{y}_i\|$;

6 $\quad \epsilon(k, b) \leftarrow \rho \cdot \text{opt}_\infty(k) + \rho\Delta(b) + \rho^2 \cdot \text{opt}_\infty(k)\Delta(b)$;

7 $(k^*, b^*) \leftarrow \arg\min \epsilon(k, b)$;

8 **return** (k^*, b^*);

4.4.3 Discussions

(1) Performance Comparison The straightforward solution EM (Sect. 4.3.3) directly minimizes the error bound (7.25a) and is thus expected to find the best configuration for CS + QT. In comparison, the two proposed algorithms (EVD-MECB and MD-MECB) only optimize approximations of the error bound. It is difficult to theoretically analyze or compare the ML errors of these algorithms since the bound may be loose and the approximations may be smaller than the bound.

7 Coreset-Based Data Reduction for Machine Learning at the Edge

Instead, we will use empirical evaluations to compare the actual ML errors achieved by these algorithms (see Sect. 4.6).

(2) Complexity Comparison In terms of complexity, EM involves executions of the k-means algorithm for all (k, b) pairs, which is thus computationally complicated. In comparison, EVD-MECB only requires one eigenvalue decomposition (EVD) and one matrix multiplication, while MD-MECB only needs to invoke Algorithm 4 once. Therefore, both of them can be implemented efficiently. As EVD can be computed with complexity $O(n^3)$ [52], EVD-MECB has a complexity of $O(n^3 + d + b_0)$. Since the computational complexity of Algorithm 4 is $O(n^2)$ (achieved at $K = n$), MD-MECB has a complexity of $O(n^2 + b_0)$. Hence, MD-MECB is expected to be more efficient than EVD-MECB, which will be further validated in Sect. 4.6.

4.5 Budget Allocation in Distributed Setting

We now describe how to construct a quantized coreset under a global communication budget in the distributed setting. Suppose that the data of interest are distributed over N nodes as $\{\mathcal{Y}_1, \ldots, \mathcal{Y}_N\}$. Our goal is to configure the construction of local quantized coresets $\{\mathcal{S}_1, \ldots, \mathcal{S}_N\}$ such that $\bigcup_{i=1}^N \mathcal{S}_i$ can be represented by no more than B bits and is an ϵ-coreset for $\bigcup_{i=1}^N \mathcal{Y}_i$ for the smallest ϵ. Given an allocation of the global budget B to each node, we can use the algorithms in Sect. 4.4 to make each $\mathcal{S}_i$ ($i \in [N]$) an ϵ_i-coreset of the local dataset $\mathcal{Y}_i$ for the smallest ϵ_i subject to the allocated local budget. However, the following questions remain: (1) How is ϵ related to $\epsilon_1, \ldots, \epsilon_N$? (2) How can we allocate the global budget B to minimize ϵ? Below, we will answer these questions by formulating and solving the distributed version of the MECB problem.

4.5.1 Problem Formulation in Distributed Setting

In the following, we first show that $\epsilon = \max_{i \in [N]} \epsilon_i$, and then formulate the MECB problem in the distributed setting.

Lemma 8 *If C_1 and C_2 are ϵ_1-coreset and ϵ_2-coreset for datasets $\mathcal{Y}_1$ and $\mathcal{Y}_2$, respectively, w.r.t. a cost function, then $C_1 \bigcup C_2$ is a $\max\{\epsilon_1, \epsilon_2\}$-coreset for $\mathcal{Y}_1 \bigcup \mathcal{Y}_2$ w.r.t. the same cost function.*

Proof We consider both sum and maximum cost functions. Without loss of generality, we assume $\epsilon_2 \geq \epsilon_1$.

Sum cost: Given a feasible solution $\mathbf{x}$, we consider sum cost as $\text{cost}(\mathcal{Y}, \mathbf{x}) = \sum_{\mathbf{y} \in \mathcal{Y}} \text{cost}(\mathbf{y}, \mathbf{x})$. According to Definition 2, we have

$$(1 - \epsilon_1) \sum_{\mathbf{y} \in \mathcal{Y}_1} \text{cost}(\mathbf{y}, \mathbf{x}) \leq \sum_{\mathbf{c} \in C_1} \text{cost}(\mathbf{c}, \mathbf{x}) \leq (1 + \epsilon_1) \sum_{\mathbf{y} \in \mathcal{Y}_1} \text{cost}(\mathbf{y}, \mathbf{x}) \qquad (7.29)$$

$$(1 - \epsilon_2) \sum_{\mathbf{y} \in \mathcal{Y}_2} \text{cost}(\mathbf{y}, \mathbf{x}) \leq \sum_{\mathbf{c} \in C_2} \text{cost}(\mathbf{c}, \mathbf{x}) \leq (1 + \epsilon_2) \sum_{\mathbf{y} \in \mathcal{Y}_2} \text{cost}(\mathbf{y}, \mathbf{x}). \qquad (7.30)$$

Summing up these two equations and noting that $\epsilon_2 \geq \epsilon_1$, we can conclude that $C_1 \bigcup C_2$ is an ϵ_2-coreset for $\mathcal{Y}_1 \bigcup \mathcal{Y}_2$.

Maximum cost: The proof for maximum cost function is similar as above but taking the maximum instead. $\qquad\square$

We can easily extend Lemma VI.1 to multiple nodes to compute the global ϵ error as: $\epsilon = \max_i \epsilon_i$. Thus the objective of minimizing ϵ is equivalent to minimizing the largest ϵ_i for $i \in \{1, \dots, N\}$.

Let B_i denote the local budget for the i-th node. Intuitively, the larger the local budget B_i, the smaller ϵ_i. Therefore, we model ϵ_i as a non-increasing function w.r.t. the local budget B_i, denoted by $\epsilon_i(B_i)$.

Then, we can formulate *the MECB problem in the distributed setting (MECBD)* as follows:

$$\min \quad \max_{i \in \{1, \dots, N\}} \epsilon_i(B_i) \qquad (7.31a)$$

$$\text{s.t.} \quad \sum_{i=1}^{N} B_i \leq B. \qquad (7.31b)$$

Note that to compute $\epsilon_i(B_i)$ for a given B_i, we need to solve an instance of the MECB problem in (7.25) for dataset $\mathcal{Y}_i$ and budget B_i.

4.5.2 Optimal Budget Allocation Algorithm for MECBD (OBA-MECBD)

The MECBD problem in (7.31) is a *minimax knapsack problem* [53, 54] with a nonlinear non-increasing objective function. Special cases of this problem with strictly decreasing objective functions have been solved in [54]. However, the objective function of MECBD is a step function as shown below, which is not strictly decreasing. Below we will develop a polynomial-time algorithm to solve our instance of the minimax knapsack problem using the following property of $\epsilon_i(B_i)$.

We note that $\epsilon_i(B_i)$ is a non-increasing step function of B_i (see Fig. 7.8). This is because the configuration parameters k and b in the CS + QT procedure are integers. Therefore, there exist intervals $[B_{i,j}, B_{i,j+1})$ $(j = 0, 1, 2, \dots)$ such that for any $B_i, B_i' \in [B_{i,j}, B_{i,j+1})$, we have $\epsilon_i(B_i) = \epsilon_i(B_i')$, as shown in Fig. 7.8. Given a target value of ϵ_i, the minimum B_i for reaching this target is thus always within the set $\{B_{i,j}\}$.

7 Coreset-Based Data Reduction for Machine Learning at the Edge

Algorithm 6: OBA-MECBD

input : Distributed datasets $\{\mathcal{Y}_i\}_{i=1}^{N}$, Lipschitz constant ρ, communication budget B.
output: Optimal $\{(k_i^*, b_i^*)\}_{i=1}^{N}$ to configure the construction of local quantized coresets within a global budget B.

each node $i = 1, \ldots, N$:
1. $B_0 \leftarrow 1 \cdot (1 + m_e) \cdot d$;
2. compute $\epsilon_i(B_0)$ by MD-MECB or EVD-MECB;
3. $\mathcal{E}_i \leftarrow \{(B_0, \epsilon_i(B_0))\}$;
4. **foreach** integer $B_i \in [B_0 + 1, |\mathcal{Y}_i| \cdot b_0 \cdot d]$ **do**
5.   compute $\epsilon_i(B_i)$ by MD-MECB or EVD-MECB;
6.   **if** $\epsilon_i(B_i) < \min\{\epsilon_{i,j} : (B_{i,j}, \epsilon_{i,j}) \in \mathcal{E}_i\}$ **then**
7. $\mathcal{E}_i \leftarrow \mathcal{E}_i \cup \{(B_i, \epsilon_i(B_i))\}$;
8. report $\mathcal{E}_i$ to the server;

the server:
9. $\mathbf{E}$ is an ordered list of ϵ-values in $\bigcup_i \mathcal{E}_i$, sorted in descending order;
10. $I_{min} \leftarrow$ last index in $\mathbf{E}$;
11. **while** *true* **do**
12. $I \leftarrow \lfloor \frac{I_{max} + I_{min}}{2} \rfloor$;
13. $\epsilon_I \leftarrow$ the I-th element in $\mathbf{E}$;
14.   **for** $i = 1, \ldots, N$ **do**
15. $B_i(\epsilon_I) \leftarrow \min\{B_{i,j} : (B_{i,j}, \epsilon_{i,j}) \in \mathcal{E}_i, \epsilon_{i,j} \leq \epsilon_I\}$;
16.   **if** $I_{min} = I_{max} + 1$ **then**
17.   send $B_i(\epsilon_I)$ to node i for each $i = 1, \ldots, N$;
18.   break **while** loop;
19.   **else**
20.   **if** $\sum_{i=1}^{N} B_i(\epsilon_I) > B$ **then**
21. $I_{min} = I$;
22.   **else**
23. $I_{max} = I$;

each node $i = 1, \ldots, N$:
24. find local (k_i^*, b_i^*) under budget $B_i(\epsilon_I)$ given by the server by MD-MECB or EVD-MECB;

25. **return** $\{(k_i^*, b_i^*)\}_{i=1}^{N}$;

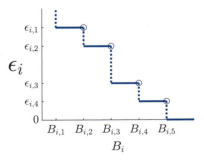

Fig. 7.8 Illustration of step objective functions

Our algorithm, shown in Algorithm 6, has three main steps. First, in lines 1–8, each node computes the set $\mathcal{E}_i$ of all pairs of $B_{i,j}$ and the corresponding $\epsilon_i(B_{i,j})$. This is achieved by evaluating $\epsilon_i(B_i)$ according to MD-MECB or EVD-MECB for gradually increasing B_i and recording all the points where $\epsilon_i(\cdot)$ decreases. After that, the set $\mathcal{E}_i$ is sent to a server.

Second, the server allocates the global budget to the nodes according to lines 9–23. To this end, it computes an ordered list $\mathbf{E}$ of all possible values of the global $\epsilon := \max_i \epsilon_i$. Let $B_i(\epsilon)$ denote the smallest value of B_i such that $\epsilon_i(B_i) \leq \epsilon$. The main idea is to perform a binary search for the target value of $\epsilon \in \mathbf{E}$ (lines 11–23). For each candidate value of ϵ, we compute $B_i(\epsilon)$ for all i. If $\sum_i B_i(\epsilon) < B$ (i.e., we are below the budget when targeting at the current choice of ϵ), we will eliminate all $\epsilon' \in \mathbf{E}$ such that $\epsilon' > \epsilon$; otherwise, we will eliminate all $\epsilon' \in \mathbf{E}$ such that $\epsilon' < \epsilon$. After finding the target value of ϵ such that $\sum_i B_i(\epsilon)$ achieves the largest value within B, the server sends the corresponding local budget $B_i(\epsilon)$ to each node.

Finally, each node uses MD-MECB or EVD-MECB to compute its local configuration (k_i^*, b_i^*) under the given budget.

Complexity We analyze the complexity step by step. First, computing $\mathcal{E}_i$ at each node (lines 1–8) has a complexity of $O((n^2 + b_0)db_0n)$ if using MD-MECB and $O((n^3 + d + b_0)db_0n)$ if using EVD-MECB, dominated by line 5. Second, computing the budget allocation at the server (lines 9–23) has a complexity of $O(db_0n \log(db_0n))$. Specifically, as $\mathbf{E}$ has $O(db_0n)$ elements, sorting it takes $O(db_0n \log(db_0n))$. The **while** loop is repeated $O(\log(db_0n))$ times, as each loop eliminates half of the candidate ϵ values in $\mathbf{E}$, and each loop takes $O(db_0n)$, dominated by lines 14–15. Finally, computing the local configuration at each node (line 24) takes $O(n^2+b_0)$ using MD-MECB and $O(n^3+d+b_0)$ using EVD-MECB.

Optimality Next, we prove the optimality of OBA-MECBD in budget allocation. Let $\epsilon_i^\pi(B_i)$ be the error bound for a given solution π of the MECB problem for dataset $\mathcal{Y}_i$ and budget B_i. We show that OBA-MECBD is optimal in the following sense.

Theorem 6 *Using a given MECB algorithm π as the subroutine called in lines 2, 5, and 24, OBA-MECBD solves MECBD optimally w.r.t. π, i.e., its budget allocation $(B_i)_{i=1}^N$ is the optimal solution to (7.31) with $\epsilon_i(B_i)$ replaced by $\epsilon_i^\pi(B_i)$.*

Proof Let $B_i^\pi(\epsilon)$ denote the smallest value of B_i such that $\epsilon_i^\pi(B_i) \leq \epsilon$. By lines 20–23 in Algorithm 6, I_{min} and I_{max} should always satisfy $\sum_{i=1}^N B_i^\pi(\epsilon_{I_{min}}) > B$ and $\sum_{i=1}^N B_i^\pi(\epsilon_{I_{max}}) \leq B$ for all nontrivial values of B. Let ϵ^* denote the value of ϵ_I at the end of budget allocation, which is the value of (7.31a) achieved by OBA-MECBD. As $I = I_{max}$ and $I_{min} = I_{max} + 1$ at this time, ϵ^* must be the smallest value of ϵ such that $\sum_{i=1}^N B_i^\pi(\epsilon) \leq B$. Therefore, for any other budget allocation $(B_i')_{i=1}^N$ such that $\sum_{i=1}^N B_i' \leq B$, we must have $\max_i \epsilon_i^\pi(B_i') \geq \epsilon^*$. $\qquad\square$

4.6 Performance Evaluation

We now evaluate our proposed algorithms using multiple real-world datasets for various ML tasks, in order to validate the efficacy and efficiency of our proposed algorithms (EVD-MECB, MD-MECB, OBA-MECBD) against benchmarks.

4.6.1 Experiment Setup

Datasets In our experiments, we use four real-world datasets to evaluate our algorithms: (1) Fisher's Iris dataset [55], with 3 classes, 50 data points in each class, 5 attributes for each data point; (2) Facebook metric dataset [56], which has 494 data points with 19 attributes; (3) Pendigits dataset [57] with 17 attributes for each data point, where we use 6000 randomly selected data points for training and the remaining 1494 data points for testing; (4) MNIST handwritten digits dataset in a 784-dimensional space [34], where 60,000 data points are for training and the remaining 10,000 data points are for testing. We leverage the approach in Sect. 3.3.1 for processing the labels, i.e., each label is mapped to a number such that distance between points with the same label is smaller than distance between points with different labels. All the data are represented in the standard IEEE 754 double-precision binary floating-point format [42]. For conciseness, we will only show the results on the Pendigits dataset, and refer to [58] for the results on the other datasets.

ML Problems We consider five ML problems as concrete examples: (1) minimum enclosing ball (MEB) [41], (2) k-means ($k = 2$ in our experiments), (3) principal component analysis (PCA), (4) support vector machine (SVM), and (5) neural network (NN) (with three layers, 100 neurons in the hidden layer). Tasks (1–3) are unsupervised, and tasks (4–5) are supervised.

Configuration Algorithms For the centralized setting, we consider five algorithms for comparison. The first two are proposed by us, i.e., EVD-MECB in Algorithm 3 (denoted by *EVD*), MD-MECB in Algorithm 5 (denoted by *MD*). The third algorithm is the straightforward solution *EM* (see Sect. 4.3.3). The fourth algorithm aims to Maximize the Precision (*MP*), i.e., using the configuration $k = \left\lfloor \frac{B}{d \cdot b_0} \right\rfloor$ and $b = b_0$ to construct coresets. The fifth algorithm aims to Maximize the Cardinality (*MC*), i.e., using $k = \min(n, \left\lfloor \frac{B}{d \cdot (1+m_e)} \right\rfloor)$ and $b = \max(1 + m_e, \left\lfloor \frac{B}{d \cdot n} \right\rfloor)$ to construct coresets, where $1 + m_e$ is the minimum number of bits required to represent a number by the rounding-based quantizer (Sect. 4.2.3). MP and MC serve as baselines, where MP only performs coreset construction, and MC mainly performs quantization.

For the distributed setting, we consider six algorithms for comparison. The first five algorithms correspond to instances of OBA-MECBD in Algorithm 6 that use EVD-MECB, MD-MECB, EM, MP, and MC as their subroutines, respectively. We denote these algorithms by OBA-EVD, OBA-MD, OBA-EM, OBA-MP and

OBA-MC, respectively. The sixth algorithm is DRCC as proposed in Sect. 3.2 that optimizes the allocation of a given coreset cardinality to individual nodes.

Performance Metrics We use the *normalized ML cost* to measure the performance of unsupervised ML tasks. The normalized ML cost is defined as $\text{cost}(\mathcal{Y}, \mathbf{x}_{\mathcal{S}})/\text{cost}(\mathcal{Y}, \mathbf{x}^*)$, where $\mathbf{x}_{\mathcal{S}}$ is the model learned from coreset $\mathcal{S}$ and $\mathbf{x}^*$ is the model learned from the original dataset $\mathcal{Y}$. For supervised ML tasks, we use *classification accuracy* to measure the performance. Furthermore, we report the running time of each algorithm. All the metrics are computed over 40 Monte Carlo runs.

4.6.2 Experiment Results

Results in Centralized Setting For unsupervised learning on the Pendigits dataset, we evaluate MEB, k-means, and PCA. Figure 7.9a–c shows the cumulative distribution function (CDF) of the normalized ML cost of each algorithm, where the budget is set to 2% of the size of the original dataset, i.e., $B = 130{,}560$ bits. Note that for MEB and k-means, i.e. Fig. 7.9a and b, MC has a much larger cost and

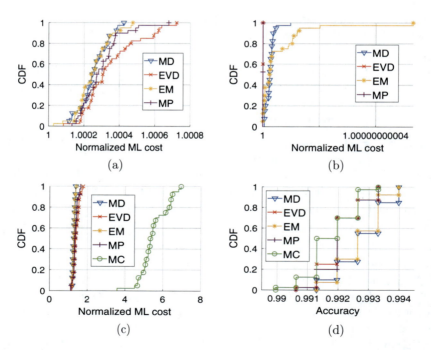

Fig. 7.9 Evaluation on Pendigits dataset in centralized setting (MC is omitted in (a–b) as its cost is much larger than the others). (**a**) MEB. (**b**) k-means ($k = 2$). (**c**) PCA (11 components). (**d**) SVM

7 Coreset-Based Data Reduction for Machine Learning at the Edge 259

is hence omitted from these plots to maintain the readability of the other results. For supervised learning on the Pendigits dataset, we only evaluate SVM-based binary classification, treating digit 0 as label '1' and the other digits as label '−1', because NN-based classification is not suitable for this dataset due to its small size. Figure 7.9d shows the CDF of the classification accuracy on testing data, under the same budget as in Fig. 7.9a–c. Note that in contrast to cost, higher accuracy means better performance. Figure 7.10a shows the average running time of each algorithm. For more insights, we also list the returned b^* values over the Monte Carlo runs for EVD-MECB (*EVD*), MD-MECB (*MD*), and EM in Table 7.6. We have the following observations: (1) Compared with MP and MC that only rely on a single operation, the algorithms jointly applying coreset construction and quantization (EM, MD-MECB, EVD-MECB) achieve better ML performance over all. (2) Although the brute-force approach EM can find a good configuration, it has a much higher running time than the proposed algorithms (MD-MECB, EVD-MECB). (3) Between the proposed algorithms, MD-MECB is not only faster but also better in approximating EM, demonstrated by the value of b^* in Table 7.6.

Remark A few remarks are in order. First, we note that all the algorithms perform close to each other in Fig. 7.9d because of the small dataset and the simple model. In our experiments with a larger dataset (MNIST) and a more complex model (a three-layer NN with 100 neurons in the hidden layer), their performances are notably different from each other, with MD-MECB performing the best and MC performing far worse than the others; see Figure 5 in [58]. Moreover, we note that although MP happens to perform well under 2% budget, its performance is highly sensitive to the budget B, while the algorithms that explicitly optimize the configuration (EM, MD-MECB, EVD-MECB) can adapt to a wide range of budgets.

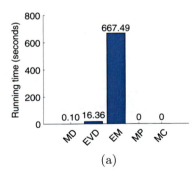 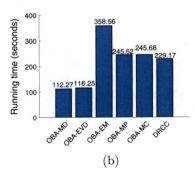

Fig. 7.10 Running times on Pendigits dataset. (**a**) Centralized setting. (**b**) Distributed setting

Table 7.6 Returned b^* for Pendigits

Algorithms	b^*	# of occurrences
EVD	[52]	[40]
MD	[11, 12, 13]	[1, 29, 10]
EM	[10, 11, 12, 13, 14, 15, 16]	[2, 9, 14, 8, 3, 3, 1]

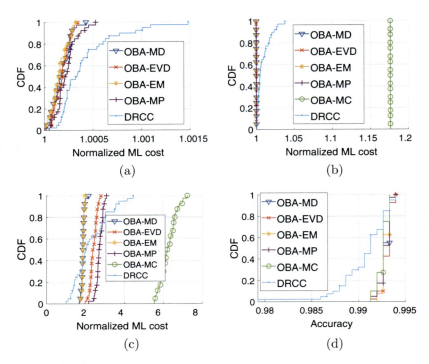

Fig. 7.11 Evaluation on Pendigits dataset in distributed setting (OBA-MC is omitted in (**a**) as its cost is much larger than the others). (**a**) MEB. (**b**) k-means ($k = 2$). (**c**) PCA (11 components). (**d**) SVM

Results in Distributed Setting In this experiment, we use the Pendigits dataset to evaluate our proposed distributed algorithm (Algorithm 6). The original data points are randomly distributed across 10 nodes. The global communication budget is set to 652, 800 bits, which amounts to 10% of the original data.

Figure 7.11 shows the performances of the learned models and Fig. 7.10b shows the average running times, both in the distributed setting. Note that for MEB, OBA-MC has a much larger cost than the other algorithms, and is hence omitted from Fig. 7.11a for better readability of the other results. We have the following observations: (1) With the exception of OBA-MC, most of the algorithms equipped with optimal budget allocation (OBA) outperform DRCC. (2) In terms of the quality of the trained model (measured by cost/accuracy), OBA-EM and OBA-MD perform the best. (3) However, OBA-MD is much more efficient than OBA-EM in terms of running time.

Summary of Experiment Results
- It is possible to achieve reasonable ML performance and substantial data reduction by combining coreset construction with quantization.

- Jointly optimizing the configurations of coreset construction and quantization can achieve a much better tradeoff between the communication cost and the quality of the trained models than relying on only one of these operations.
- The proposed algorithms, particularly MD-MECB and its distributed variant (OBA-MD), can approximate the performance of the brute-force solution (EM) with a significantly lower running time.

5 Conclusion

In this chapter, we consider supervised/unsupervised learning over distributed big data via the "sharing summary" approach (Sect. 2.1). We show, both theoretically and empirically, that coresets and their variations (e.g., quantized coresets) have the potential of significantly reducing the communication cost in offloading ML computation from data-capturing devices to edge servers, while providing guarantees on the learning results. In Sect. 3, we show that the centers of k-clustering (with suitable weights) form a coreset that provides a guaranteed approximation for a broad set of ML problems with sufficiently continuous cost functions, which allows us to leverage existing k-clustering algorithms for robust coreset construction. In Sect. 4, we further extend the solution space by combining coreset construction with quantization, where we present a general framework as well as efficient algorithms to jointly configure these operations in order to minimize the ML error under a given communication budget. The performances of all the proposed algorithms are validated through both theoretical analysis and empirical evaluations on real-world datasets. As our solutions only depend on a smoothness parameter (Lipschitz constant) of the ML cost function, our results have broad applicability to ML tasks arising in AI applications.

Remark In comparison with the popular approach of "sharing model" (e.g., federated/decentralized learning), the approach of "sharing summary" (e.g., coreset-based learning) has fundamentally different design objectives. In the spectrum of tradeoffs among communication cost, computation cost, and accuracy (of the learned models), federated/decentralized learning generally aims at achieving the accuracy of centralized learning at the cost of heavy communication/computation at data sources, while coreset-based learning and its variations generally aim at relieving data sources from heavy communication/computation at the cost of sacrificing some accuracy compared with centralized learning. Thus, instead of blindly following a fixed approach, application developers need to carefully weigh the pros and cons of different approaches in each application scenario to make a proper selection.

References

1. K. Sayood, *Introduction to Data Compression, Fourth Edition*, 4th ed. (Morgan Kaufmann Publishers Inc., San Francisco, CA, 2012)
2. D. Wolpert, Stacked generalization. Neural Netw. **5**(2), 241–259 (1992)
3. G. Tsoumakas, I. Vlahavas, Effective stacking of distributed classifiers, in *ECAI* (2002)
4. J. Kittler, M. Hatef, R.P. Duin, J. Matas, On combining classifiers. IEEE Trans. Pattern Anal. Mach. Intell. **20**(3), 226–239 (1998)
5. H. McMahan, E. Moore, D. Ramage, S. Hampson, B.A. y Arcas, Communication-efficient learning of deep networks from decentralized data, in *AISTATS* (2017)
6. V. Smith, C.-K. Chiang, M. Sanjabi, A.S. Talwalkar, Federated multi-task learning, in *Advances in Neural Information Processing Systems* (2017), pp. 4424–4434
7. X. Lian, C. Zhang, H. Zhang, C.-J. Hsieh, W. Zhang, J. Liu, Can decentralized algorithms outperform centralized algorithms? a case study for decentralized parallel stochastic gradient descent, in *Proceedings of the 31st International Conference on Neural Information Processing Systems* (2017), pp. 5336–5346
8. J.M. Phillips, Coresets and sketches, *CoRR*, vol. abs/1601.00617 (2016)
9. A. Munteanu, C. Schwiegelshohn, Coresets-methods and history: a theoreticians design pattern for approximation and streaming algorithms. KI - Künstliche Intelligenz **32**(1), 37–53 (2018)
10. J. Wang, J.D. Lee, M. Mahdavi, M. Kolar, N. Srebro et al., Sketching meets random projection in the dual: A provable recovery algorithm for big and high-dimensional data. Electron. J. Stat. **11**(2), 4896–4944 (2017)
11. M. Bādoiu, S. Har-Peled, P. Indyk, Approximate clustering via core-sets, in *ACM STOC* (2002)
12. D. Feldman, M. Langberg, A unified framework for approximating and clustering data, in *STOC*, June (2011)
13. M. Bādoiu, K.L. Clarkson, Smaller core-sets for balls, in *SODA* (2003)
14. D. Feldman, A. Munteanu, C. Sohler, Smallest enclosing ball for probabilistic data, in *SOCG* (2014)
15. I.W. Tsang, J.T. Kwok, P.-M. Cheung, Core vector machines: Fast SVM training on very large data sets. J. Mach. Learn. Res. **6**, 363–392 (2005)
16. S. Har-Peled, D. Roth, D. Zimak, Maximum margin coresets for active and noise tolerant learning, in *IJCAI* (2007)
17. D. Feldman, M. Volkov, D. Rus, Dimensionality reduction of massive sparse datasets using coresets, in *NIPS* (2016)
18. M. Langberg, L.J. Schulman, Universal ϵ approximators for integrals, in *SODA* (2010)
19. D. Feldman, M. Feigin, N. Sochen, Learning big (image) data via coresets for dictionaries. J. Math. Imaging Vis. **46**(3), 276–291 (2013)
20. A. Molina, A. Munteanu, K. Kersting, Core dependency networks, in *AAAI* (2018)
21. D. Feldman, A. Fiat, M. Sharir, Coresets for weighted facilities and their applications, in *FOCS* (2006)
22. G. Frahling, C. Sohler, Coresets in dynamic geometric data streams, in *STOC* (2005)
23. A. Barger, D. Feldman, k-means for streaming and distributed big sparse data, in *SDM* (2016)
24. S. Har-Peled, S. Mazumdar, On coresets for k-means and k-median clustering, in *STOC* (2004)
25. D. Feldman, M. Schmidt, C. Sohler, Turning big data into tiny data: Constant-size coresets for k-means, PCA, and projective clustering, in *SODA* (2013)
26. C. Boutsidis, P. Drineas, M. Magdon-Ismail, Near-optimal coresets for least-squares regression. IEEE. Trans. IT **59**(10), 6880–6892, October (2013)
27. D. Feldman, A. Krause, M. Faulkner, Scalable training of mixture models via coresets, in *NIPS* (2011)
28. H. Lu, M.-J. Li, T. He, S. Wang, V. Narayanan, K.S. Chan, Robust coreset construction for distributed machine learning. IEEE J. Sel. Areas Commun. **38**(10), 2400–2417 (2020)
29. D. Aloise, A. Deshpande, P. Hansen, P. Popat, NP-hardness of Euclidean sum-of-squares clustering. Mach. Learn. **75**(2), 245–248 (2009)

30. N. Megiddo, K.J. Supowit, On the complexity of some common geometric location problems. SIAM J. Comput. **13**(1), 182–196 (1984)
31. M. Cohen, Y.T. Lee, G. Miller, J. Pachocki, A. Sidford, Geometric median in nearly linear time, in *STOC* (2016)
32. M.F. Balcan, S. Ehrlich, Y. Liang, Distributed k-means and k-median clustering on general topologies, in *NIPS* (2013)
33. A. Virmaux, K. Scaman, Lipschitz regularity of deep neural networks: analysis and efficient estimation, in *Advances in Neural Information Processing Systems* (2018), pp. 3835–3844
34. Y. LeCun, C. Cortes, C. Burges, The MNIST database of handwritten digits," http://yann.lecun.com/exdb/mnist/ (1998) [Online]. Available: http://yann.lecun.com/exdb/mnist/
35. D. Anguita, A. Ghio, L. Oneto, X. Parra, J.L. Reyes-Ortiz, A public domain dataset for human activity recognition using smartphones, in *ESANN* (2013)
36. H. Lu, C. Liu, S. Wang, T. He, V. Narayanan, K.S. Chan, S. Pasteris, Joint coreset construction and quantization for distributed machine learning, in *IFIP Networking*, June (2020)
37. S. Han, H. Mao, W.J. Dally, Deep compression: Compressing deep neural networks with pruning, trained quantization and Huffman coding, preprint arXiv:1510.00149 (2015)
38. A. Zhou, A. Yao, Y. Guo, L. Xu, Y. Chen, Incremental network quantization: towards lossless CNNs with low-precision weights, preprint arXiv:1702.03044 (2017)
39. D. Lin, S. Talathi, S. Annapureddy, Fixed point quantization of deep convolutional networks, in *International Conference on Machine Learning* (2016), pp. 2849–2858
40. A. Gersho, R.M. Gray, *Vector Quantization and Signal Compression*, vol. 159 (Springer Science & Business Media, 2012)
41. K.L. Clarkson, Coresets, sparse greedy approximation, and the frank-wolfe algorithm. ACM Trans. Algorithms (TALG) **6**(4), 63 (2010)
42. IEEE, 754–2019 - ieee standard for floating-point arithmetic (2019) [Online]. Available: https://ieeexplore.ieee.org/servlet/opac?punumber=8766227
43. H. Lu, M.-J. Li, T. He, S. Wang, V. Narayanan, K.S. Chan, Robust coreset construction for distributed machine learning, in *IEEE Globecom*, December (2019)
44. D. Arthur, S. Vassilvitskii, k-means++: The advantages of careful seeding, in *SODA*, January (2007)
45. C. Ding, X. He, K-means clustering via principal component analysis, in *Proceedings of the Twenty-first International Conference on Machine Learning* (ACM, 2004), p. 29
46. G.W. Stewart, A Krylov–Schur algorithm for large eigenproblems. SIAM J. Matrix Anal. Appl. **23**(3), 601–614 2002)
47. A. Lim, B. Rodrigues, F. Wang, Z. Xu, k-center problems with minimum coverage. Theor. Comput. Sci. **332**(1–3), 1–17 (2005)
48. S. Khuller, Y.J. Sussmann, The capacitated k-center problem. SIAM J. Discrete Math. **13**(3), 403–418 (2000)
49. S. Khuller, R. Pless, Y.J. Sussmann, Fault tolerant k-center problems. Theor. Comput. Sci. **242**(1–2), 237–245 (2000)
50. D.S. Hochbaum, D.B. Shmoys, A best possible heuristic for the k-center problem. Math. Oper. Res. **10**(2), 180–184 (1985)
51. J. Kleinberg, E. Tardos, *Algorithm Design* (Pearson Education India, 2006)
52. J. Demmel, I. Dumitriu, O. Holtz, Fast linear algebra is stable. Numer. Math. **108**(1), 59–91 (2007)
53. H. Luss, A nonlinear minimax allocation problem with multiple knapsack constraints. Oper. Res. Lett. **10**(4), 183–187 (1991)
54. H. Luss, An algorithm for separable non-linear minimax problems. Oper. Res. Lett. **6**(4), 159–162 (1987)
55. R. Fisher, Iris data set. https://archive.ics.uci.edu/ml/datasets/iris (1936) [Online]. Available: https://archive.ics.uci.edu/ml/datasets/iris
56. S. Moro, P. Rita, and B. Vala, "Facebook metrics data set," https://archive.ics.uci.edu/ml/datasets/Facebook+metrics, 2016. [Online]. Available: https://archive.ics.uci.edu/ml/datasets/Facebook+metrics

57. E. Alpaydin, F. Alimoglu, Pen-based recognition of handwritten digits data set. https://archive.ics.uci.edu/ml/datasets/Pen-Based+Recognition+of+Handwritten+Digits (1996) [Online]. Available: https://archive.ics.uci.edu/ml/datasets/Pen-Based+Recognition+of+Handwritten+Digits
58. H. Lu, C. Liu, S. Wang, T. He, V. Narayanan, K.S. Chan, S. Pasteris, Joint coreset construction and quantization for distributed machine learning. Preprint. arXiv:2204.06652 (2022)

Chapter 8
Lightweight Collaborative Perception at the Edge

Ila Gokarn, Kasthuri Jayarajah, and Archan Misra

Abstract This chapter introduces the paradigm of *collaborative edge perception*, whereby the sensing and DNN-based inference pipelines associated with multiple sensor nodes and edge devices are optimized jointly to overcome both computational and communication resource constraints. Collaborative Edge Perception exploits the fact that multiple sensor nodes often observe the same physical phenomena and/or the same objects, but from different spatial perspectives and/or at different instants of time. Intuitively, such observations provide a degree of redundancy or hints, which can be used to eliminate or reduce unnecessary computation without sacrificing perception accuracy. The chapter describes a core set of techniques (for both RGB and LIDAR sensing data), including DNN state sharing, adaptive attention and dynamic scheduling, that exploit such spatiotemporal correlation to significantly reduce both the communication overheads at the sensor node and the inference overheads at either the edge device or the sensor node. We also describe how such collaboration can help optimize the concurrent execution of multiple perception tasks and outline a set of open problems and prospective approaches.

1 Introduction

Mobile edge computing has rapidly become the new paradigm for information processing, especially for the upcoming Internet of Things (IoT) era, where billions of connected devices (often embedded in everyday objects such as cars, coffee makers and toys) will produce new streams of data capturing our physical environment. While edge computing was initially conceptualized [1] fundamentally

---

I. Gokarn · A. Misra (✉)
Singapore Management University, Singapore, Singapore
e-mail: ingokarn2019@phdcs.smu.edu.sg; archanm@smu.edu.sg

K. Jayarajah
University of Maryland, Catonsville, MD, USA
e-mail: kasthuri@umbc.edu

© The Author(s), under exclusive license to Springer Nature Switzerland AG 2023
M. Srivatsa et al. (eds.), *Artificial Intelligence for Edge Computing*,
https://doi.org/10.1007/978-3-031-40787-1_8

as a way to offload computation for resource-intensive applications to more capable devices, edge-based processing has increasingly become [2] the de-facto approach to overcome both (a) the high latency of cloud-based processing and support a variety of interactive, latency-sensitive applications (such as augmented reality) and (b) reduce the significant bandwidth load that would be incurred to provide wide-area transfer for high bandwidth applications (such as streaming of UHD video and LIDAR pointcloud streams).

Many emerging pervasive applications, such as automated roadside monitoring, drone-based crowd surveillance and shopfloor defect analysis, rely on the use of AI models (typically DNN models) to extract real-time intelligence from a stream of increasingly high-bandwidth sensor streams, including audio, video and LIDAR data. Not surprising, there has been a growth in the number of edge device products (e.g., Google Coral Board™, Nvidia Jetson products such as the Nano and TX2) that include relatively low-end GPUs or TPUs to permit accelerated execution of DNNs, both for training and inference. Most DNN models, which embody the machine intelligence needed to automate a variety of perception and decision making tasks, are simply too heavyweight and complex to be executed efficiently, if at all, on such edge devices—constraints include memory, processing latency (which indirectly restricts throughput) and energy (especially relevant for battery-powered and more futuristic energy harvesting based edge platforms). A variety of approaches and techniques have thus been devised to support the lightweight, real-time execution of AI inference models, driven by pervasive sensing data, on such resource-limited platforms—such approaches include both static DNN model compression techniques and dynamic (runtime) DNN optimization techniques that have been covered elsewhere in this book. The vast majority of such compression and optimization techniques focus on a single task (e.g., object detection over RGB vision sensor data), embodied through a specific DNN model, and executing on a single sensor stream *in isolation*.

In this chapter, we explore another distinct approach to reduce the complexity of DNN-based inference execution—one based on the notion of *collaborative edge intelligence*. Such collaborative intelligence hinges on the central observation that many AI-driven pervasive applications involve the use of not just a single sensor node, but a collection of such sensor nodes (whether static or mobile) or at least multiple sensor streams of different modalities (e.g., RGB and LIDAR) from a common sensor node. In addition, for many applications and deployments, the same sensor data stream is used to execute multiple inference tasks, each using a distinct custom DNN model, on the same edge device. In such applications, the intelligence is gleaned not just by processing data from one sensor stream, but by fusing together insights generated from different sensor nodes or streams. It is worth noting that the collaborative edge intelligence approach is complementary to, and can be combined with, other static and dynamic optimization approaches mentioned in other chapters.

These sensor nodes effectively provide a distributed sensing substrate that captures underlying physical phenomena with significant *spatiotemporal overlap or correlation*. Such spatiotemporal overlap or correlation implies that an individual sensor data stream is highly likely to exhibit some degree of *redundancy*—i.e., at

8 Lightweight Collaborative Perception at the Edge 267

least parts of its sensed data will conform to some predictable spatial or temporal evolution of data that is also sensed by some other sensor modality or node. The central objective in collaborative edge intelligence is thus to find inexpensive ways to isolate such redundant portions of sensor data, across a set of spatiotemporally disparate sensors streams, and modify the computation pipeline so that it avoids (or minimizes) computation on such redundant data. Phrased another way, the computational resources are preferentially redirected only towards data with *high criticality*—e.g., on RGB image data, DNN operations are executed at higher fidelity on portions of the image that are judged to more likely have relevant information content. While such selective computation effectively reduces the total computational load of DNN inference pipelines, it is important to note that such computational reduction should, ideally, not result in any loss of inference accuracy as it is applied judiciously only on data segments that do not contain any unique information content. Of course, such selection computation will also reduce system-level overheads, such as memory, latency and energy consumption, as long as the pre-processing steps for prioritizing the portions of the data streams is itself lightweight and low-overhead. Besides such preferential computation redirection, the proposed approaches discussed here also attempt to find common computational elements across multiple DNN models (each corresponding to a different task) or to better schedule the execution of such concurrent tasks.

The chapter will introduce the concepts mentioned above through the following sections:

- Section 2 will introduce, using relevant examples, three primary approaches in collaborative edge perception—(a) *state sharing across different sensors and DNN pipelines*, (b) *adaptive attention* and (c) *dynamic scheduling*—as applied to the common use cases involving vision tasks performed using either 2D RGB image data or 3D point cloud data.
- Section 3 will focus on the challenge of executing multiple independent tasks concurrently on an edge device, and describe the use of two primary approaches—runtime dynamic scheduling and model merging—to reduce the associated runtime overheads.

2 Collaboration Between Sensors and Edge Nodes

2.1 Understanding the 2D Scene

Driven by the proliferation of low-cost, high-definition, intelligent cameras [3], and the evolution of deep neural networks (DNNs) as the state-of-the-art for visual understanding, individuals and organizations are deploying cameras for a range of vision-based tasks. These include applications in Industry 4.0 (such as machine diagnostics and supply chain logistics monitoring), surveillance and homeland defense, and high-definition sports telecasting among many others. Edge supported

HD, UD, and 4K cameras have now reached economies of scale. For example, a QualComm Vision AI DevKit available with a 4K high frame-rate camera is available for USD \$100 [3], a significant reduction in cost as compared to 2018, while the latest Samsung S22 smartphone in 2021 boasts of a 108 mega-pixel camera with 8K@24 fps video quality [4]. "AI cameras", or high definition cameras with high framerates, onboard memory and computing resources, are also rapidly appearing in the marketplace (e.g., [3, 5]). The goal of such an AI camera is to reduce the network bandwidth used to transfer raw frames to the edge node for processing and instead enable intelligence at the point of data creation.

Resource-efficient and accurate visual understanding is key to the success of such deployments, both at the edge and the cloud. At the time of inception of real-time video processing at the edge, research works focused on supporting continuous vision tasks such as object detection on a resource-constrained mobile device. With limited on-device computing resources, these tasks were was supported by:

1. **Execution-Aware Offloading:** workload sharing or the offloading DNN feature maps or weights from a partially executed visual pipeline to the cloud or a co-located edge computing device [6–8] to complete the DNN pipeline execution
2. **Content-Aware Offloading:** offloading partial or entire portions of the captured camera frames in an intelligent manner to improve the cost-accuracy tradeoff for the specific DNN task [9, 10]

Offloading any part of the visual computing pipeline, either partially or in its entirety, to the edge, however, generates challenges such as high communication latency, bandwidth overheads and privacy concerns. Recent works have thus focused on reducing the overhead of video processing at the network edge, resulting in a research area rich with interesting solutions to address different aspects of deployment [11–14]. Such solutions, however, tackle the data stream generated by an individual vision sensor (camera) in isolation, and do not consider the additional challenges and opportunities that arise in many real-world deployments, where a *collection* of camera nodes collectively monitor a specified area, region or object.

The overheads and drawbacks associated with such offloading of parts of the processing pipeline to an edge node are amplified in three distinct ways when a deployment consists of multiple such cameras [15]. First, multiple high-definition cameras (i.e. 4K/8K resolution) require prohibitively high network bandwidth to transfer the frame to the edge node for processing. Second, the resolution and frame rate of the video received at the edge also influences the computation requirements of the model chosen to process such video, and the resulting inference latency and accuracy of the DNN pipeline. Effectively, the higher the resolution and frame rate of the video and the larger the model, the better the inference accuracy but the greater the resource demand on a resource-constrained edge computing device. Lastly, the reality that state-of-the-art DNNs (which now contain 200+ layers and one trillion+ parameters) adopted by vision and perception tasks are not designed to conform to processing constraints imposed by the resource constrained edge is especially acute when considering resource contention from multiple camera sensors.

With these three challenges in mind, this section introduces the *Lightweight Collaborative Perception* paradigm for RGB camera sensors that builds on the methodologies of offloading content or partial execution of a vision pipeline to the edge node to improve the cost-accuracy tradeoff for vision pipelines operating on multiple camera sensors. We first describe the pre-requisites for collaborative perception in Sect. 2.1.1, with collaboration between the sensors and/or between the sensor and the edge being moderated by the edge node. In Sect. 2.1.2 we introduce the concepts underlying lightweight execution-aware collaboration or DNN state sharing between perception pipelines of different camera nodes. On the other hand, Sect. 2.1.3 describes methodologies that focus on content-aware processing of multiple concurrent camera streams by paying selective attention to only critical portions of the camera frames, and scheduling salient portions of the sensing field for DNN inference on the resource-constrained edge.

2.1.1 Opportunities for Collaboration in Multi-Camera Deployments

High-density camera deployments very often have groups of cameras that have shared sections of overlap in their individual fields of view (FoV) with other cameras in the group. It stands to reason that all the cameras within a single group then observe high correlated events—both spatial and temporal - arising from observing a shared physical phenomenon [15]. Spatial correlations arise from the probability that an object observed by one camera may also be simultaneously observable within the FoV of another adjacent camera. Temporal correlations arise from the probability that an object observed in one camera might move into the FoV of a neighboring deployed camera after a specific time interval. For example, a group of cameras monitoring a busy intersection may all observe the same pedestrians and cars, but all from different angles and/or perspectives. Alternatively, the camera at a downstream interesection is highly likely to observe the *same set* of vehicles observed by an upstream camera, with a time lag between 60–120 s. The fundamental idea then is that these cameras could be considered collectively to take advantage of these cross-camera spatiotemporal correlations to reduce redundant computations and enable more efficient use of computation resources. Combining the information embedded across multiple such perspectives will help maintain (or even increase) the task accuracy, compared to an independent task execution by each camera node. For example, in a deployment of static cameras whose extrinsic qualities are known ahead of processing time, it could be determined that if person A is visible to camera 1, due to the nature of the camera placements, the same person A would be visible to camera 2 in a matter of frames, as seen in Fig. 8.1a. Here it would behove the edge device to share some features of the person as seen in camera 1 to camera 2 (effectively, a set of 'priors' in the language of statistical inference) to reduce the computation required of camera 2's pipeline. In another example, if person A is occluded from view in camera 1's FoV but is visible in camera 3's FOV from another angle, it might be useful for camera 1 to receive such 'inference hints' to increase the inference accuracy of its vision pipeline, as seen in Fig. 8.1b.

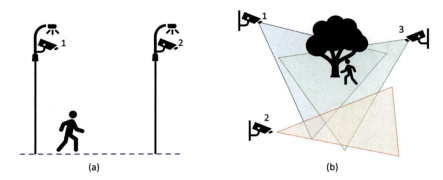

Fig. 8.1 (**a**) Person visible to Camera 1 will be visible to Camera 2 in a matter of frames. (**b**) Camera 1 has an occluded view of the person and can receive intelligence from Camera 3 to increase its visual inference accuracy

A canonical application that would rely on spatiotemporal correlations among multiple cameras deployed at an edge node is "spotlight search" which involves detection of an object or an activity, and subsequently tracking the object as it moves in physical space as observed over time by different cameras in the network [15]. In the context of surveillance for public safety, vehicle tracking or retail store security, such spotlight search will be required in both forward-direction i.e. tracking a speeding vehicle, suspect, or shoplifter as they move through the camera network over time, as well as the backward direction i.e. replaying or investigating where the vehicle, suspect, or shoplifter traversed through before an activity of interest like speeding or theft occurred. Recent research works focus on a myriad of ways to take advantage of spatiotemporal correlations to improve DNN inference accuracy while reducing the computation load imposed by multiple cameras on a single edge node. We now detail a few of the proposed approaches that address bandwidth, computation, and inference accuracy challenges to such deployments.

2.1.2 Lightweight State Sharing for Improved Perception

The benefits of collaborative perception have been well documented in recent Computer Vision literature—e.g., Deep-Occlusion [17], DeepMCD [18], MVDet [19]. The majority of these works rely on designing custom deep neural network models which assume that collaborating camera peers share the *raw images* with a single node that executes the custom model (the model being customized to handle multiple input images simultaneously). The drawback of this design is that it does not consider constraints in the bandwidth availability, especially, as the number of nodes in the network increases. Another key problem with this design is the tight coupling of the visual perception pipeline with the collaboration logic—custom training of DNN models means that the number and spatial configuration of the collaborators need to be known a-priori, and any configuration change effectively requires re-training of the network for reliable inferencing. To overcome these challenges, recent works have looked at devising strategies to be selective in terms of *what information* is exchanged while trading off on the network consumption and the processing required at the edge nodes themselves [16, 20] as well as decoupling the perception and collaboration logic so as to remain resilient to churn in the network [20]. Such approaches fundamentally involve the transfer or exchange of *internal DNN state*—i.e., different features or digests that capture intermediate values of the DNN layers that operate on a camera's image, and that arguably provide succinct summaries or hints of the objects being observed by the corresponding camera. Approaches differ in terms of what layers generate such state: *late* state sharing approaches share outputs generated at the last 1–2 layers of a DNN pipeline and effectively correspond to late Bayesian fusion, whereas *intermediate/early* state sharing approaches extract hints from the features in the first 1–5 layers of the DNN pipeline. However, they all assume that the cameras are all equipped with onboard compute resources that allow them to execute deep neural network (DNN) models for perception tasks (such as object detection and occupancy map generation), either end-to-end or at least partially.

The following techniques describe modifications to a class of state-of-the-art object detection models known as *single shot detectors*[21] that operate on the basis of *prior* or *anchor* boxes. These types of detectors combine the region proposal and object localization steps into a single deep neural network (DNN) leading to lower runtimes, in comparison to search-based approaches [22], that make them suitable for real time applications. A predefined set of anchor boxes, of varying sizes and aspect ratios, tiled across an image of fixed size represent *likely* positions/sizes of different objects. The runtime inference phase consists of detecting whether objects are present in each of the anchor boxes and if so, computing (a) probabilities for each object class through a classification step and (b) positional adjustments/offsets of these boxes through a regression step. These class probability and adjusted boxes are then used to output a set of detections in the forms of bounding boxes, each associated with an object class, class likelihood, location and size. Each image passes through several convolutional layers (feature extractors), followed by a SoftMax layer (where the highest probability class is selected, per anchor box) followed by a Non-Maximum Suppression (NMS) step. NMS is a post-processing step which assumes that bounding boxes with significant overlap (measured as the intersection over union, IoU, higher than a threshold) correspond to the same object, and thus eliminates redundant boxes (thereby improving overall precision).

Early Vs. Late Fusion Weerakoon et al. [16] propose two alternate techniques, the Collaborative Non Maximum Suppression (CNMS) and the Collaborative Single Shot Detection (CSSD). In both approaches, individual cameras run their DNN models till *later* layers of their pipeline (giving rise to the term *late fusion*), and exchange processed outputs which are then consumed by the peer cameras at various stages of their own pipelines.

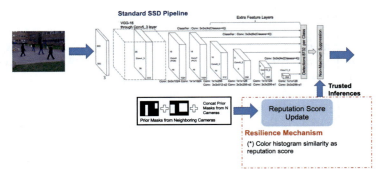

Fig. 8.2 In the Collaborative Non Maximum Suppression (CNMS) approach, individual inferences of participating cameras are fused at the last stages of the perception pipeline [16]

In the **CNMS** approach (depicted in Fig. 8.2), the object detector DNN is executed, during the inference stage, until the NMS step to output object bounding boxes in a similar fashion to the baseline detector. Prior to the NMS stage, a camera running CNMS *fuses* both its own outputs as well as peer cameras' outputs. More specifically, bounding boxes from peers are first converted to the reference camera's coordinate system. Suppose a peer camera generates a set of m bounding boxes $P = \{p_1, p_2, \ldots, p_m\}$, and the reference camera generates a set of n bounding boxes $R = \{r_1, r_2, \ldots, r_n\}$ with n corresponding confidence scores represented by $C = \{c_1, c_2, \ldots, c_n\}$. Then, overlaps between the two sets of bounding boxes are computed as $IoU(p_i, r_j)$ for all $i \in [1, m]$ and $j \in [1, n]$. A higher $IoU(p_i, r_j)$ (e.g., $IoU > 0.8$) implies that both the peer and reference cameras have detected an object at the same location. This reinforces the belief that there is indeed an object present leading to an update of the confidence score as $c_j^{new} = (1 - \alpha) * c_j + \alpha * IoU(c_i, r_j)$. α is a constant for a given peer camera, set based on its spatial overlap and reputation (for e.g., $\alpha = 0.2$). The confidence update procedure is then repeated in a round robin fashion for every peer camera, Finally, the NMS step is run on the *modified* bounding boxes. It is worth highlighting that the CNMS approach performs fusion only by modifying this last step of the detection pipeline–i.e., it only performs statistical fusion of the DNN's output (the set of bounding boxes), and as such, does not require any re-training of the DNN itself.

On the contrary, the **CSSD** approach, achieves improved perception through structural changes to the DNN pipeline itself. While the *state* exchanged between the cameras remains the same (i.e., the bounding boxes from inferences), instead of fusing them through a post-processing step, CSSD augments input images from a reference camera through a pre-processing step. More specifically, peer bounding boxes are converted to a binary input mask (as illustrated in Fig. 8.3, with pixel values= '1' for areas where objects were detected and '0' elsewhere) after being appropriately spatially-transformed to the reference camera's coordinate frame. Masks from N peers are the concatenated as additional inputs to the object detector

8 Lightweight Collaborative Perception at the Edge

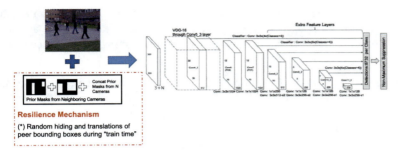

Fig. 8.3 The Collaborative Single Shot Detector (CSSD) approach, individual inferences of participating cameras are fused at the input stage of the perception pipelines of the receiving cameras, with a time delay [16]

DNN as additional N channels to the reference camera's 3-channel RGB input. We highlight to the reader that while the CSSD approach requires end-to-end re-training of the DNN unlike CNMS, as experiments on the PETS2009 multi-view camera data [23] show, it can provide significantly improved detector performance. For instance, while both approaches outperform the non-collaborative baseline detector, the CSSD detector achieves an impressive mAP improvement of $\approx 14\%$, compared to only a $\approx 7.5\%$ improvement for CNMS.

Intermediate State Sharing While the above *late fusion* approaches improve the perception accuracy, they are fundamentally limited by the accuracy of the baseline detector itself; for instance, without collaboration, the SSD detector has an average F-score of only $\approx 68\%$ on the PETS2009 dataset [23] as reported by Weerakoon et al. [16]. As an alternative, **ComAI** [20] proposes using state information extracted from intermediate features of the detector DNN. Such techniques can be advantageous in two principal ways: processed states or features can (a) be considerably less bandwidth consuming than sharing raw video streams and (b) embed more useful information and nuanced hints than fully processed outputs from DNNs of subpar performance.

The design of ComAI is motivated by two main observations. Firstly, objects that appear small or occluded account for a large percentage of failed detections in the DNN (owing to lower output confidence values) and such objects may appear larger or un-occluded to other cameras in the network just by the virtue of those cameras being placed in a different position or orientation. Secondly, intermediate outputs (i.e., outputs from convolutional kernels called feature maps) from even the very first layers of a DNN can be reasonably discriminating between object classes and can be leveraged to extract *early* hints about a scene. ComAI operates based on two key ideas: (a) as part of a two-step process, it identifies the most effectual, bandwidth-parsimonious scene summaries that peer cameras can share, and (b) then modifies the runtime execution of the DNN pipeline to ingest those summaries for improved perception, in a manner that does not require re-training of the model.

Stage 1: Lightweight Scene Summarization During the initial training stage, ComAI first determines the most *discriminative* filters (i.e., kernels at each of the convolutional layers) that are most effective in detecting the presence of target objects such as "persons" as follows (see Fig. 8.4):

1. As each image frame f is processed by the DNN, intermediate feature maps, $F_{f,i,j}$, output by filter j, at every convolutional layer i, as well as the final inference, I_f (i.e., a set of bounding boxes), are captured.
2. Then, for each frame f, the output $F_{f,i,j}$ is used to create a corresponding binary mask, $B_{f,i,j}$, where activation values beyond a threshold are marked as "1" and others as "0" and appropriately resized to match the dimensions of the inference image. The inference image, I_f, is also binarized with regions containing target objects marked with "1" and "0" elsewhere.
3. For each such **binary** fmap, $B_{f,i,j}$, a similarity score (e.g., intersection over union) is computed against the binary version of the inference image. Finally, filters are ranked based on the average similarity score, in descending order, over all frames where filters: filters with the highest scores are deemed to be the most effectual in picking out pixels corresponding to those target classes.

Next, as illustrated in Fig. 8.5, a shallow machine learning-based classifier is trained using the top-k *discriminative* feature maps ($\hat{F}_i$) from the previous step to *estimate, in advance,* the class-specific probabilities that will be eventually output

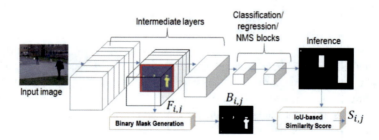

Fig. 8.4 Intermediate feature map selection process [20]

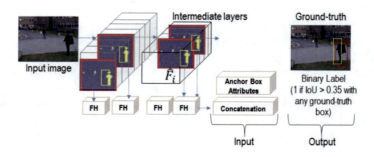

Fig. 8.5 ComAI's shallow classification process [20]

8 Lightweight Collaborative Perception at the Edge 275

by the DNN's predictor layer. The problem is posed as a binary classification task where the model predicts whether an object of a particular class is either present or not, in a region (such as an anchor box) within the image. For an image frame (f), the classifier predicts the target-class probability for each anchor box a in the anchor box space (A) using both (1) activation-value based features extracted from the feature maps, $\hat{F}_i$, (represented as histograms of predetermined lengths and concatenated) as well as (2) properties of the anchor boxes such as the location within the image view and the aspect ratio of the box.

Stage 2: Modified Runtime Execution During runtime, ComAI nodes run the shallow classifier to estimate class confidence values based on intermediate outputs from the initial layers and share them with peer nodes while continuing to execute the remainder of the DNN pipeline in parallel. On receiving such *perspective summaries*, or states, in the form of class-specific probabilities, the receiving (reference) nodes then *bias* the confidence values computed by their own DNN, and thereby improve inference accuracy. We emphasize again that this process where the biased confidence value is simply a cumulative sum of all the confidence values received from all cameras (for regions with overlaps with at least one other camera) occurs at runtime and does not require any retraining or structural modification of the DNN model. The biased class confidence values are then forwarded to the subsequent (Soft max) layer for object classification and localization, similar to a non-collaborative baseline detector.

Experiments on a ComAI deployment consisting of multiple camera nodes (each supported by a Jetson Nano board with 128-core NVIDIA Maxwell GPU and 4 GB of memory) demonstrates the significant performance gains of an intermediate state sharing based collaborative approach. Across two benchmark datasets, PETS2009 [23] and WILDTRACK [24], ComAI's collaborative confidence boosting helps increase object detection recall by 20–50% within overlapping regions outperforming both non-collaborative as well as other *early* and *late* fusion baselines such as CNMS and CSSD. At the same time, due to its lightweight scene summarization technique and fusion approaches, ComAI only incurs a marginal increase in the inference latency ($\leq$ 10–14%) and memory overhead ($\leq$ 0.017%). Furthermore, sharing states in the form of highly selective confidence values (as opposed to bounding box outputs as binary masks [16] or feature maps [19]) also means that ComAI only expends a minimal communication overhead ($\leq$ 1 KB/frame/pair).

2.1.3 Content-Aware Collaboration: Attention and Scheduling

Early works on continuous vision tasks at the resource-constrained edge have showcased the benefits of content-awareness in the processing pipeline. Camera-directed methodologies, such as sampling and filtering whole input frames [11, 25, 26], using microclassifiers or cascading filters for discarding frames which contain no new information [26, 27], are all based on the principle of using temporal locality of objects of interest between consecutive frames. This temporal understanding acts as a guide for camera-directed moderation of the amount of data to be transmitted to the edge for further processing to reduce bandwidth consumption. Edge-directed methodologies such as caching of input frames described by DeepCache [9] show the use of spatiotemporal locality or objects over multiple frames from the same camera stream to improve the DNN inference pipeline and resulting reduced computation load. Recently, some works have argued that processing a single whole input frame on the edge imposes the notion of *priority-inversion* on the individual objects/regions of interest in the frame, where high-priority objects of interest (e.g. pedestrians or vehicles) are processed alongside low-priority regions of the frame (e.g. the sky or a static background) [28, 29]. Intuitively, this is a wholly disadvantageous philosophy which also imposes harsh cost-accuracy constraints on the edge device. These works assert that individual camera input frames can be decomposed into cropped regions containing high-priority objects of interest. By ascribing higher attention to these high-priority regions containing objects of interest, the system can simultaneously (1) reduce bandwidth and energy consumption at the camera, and (2) reduce the computation load at the edge by generalising to regions of interest across multiple spatio-temporally correlated cameras. In this section we introduce the separate but related concepts of *Attention* and *Scheduling* which focus on spatio-temporally correlated content-aware offloading and processing.

Filtering Camera Streams with Spatio-Temporal Correlations One way to ease the network bandwidth costs associated with multicamera deployments is to perform input filtering (i.e., reducing the number of frames transmitted to the edge) by taking advantage of spatiotemporal kinematics or the movement of objects in time and space across a frame.

Jain et al proposed **Spatula** [30], a cost-efficient system that takes into consideration the spatiotemporal correlations across the multi-camera deployment for detecting and tracking objects of interest across frames from multiple cameras. The key methodology focuses on pruning the search space of a query identity (i.e. limiting an object to be searched and tracked only to a subset of frames across cameras that are more likely to contain the object), thereby drastically reducing computation and communication costs from the baseline of transferring and processing all frames. Significant challenges to such a methodology lie in the fact that it is non-trivial to derive cross-camera spatiotemporal correlations on unlabelled video data, and that additional system-level software support is needed to facilitate online filtering which is bound to contain errors from missed objects that were precluded from the search space. To address these challenges, Spatula adopts three key phases in its operation—(1) An offline phase constructs the spatiotemporal correlation model from unlabelled video data, (2) At runtime for inference on a particular query, this correlation model based on cross-camera relationships, as seen in Fig. 8.6, is used to filter out those cameras that are not correlated to the query identity's current position and therefore have low likelihood of observing the object in their FoV or after a particular interval of time, and (3) Spatula maintains a cache of recently filtered frames and is able to perform a fast-replay search in

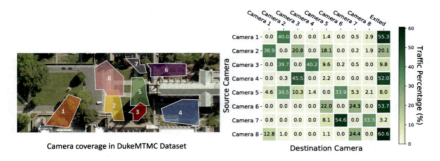

Fig. 8.6 (**a**) Camera coverage and regions in the DukeMTMC dataset [31] (**b**) Cross-camera correlations—each cell shows the percentage of outbound traffic from one camera is seen in the other cameras [30]

case of missed query instances during the live search. Spatula reports savings in compute cost by a factor of 8.3× and network cost of 5.5× (as compared to the baseline) in an 8-camera deployment and reports a 39% improvement in precision with a 1.6% drop in recall due to the reduced number of false positives that the correlation model filters out. In larger deployments of 130/600 simulated cameras Spatula reports dramatic savings of 23× and 86× respectively.

Another approach to input filtering is shown by Li and Padmanabhan in their work **Reducto**, which operates at the camera-level rather than filtering at the edge at inference time [10]. Although this approach leverages only the video stream from individual stateless cameras, it shows a trend where multiple cameras can be moderated or directed by the edge as to their input frame filtering to benefit the overall system. At the camera level, frame filtering is possible on commodity cameras through purely low-level feature analysis of differences between subsequent frames. However, the reliance on low level features can lead to significant loss in query accuracy if the feature analysis is implemented incorrectly and a required frame is filtered out. On the other hand, such filtering has the potential to reduce some downstream computation cost by reducing the number of frames to be processed in relation to a specific query identity. Reducto overcomes this challenge by building dynamic adaptive online filtering decisions at runtime based on temporally-variant correlations between the low level feature types, the associated feature thresholds, query inference accuracy, and the video content itself. The key point in this system is that a Reducto camera does not transmit any frames until a specific user-given query, desired inference accuracy level, and best features for the query are sent to the camera. With this system, Reducto reports 51–97% savings in filtering benefits while achieving the visual inference accuracy. In a multi-camera deployment, applications might focus on overall scene representations rather than detecting and tracking an object across multiple cameras simultaneously, as posited by Guo et al. in their work **CrossRoI** [32]. The methodology in this work takes advantage of spatiotemporal correlations across multiple cameras to remove repetitive appearances of the same object from different FoV without impacting the overall scene representation and understanding. While this work also focuses on object detection and tracking, they assert that detection of an object in any one of the overlapping views is sufficient to achieve adequate inference accuracy. With this aggressive methodology of filtering frames, CrossRoI achieves significant savings in network overhead up to 42–65% and processing latency savings up to 25–34%. In conjunction with state of the art camera-based filtering methodologies like Reducto, CrossRoI reports savings in network overhead up to 50–80% and end to end processing latency savings up to 33–61%.

Attention to Regions and Resolution of a Frame As established earlier, larger and more complex DNN models are required to accurately infer visual tasks on frames produced by high resolutions and framerates of cameras available today which can incur significant computation costs at the edge. One way to address the resource-accuracy tradeoff is by tuning the DNN configuration and camera characteristics configuration with respect to framerate and resolution to achieve inference accuracy while using available computation resources at the edge efficiently.

In this respect, Jiang et al. propose **Chameleon** [33], a system controller that dynamically adjusts the resolution and framerate of the input video for existing DNN pipelines to reduce computation costs without significant loss in inference accuracy. Chief among the motivators for such a dynamic controller is that scenes observed by a set of cameras can vary over time—for example, rush hour traffic may require frames of higher resolution and framerate to accurately detect multiple closely-packed vehicles, while non-rush hour traffic may be better serviced by a lower framerate that could save on computing resources. A key challenge with such dynamic adjustment of DNN characteristics is that the search space of permissible configurations is exponentially large in the number of configurations and their values, and periodically traversing through such a search space to ascertain the best configuration can itself stand to incur significant computation overheads. Chameleon posits that leveraging spatiotemporal correlations of objects across cameras can amortise the search cost over time to achieve a resource-accuracy tradeoff, as the shared characteristics among proximal cameras observing the same physical space will allow the configuration of one camera to be applied to all cameras in that group. The key idea is to profile different pipeline configurations as tuneable knobs—i.e. models of choice, input video resolution, and framerate—and periodically tune or adjust those knobs that are predicted to either favour accuracy or resource utilization to meet the task and user requirements. Furthermore, patterns of usage among top-k configurations will emerge over time and such patterns can be used to prune the search space for optimal configurations for large camera deployments over time. Chameleon takes advantage of the observation that all the knobs are independent from each other in their impact on the resource-accuracy trade-off. Chameleon then uses a greedy hill climbing-inspired profiling method to learn the persistence of the top-k configurations. Chameleon identifies a leader camera that is profiled at the start of the profiling window or an interval; in addition, a set of good top-k configurations are identified and used by all follower cameras that are grouped under the leader camera. Compared to a baseline system that

8 Lightweight Collaborative Perception at the Edge 279

picks a single configuration offline for the DNN pipeline, Chameleon reports 20–50% accuracy gains while using the same amount of computational resources or achieve a speedup of 2–3× by using only 30–50% of the resources to achieve the same accuracy. Continuously adapting the configurations of the DNN pipeline show savings in computation cost up to 10× and an increase in accuracy by 2× when compared to selecting a single spatiotemporal invariant configuration at the start of the processing pipeline.

Chameleon applies configuration knobs to the entire frame received within a profiling window, but a complementary area of research is in differential treatment or application of different configuration knobs to different regions from a single high-resolution frame to moderate the resource-accuracy-processing latency trade-off. Jiang et al. propose **REMIX** [34], a framework that takes a latency budget as an input and creates a set of non-uniform frame partitions based on regions of interest and a model execution plan to run off-the-shelf DNN pipelines on different partitions. Effectively, REMIX selectively focuses the models with larger computation costs on those partitions that have more regions of interest while applying a cheaper model to the partition with no areas of interest or discarding them entirely, thus achieving overall higher inference accuracy on the objects of interest while saving on computation cycles of the edge device. The authors identify two key challenges that arise from such a framework—(1) the search space of the optimal partition of an image that best balances the accuracy-processing latency trade-off is non-trivial, and (2) spatiotemporal variations in the input video over time influences the choice of model allocated to each partition and the effective execution of the model on each partition, which in-turn is also based on available computation resources on the edge node. REMIX addresses the first challenge by learning the spatial distribution patterns of different objects observed over time and builds a profiler that estimates the inference accuracy of all available DNNs on objects of different sizes. With this profiler, REMIX can reduce the search space by 35×. To address the second challenge, given a set of partitions, REMIX utilizes previous detection results as feedback to determine which partitions can need more attention from stronger models, which partition can achieve high accuracy with weaker/smaller DNN models, and which partition can be discarded entirely. REMIX reports 1.7× to 8.1× of processing latency speedup with an accuracy drop of only 0.2% when compared to the baseline of the state-of-the-art models. On the other hand, with frame processing latency as constant, REMIX reports an improvement of 65.3% in inference accuracy on average.

In a parallel research thread, low-power vision systems battle the same resource-accuracy-latency trade-offs but with the additional consideration of energy costs to transfer images from battery-less or low-energy camera systems to the edge. In this line of work, cameras simply capture and offload low-resolution frames as determined by the available energy budget to a GPU enabled edge node with more resources for visual inference. However, images captured by low-energy cameras would either sacrifice resolution or framerate to operate within the limited energy budget, thus adversely impacting the inference accuracy for the vision task. Wu et al. propose **MRIM** [35], a framework that tackles the fidelity-vs-accuracy tradeoff.

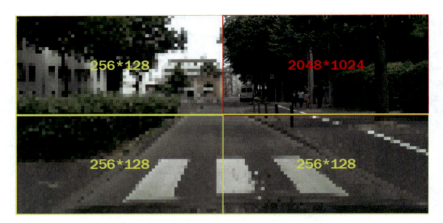

Fig. 8.7 Uniform partitions and non-uniform resolution adjustment based on regions of interest as determined by MRIM [35]

MRIM executes a lightweight pre-processing step to accurately determine regions of interest in a low-resolution frame and then perform non-uniform downscaling of the resolution of the image so as to preserve saliency in those regions of the frame to achieve the required DNN inference accuracy, while reducing attention given to the background of uninteresting regions of the frame, as illustrated in Fig. 8.7. The amount of resolution adjustment for each region is calculated based on offline profiling of the impact of different resolutions on object detection confidence and the resulting DNN inference accuracy. MRIM effectively creates a *mixed resolution image* with regions of interest having a higher resolution, thereby preserving pixel-level details preferentially in the regions of interest, which then leads to higher inference accuracy. MRIM proposes two algorithms—min-max optimisation and water filling method—to efficiently compute appropriate resolution choices for different regions of the frame constrained by the energy budget and accuracy requirements. MRIM reports high throughput and energy savings up to 35% or inference accuracy by 8% over the baseline method of uniformly scaling an image to fit the energy constraints. While this work focuses on discreet camera streams, further optimisation to the offline profiler could use spatio-temporal correlations between static cameras to further fine-tune the resolutions of those camera frame regions that fall in the shared FOV between cameras.

Attention in Multi-Target Tracking So far, the works introduced in this section focus on single object query, detection, and tracking across a multi-camera deployment and tuning of configurations within a specified resource-accuracy-latency budget. The challenge becomes much greater if we attempt to track multiple objects as they move in space and time across a camera deployment supported by the same edge with the same tradeoff matrices.

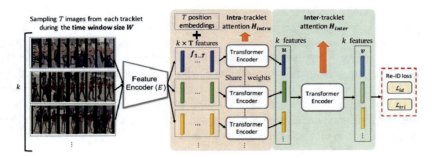

Fig. 8.8 Proposed architecture to pay attention to inter-tracklet differences and intra-tracklet discriminators Li et al. [36]

Li et al. address such a multi-target multi-camera tracking (MTMC) mechanism [36] which enables the detection and tracking of multiple targets over time-synchronised cameras whose pose or calibration is unknown at runtime. Using spatiotemporal correlations of the multiple detected objects, multiple targets or objects are detected and their trajectories are associated across the cameras. This process is simplified as only those objects that appear in the overlapping sections between the cameras within the same time window need to be considered. The proposed **MTMC** approach, a variation of the re-id algorithm, models visual features from a sequence of object images or tracklet to encode a similarity matrix in relation to all other object tracklets in the scene observed from uncalibrated cameras. The proposed approach focuses on learning these differences between tracklets using an inter-tracklet attention module to better learn the representation of each tracklet. To bolster the representation of each tracklet, Li et al. pay attention to and encode intra-tracklet discriminative features as well, namely, the gait and specific motion of each person by using a transformer encoding into the individual tracklet's embedding, as illustrated in Fig. 8.8. The authors compare the performance of this modified re-id methodology against state of the art video re-id approaches like NVAN [37], VKD [38], and AP3D [39], and report a 5% improvement on average in the performance of re-id over the aforementioned methods.

Scheduling and Removing Redundant DNN Inspections In the previously discussed works, framerate was a tuneable component in the overall framework. However, for certain time-sensitive applications like surveillance and automatic braking systems in cars, a high framerate is a mandatory requirement and cannot be 'tuned', so as to not miss any events of interest. In this case a multi-camera system must look at other ways of tuning the resource-accuracy-processing latency trade-off with framerate and resolution at the best possible configurations while minimizing frame processing latency to support high framerate input videos.

Kang et al. suggest that the priority-inversion suffered by critical objects of interest in a frame can be mitigated by decomposing the input frame into smaller cropped regions for DNN inference [29]. The authors introduce the **DNN Split-**

and-Merge (DNN-SAM) framework that leverages on an unmodified DNN in two parallel sub-tasks and merges results from both. The primary (mandatory) subtask crops salient regions containing objects of interest and schedules their processing within a strict processing deadline. The second optional sub-task downsamples the entire image and schedules it for inference within an overall processing deadline to not miss those objects that might not have been included in cropped regions processed by the mandatory sub-task. The optional sub-task also employs adaptive scaling to use higher image size for inference if there is no other mandatory part being computed. With such mixed-priority scheduling of cropped regions of the input frame, DNN-SAM does not violate the deadline constraints and achieves an improvement in the detection accuracy in the critical regions by 2.0 to 3.7× and a lower average inference latency by 4.8 to 9.7× over baseline approaches.

To extend to spatio-temporally correlated multi-camera deployments, Liu et al. suggest a real-time multi-camera scheduling framework where cameras with overlaps in their FoV collaborate with each other to share the workload of object detection and tracking [40]. In this paradigm, spatiotemporal correlations across the cameras inform an object-to-camera mapping for all unique objects visible across the multiple FoV, such that the total frame processing latency is minimized across all cameras. The framework ensures that objects appearing in the overlapping areas of different cameras are assigned to a single camera for detection and tracking, and redundant computations for the same object in multiple views is avoided thereby reducing frame processing latency without missing any objects. In addition, this object-to-camera mapping would need to be dynamic owing to the changes in the object's position relative to the assigned camera over time. The proposed approach includes a fine-grained "object aware, workload-aware, latency balancing approach" where the responsibility to detect and track an object over the shared region of coverage of the multiple cameras is dynamically redistributed among the collaborating cameras. This approach has multiple components working in tandem to achieve low processing latencies with high framerates and resolutions of the input video. The first design component deals with reducing the DNN processing latency by running a full-frame DNN inference only intermittently (at predefined intervals), and using flow-tracking algorithms to determine the approximate spatial position of the detected objects over time with the end goal of restricting the DNN inference only to an approximate region of the frame where the object is likely to be found. This method reduces redundant DNN computation on background regions and reduces the frame processing latency relative to the number of objects being tracked. Another way to reduce DNN processing latency is by batching frame partials or regions of the same size for parallel DNN inferencing to achieve lower processing latency. The second design component of this approach deals with dynamic latency aware partitioning and scheduling of the object detection and tracking workloads across cameras in a way that the overall frame processing latency across all cameras is minimized when executed in a batched fashion. The proposed scheduling algorithm is aware of both the available resources at the edge and relative object tracking workload, and redistributes the responsibility of tracking of objects in the shared overlap region from a a heavy-loaded camera to

8 Lightweight Collaborative Perception at the Edge   283

a light-loaded peer camera nodes. With these design components, the approach reports an improvement in frame processing latency of 2.45× to 6.85× with only minor degradation of inference accuracy.

Selective Activation of Static and Mobile Vision Sensors While most work has focused on collaborative execution of the DNN pipelines between one or more pervasive sensing nodes and a common edge device, researchers have also investigated mechanisms for collaboratively executing vision tasks across a group of mobile and static camera sensors. **Kestrel** [41] considers such collaborative execution for vehicle tracking, performed using a hybrid fixed/mobile camera network. Given the signiicant processing and bandwidth limitations on mobile devices, Kestrel dispenses with the common paradigm of collecting the corpus of images on a central cloud platform and subsequently executing the tracking query on the cloud. Instead, Kestrel requires the static set of cameras to upload their images to a common edge device, which then performs suitable detection and tracking to identify suitable segments of a target vehicle's trajectory. In addition, to further obtain the *missing* trajectory fragments, which may have been observed by one or more mobile sensors, the cloud also extracts suitable object feature descriptors and shares them with designated mobile nodes that *may* have observed the vehicle. Kestrel also optimizes the subsequent object detection pipeline, performed on board the mobile platforms, by (a) using the (feature, time) hints provided by the cloud to reduce the set of trajectories to be searched and (b) implementing a computationally efficient optical flow filter to further reduce the set of frames on which the object detector is executed.

2.2   Collaboration for 3D Sensing

The chapter has thus far discussed collaboration amongst pervasive and edge devices primarily to process 2-D video frames. We now discuss techniques that extends the notion of collaboration to make sense of the three-dimensional world, typically represented by point clouds; such collaboration is commonly associated with cooperation among devices such as connected autonomous vehicles (CAVs) and AR/VR platforms.

2.2.1   V2V Lidar 3D Point Cloud State Sharing

We describe techniques that leverage cooperation at the raw data level (i.e., early fusion), feature-level (i.e., intermediate representations) and high level (i.e., late fusion) [42].

Early Fusion Chen et al. [43] develop **Cooper**, one of the earliest systems leveraging cooperative perception for CAVs, designed to enhance vehicles' perception by both extending their FoV and ensuring higher object detection accuracy within their view. Cooper addresses a fundamental challenge in data sharing in V2V (vehicle-to-vehicle) networks; while running inferences over fused raw data has benefits over fusing end results from multiple vehicles, sharing such raw data imposes a significant bandwidth and latency bottleneck—for instance, self-driving cars are reported to generate over 1000 GB of point cloud data over the course of a day. To overcome this, Cooper (a) selectively shares raw data in compressed form that provides only positional coordinates and reflectance values per scan—this considerably reduces the transmission cost (e.g., to only ~200 kB per scan), and (b) runs a 3D Sparse Point-cloud object detection (SPOD) model, that is custom designed to work with *sparse* LIDAR data. In SPOD, during the preprocessing stage, the received *sparse* point clouds are first projected onto a sphere using the technique presented by Wu et al. [44] to produce a more dense, grid-based representation. This is owing to the fact that unlike RGB pixels found in images, LiDAR point clouds are typically sparse and irregular, and discretizing the entire 3D space into voxels leads to many empty voxels that are wasteful to compute. Then voxel-wise features are extracted by a feature encoding layer (similar to VoxelNet [45]) followed by sparse convolutional middle layers and then finally a region proposal network [46] constructed based on the single shot multibox detector [21]. In experiments conducted over the KITTI benchmark [47] as well as an in-house dataset, Cooper only requires an exchange of 0.5–4 Mb/s bandwidth which is shown to be feasible over the Dedicated Short Range Communications (DSRC) [48] channel common in V2V networks and incurs only a 5 ms overhead in processing time per frame.

Qiu et al. [49] propose **Augmented Vehicular Realiy (AVR)**, a capability that virtually extends the visual horizon of CAVs beyond line-of-sight through the exchange of raw point clouds, similar to Cooper. Specifically, AVR applies to risky road scenarios that typically involve a small number of CAVs—for instance, a car ("follower") following a slow moving truck ("leader") on single lane highway may decide to overtake the truck; however, due to limited visibility (occluded by the truck), the car is oblivious to the oncoming traffic from the other side. In such a scenario, the truck which does have full visibility can share its sensed data with the car trailing behind such that the car's path planner avoids risky maneuvers. In realizing this vision, AVR overcomes several technical challenges. AVR devises a novel relative positioning technique to suitably transform the 3D point clouds between the sending and receiving vehicles by adapting a feature-based SLAM technique [50]. As transmitting full 3D point clouds per scan be infeasible in bandwidth-constrained environments, AVR processes the 3D point clouds in order to isolate the dynamic objects in the surroundings using a homography-based transformation and shares point cloud data pertaining to only those objects, and applies lossless compression algorithms to further downsize the payload. AVR also actively eliminates redundant data transmission, for example, by avoiding repeated sharing of dynamic detected objects that lie within the FoV of more than one vehicle. AVR also implements a range of optimizations such as carefully planned pipelining to improve the overall frame rate. Extensive trace-based simulations demonstrate that AVR is optimized to process frame at a rate of 30 frames per second with each frame transmitted with a nominal latency of only 150–200 ms using 60 GHz radio.

While the cooperation benefits of AVR are convincing, its use is limited to only two vehicles at a time. **EMP [51]** is an edge-assisted platform which relies on vehicle-to-infrastructure (V2I) communication to enable cooperation and extends support to up to 6 vehicles. Similar to Cooper and AVR, in EMP, proximal CAVs share their raw sensed data, but with an edge server which then merges individual perspectives into high-resolution, high-fidelity views. EMP enables reliable and scalable data sharing over highly fluctuating wireless links by implementing a series of novel algorithms. In EMP, CAVs create a *disjoint spatial partition* of the environment where each vehicle only shares data pertaining to its immediate proximity. This ensures that the shared data is of the highest quality (as quality degrades with increasing distance) and that it also avoid redundant data being shared (e.g., overlapping regions of more than one vehicle) and hence ensures that no bandwidth is wasted. EMP creates such partitions using Voronoi diagrams with the vertices representing the vehicles and the area they belong to represents the extent of the data that would be shared by each of the vehicles. Additionally, EMP adjusts the partition boundaries at real-time depending on the network bandwidth available to the individual vehicles, effectively, allowing vehicles with slower links to *redistribute* some of their sensing load to neighboring vehicles with faster links. Evaluations show that EMP's cooperative perception can detect blind spots faster (i.e., by 0.5–1.1 s) compared to a non-cooperative baseline. EMP achieves processing rates as high as 24 FPS while reducing the end-to-end latency by 49–65% compared to a purely peer-to-peer approach over V2V.

While EMP extends support to up to 6 vehicles, its reliance on infrastructure is problematic. Qiu et al. [52] propose **Autocast** that scales up to 40 vehicles without requiring high bandwidth, infrastructural support, for cooperative perception in scenarios such as sensemaking at an intersection. This poses two central challenges: (a) enabling sharing of data from a large number of vehicles over a shared, narrower V2V bandwidth, and (b) avoiding packet collisions by intelligent coordination and scheduling. Instead of naively tagging all *dynamic objects* for sharing, as in AVR, AutoCast implements a suite of *spatial reasoning* algorithms to determine the *visibility* and *relevancy* to evaluate the quality of the data being sent. For instance, if an object is already visible in the receiver's view, the sender can discard those objects saving on bandwidth. Similarly, occluded objects from farther away that are not relevant for near term path planning need not be sent. Furthermore, AutoCast implements an efficient scheduling algorithm, based on the application of a Markov decision process (MDP) framework, that prioritizes the most relevant information. Experiments conducted on a CARLA-based simulation environment show that AutoCast scales gracefully up to 40 vehicles whilst raising the visibility of safety critical objects such as *humans* by 2–4× of the times. AutoCast also avoids *all* collisions that are simply inevitable with other cooperative and non-cooperative baselines. Network benchmarking experiments validate that AutoCast meets all its scheduling requirements within the bounds of DSRC's channel capacity while reaching a frame rate of ≈ 10 FPS.

Intermediate State Sharing Extending their earlier work, Chen et al. design **F-Cooper** [53] which selectively shares point cloud as feature map representations that are generated during the 3D object detection pipeline. The F-Cooper framework supports two fusion schemes: (a) voxel feature fusion (VFF, for superior accuracy) and (b) spatial feature fusion (SFF, that supports dynamic resizing of the feature maps to be transmitted to other vehicles). In both settings, the CAVs process their object detection pipeline until after the voxel encoding layer (see Cooper [43]) and share *selected* feature maps with cooperating CAVs. In the VFF scheme, the voxel feature maps are directly *fused* where as in the SFF scheme, the individual maps are further processed through sparse convolutional layers to generate spatial maps which are then eventually *fused*. The remainder of the pipeline follows the baseline system, Cooper, on which F-Cooper is built. Due to the parallel execution of N sparse CNN layers ($N - 1$ cooperating CAVs and the receiving CAV), the SFF scheme is computationally more expensive. In both cases, F-Cooper uses the *maxout strategy* [54] for fusion which highlights important features and filters out others. As a straightforward floating point operator, the maxout operation incurs minimal processing overhead. Experimental results show that by fusing features, F-Cooper achieves improved object detection ($\approx 10\%$ increase for detection within 20 meters and 30% increase for distances beyond 20 meters), over non-collaborative baselines, and achieves performance comparable to Cooper [43] (which transmits raw data), but at a fraction of the bandwidth required (< 1 MB/s for most cases). F-Cooper incurs only a minimal latency—for e.g., 71 ms in certain feature selections. Furthermore, results show that the data transmitted for feature fusion is only a fraction of the size of the original data (from 10,000 points in a typical LIDAR scan amounting to 4 MB down to only 200 Kb of processed feature map data).

2.2.2 Physical Navigation in Virtual Reality

Commercial applications of AR/VR games already see multiple users in the same game and scene interacting with each other and the game. 3D sensing pipelines are often used in such games to support location-awareness and navigation, across a mix of real and virtual worlds. One of the areas of research in the AR/VR space addresses the challenges related to real walking within a bounded physical physical space while experiencing walking in an unbounded virtual space. Real walking encourages extremely high levels of immersion in the AR/VR world but is impractical to execute as this requires very large areas of physical space. Simulated walking such as walking on treadmills, teleportation, and even walking in place have been proposed as solutions to this problem but they all sacrifice user experience to achieve some form of continuous walking. Some works present redirected walking tries to transform long walking paths into smaller folded segments, but they still require large physical spaces.

To this end, Marwecki et al. present **VirtualSpace** [55], which allows for the overloading of multiple walking users donning mobile AR headsets within the same restricted physical space such that they don't feel confined within the physical space.

A key point to note in this work is that all users are independent of each other and immersed in their own distinct games/worlds, completely unaware of others who might be sharing the same physical space. The game/world is also mapped to the entirety of the physical space, giving the user the experience and illusion of being the sole inhabitor of the physical space. VirtualSpace divides the available physical area into a subset of regions or tiles and the edge coordinator ensures that a user is contained within a subset of tiles at all points in time, while manoeuvring and shuffling all users across the entire physical space using different sets of tiles and ensuring that no collisions occur. Using a variety of maneuvers such as rotation, switching, focus, VirtualSpace can effect a change in the tile that a particular user is on by triggering risks and rewards within the game or world which motivate the user to move across the physical space to control the narrative of their game or world. For example, a user playing PacMan might encounter a cherry (reward) or a ghost (risk) which will motivate the user to physically move in a different direction effectively placing them on a different VirtualSpace tile. Of course, this strategy relies heavily on the assumption that VirtualSpace correctly estimates the cognitive load and reaction time of the user to the app-presented risks and rewards for VirtualSpace to effect the users' positional manoeuvres in physical space. VirtualSpace manages the manoeuvres across multiple users by attaching value in the form of credits or money to each manoeuvre request, thereby creating a manoeuvre economy where some selfish manoeuvre might be allowed to occur if they can operate within the economic model. VirtualSpace is shown to be very effective in managing 4 users wearing GearVRs with Samsung S6 mobile phones playing 4 different immersive Unity3D games—Badminton, PanMac, Space Invaders, and Whack-a-Mole in a restricted 4×4 metres space.

2.2.3 Localisation and Wayfinding in Robotics and Autonomous Vehicles (AV)

Robotic devices and AVs typically utilise RGB camera and LiDAR sensors to perform localisation of objects in their vicinity and pathfinding for navigation in physical spaces, with the LiDAR sensor focused on motion estimation and the RGB camera sensor focused on feature extraction. However, LiDAR-RGB sensor fusion is challenging on many levels due to inherent dissimilarities in the sensor modalities, spatial and temporal resolutions, and geometric properties and alignment. Recent works in Visual-LiDAR Simultaneous Localisation and Mapping (Visual-LiDAR SLAM) leverage the joint optimisation of both sensor streams to address shortcomings such as lack of depth awareness, range, occlusion, and noise in the individual sensor streams [56, 57]. Graeter et al. introduced **LIMO** [58] - a framework that jointly leverages depth estimation from the LiDAR sensor data and the object feature tracking capabilities from the RGB camera stream to enable more nuanced and accurate visual odometry in autonomous cars. LIMO leverages a depth extraction algorithm that matches depth data from the LiDAR sensor to keypoints and features detected in an RGB camera frame to enable motion estimation of the

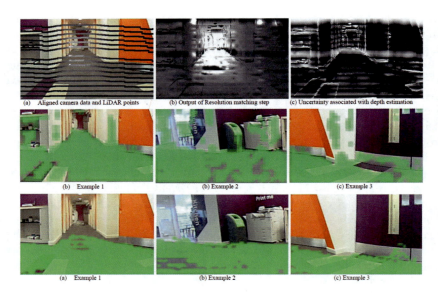

Fig. 8.9 Row 1: Illustration of the outputs of the sensor fusion. **Row 2:** Visualising the free space detection results from a baseline algorithm (SVM: HoG features+histograms). **Row 3:** Visualisation of the free space detection results from the proposed framework [59]

surrounding objects. De Silva et al. explore LiDAR-monocular camera fusion [59] in the context of driverless vehicles and address the spatial resolution misalignment in LiDAR and monocular camera sensor data. The proposed methodology describes a geometric model that spatially aligns data from the two sensor streams. As depth information may not be available for every pixel in the camera frame, the methodology also proposes a Gaussian Process (GP) regression to interpolate the missing depth information for individual pixels. The authors assert that such sensor fusion is beneficial to the overall perception pipeline as evidenced in the 62%+ improvement from the baseline in a free space detection application calculated as the perceptage of pixels wrongly classified as free space across 3 separate scenarios illustrated in Fig. 8.9.

Most existing work utilize the collaboration between distinct visual sensing modes (e.g., RGB & LIDAR) to achieve the necessary navigation and movement decisions, while some threads of research, such as the one explored by Fankhauser et al., focus on the collaboration between robots—i.e collaboration between aerial and ground robots to perform navigation and pathfinding on an unknown terrain for the ground robot with aerial assistance [60]. In the context of a 3D space, recent works focus on the fusion of spatial information from cameras and LIDARs with acoustic information from microphone arrays on the robot to achieve fine-grained audio-visual localisation and pathfinding [61].

In another vein, Chen and Jain et al. posit that adding acoustic understanding to embodied agents or robots can enable new possibilities for faster and easier localization and navigation [62], by exploiting the distinct acoustic signatures associated with various object interactions and activities. In their work **SoundSpaces**, they fuse vision and sound so that robots and robot swarms can learn the intrinsic geometries of the physical space in a method similar to sonar, and can also detect and follow objects or targets that make sound. Leveraging acoustic localisation as part of the sensing and pathfinding pipeline also expands the sensing range to objects that are audible but outside of the visual range, and provides sensing capabilities in scenarios where visual sensors fall short. Some of the challenges include (a) generating distinct and useful aural observations with respect to both the robot's current position and orientation, and the physical 3D environment it is participating in, (b) choosing movements in an unmapped unknown 3D environment to bring the robot to its target efficiently while also learning relevant details about the environment, and (c) modeling a robot's behaviour in environments with unseen scenes and unheard sounds. SoundSpaces incorporates aural embeddings into the reinforcement learning pipeline and reports encouraging results for acoustic understanding of a 3D space to aid in visual understanding.

3 Cross-Model Collaborative Execution

Lightweight collaboration at the edge can come in different forms: (a) collaboration between sensors at the edge, as we have covered in Sect. 2, (b) collaboration between edge nodes in terms of workload sharing/stealing, and (c) collaboration between multiple resource-intensive tasks that are contending for the same resource pool available at the edge. Most work in edge AI typically does not consider the scenario of multiple *concurrent* contending workloads on edge GPUs, and instead focus on optimizing system metrics (such as inference accuracy, processing latency and throughput) for individual tasks. Yi et al. [63] claim that such multiple concurrent contending workloads on a single DNN can increase the processing latency $19.7\times$ (from 59.93 ms to 1181 ms) and degrade processing throughput from 30 fps to 12 fps, primarily because as the overloaded GPU fails to process multiple tasks with the same efficiency due to high switching overheads.

In the **Heimdall** framework [63], Yi et al. focus solely on concurrent augmented reality or virtual reality (AR/VR) application apps running on a mobile device. However, the broad principles and constraints are valid in other edge environments as well. A typical AR/VR app applies multiple constraints and conditions for seamless real-time immersive user-experience. First, to decide which visual components need to be displayed to the user and where within their FoV, the app needs to analyse and synthesize physical environment and relative user behaviours, using data from multiple sensors that are processed by multiple distinct task-specific DNNs (e.g. gesture recognition, object recognition). Second, the app must seamlessly render a variety of virtual objects, gestures, and/or motions in the user's field of view for a real-time immersive experience; such rendering itself requires use of GPU resources. Third, various tasks, such as sensing, synthesizing and foreground rendering, must all be completed using the limited resources available to the app. They also observe that desktop techniques such as hyper-threading and parallelization cannot be applied to a mobile or edge context due to the limited architectural support for parallel execution. Also, techniques such as preemption would overwhelm the available memory with large context switch costs owing to large state sizes. They propose a GPU coordination framework which pseudo-preemptively breaks down all the concurrent DNNs into smaller units and schedules them for processing on the GPU. Using a time-sharing approach, Heimdall only allows the switching of contexts whenever a single semantic unit of the GPU task is complete, thus enabling time-sharing of the GPU by multiple perception and rendering tasks with minimal scheduling and memory overhead. Additionally, with processing deadlines incorporated into the scheduling mechanism, Heimdall is able to meet the strict latency requirements imposed by different components of the AR app. Heimdall reports that with their pseudo-preemption technique, processing throughput is improved from 12 fps to 30 fps, while the inference latency is reduced by 15× in the worst case (when compared to a baseline multi-threading approach).

While scheduling innovations promote run-time sharing and resource sharing among multiple DNN pipelines on the same edge device, other strategies have focused on suitable modification of DNN models (at training time) to enable more efficient concurrent DNN execution on such edge devices. Broadly speaking, the approaches utilize the concepts of *model merging* and *weights sharing* to condense multiple DNN pipelines for disparate tasks into a single pipeline that ideally preserves inference accuracy while avoiding an increase in processing latency. Chou et al. determine that resource sharing among multiple models is possible by combining multiple well-trained DNNs with known weights trained for different individual tasks and unifying them into a single but compact model [64]. The key challenge comes from merging distinct disparate DNN networks which may not have similar architectures, may not share the same input or features, and may even have multiple inputs. They propose a **NeuralMerger** framework that consists of two stages. In the first stage, visually illustrated in Fig. 8.10a, the NeuralManager will align the architectures of all the DNN pipelines and encode the weights such that they are shared among the DNN models. The purpose is to unify the weights

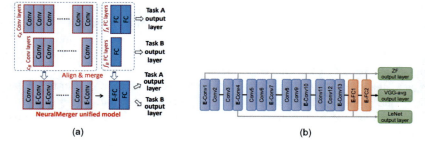

Fig. 8.10 (a) NeuralMerger framework [64] (b) Resulting Model when merging ZF, VGG-avg, and LeNet into a single model for inference

so that the filters and weights of different neural networks can be co-used, therefore reducing computation needs. The second stage is to fine-tune the unified model with partial or all training data to calibrate the unified model to achieve reasonable accuracy across all the tasks. NeuralMerger shows (see Fig. 8.10b) that this approach can be used to merge three distinct pipelines—VGG-avg for clothing classification, ZF-net for gender classification, and LeNet for sound classification—into a single unified model, with <1% accuracy loss across all three tasks.

4 Conclusion

Many AIoT deployments, especially for a variety of perception tasks, are characterized by the use of (a) multiple sensor nodes, often mobile, interacting with one or more static, better-resourced edge devices, and (b) the execution of computationally-heavy DNN pipelines that need to be optimized for such edge and pervasive devices. This chapter has introduced the concept of *collaborative edge perception*, which seeks to exploit two properties of such deployments. First, the same set of objects and scenes are observed either concurrently, or within a predictable time interval, from different perspectives by multiple such sensor devices. Second, the sensor nodes often contain multiple different sensor modalities (e.g., RGB, LIDAR, audio) that enable the same physical phenomena to be captured at different levels of sensitivity and resolution. Collaborative edge perception seeks to optimize the execution of the DNN pipelines to take advantage of the redundancy and correlation arising from such multiple sensing modalities and the spatitemporal correlation across multiple sensor nodes. We have described approaches that involve either state sharing (of various stages from the underlying DNN pipelines), modified attention or dynamic scheduling mechanisms, especially for perception tasks utilizing RGB sensor data. We have also outlined various approaches that apply similar principles to 3D point cloud data, often generated by LIDAR sensors, to not only improve the accuracy of perception tasks but also aggressively reduce the high volumes of 3D point cloud data that is generated during such sensing. Besides such collaborative execution of the *same task* across nodes, we have also described approaches that apply such principles of collaboration, both at runtime and at model development/training time, to tackle concurrent execution of *multiple tasks* on edge devices.

In spite of the significant technical advances, many open problems remain to be tackled in future work. Additional approaches, currently under active investigation, to utilize the shared, combined sensing capabilities of multiple vision sensors include the possibility of combining the mechanisms of mixed-resolution imaging (as suggested by MRIM) and intermediate state sharing (as suggested by COMAI). In addition, the processing of multiple image/LIDAR sensing streams on a single edge device can be significantly optimized by combining spatial attention mechanisms (which effectively restrict the DNN computation to only selective portions of an individual image/LIDAR frame) with temporal scheduling techniques (which effectively apply the DNN computation selectively to dynamically chosen subsets of consecutive image/LIDAR frames). There is also incipient interest in harnessing the power of event-based vision sensing (EVS) devices (such as Prophesee's Metavision™sensor) for energy-efficient collaborative inferencing. The core idea is to exploit the complementary strengths and weaknesses of event-based vs. frame-based vision sensors: EVS streams support hyper-fast tracking of dynamic events (over 10,000 'frames' per sec) with low power but do not provide color or texture cues, whereas frame-based sensor data operates continuously with lower sampling frequency (typically 100–200 fps) and independently of event dynamics (thus incurring much higher average power) but provides a richer set of visual features.

For AR/VR tasks, the current DNN pipelines that process LIDAR data can also be arguably optimized, using the core principle of selective attention, to use cues provided by additional modalities of human instructional input—e.g., using verbal cues about objects or regions of special interest to a human participant to reduce both the volume of 3D point cloud data that must be transmitted and the portions of a sequence of 3D point cloud streams that need to be processed at higher resolution. More broadly, many of the existing techniques and approaches are focused on short-term optimization of the task inference pipelines and may not be suitable for future energy-harvesting based sensor platforms, where the available energy is significantly more constrained and may be subject to environment-dependent uncertainty and fluctuations. For such environments, the collaborative execution of perception tasks need to consider not just the relationship between *instantaneous* task accuracy and system overheads (such as bandwidth, energy and latency), but also the overall task accuracy that may be achieved over a finite future horizon.

References

1. M. Satyanarayanan, V. Bahl R. Caceres, Davies, The case for VM-based cloudlets in mobile computing. IEEE Pervasive Comput. (2009)
2. S. Noghabi L. Cox S. Agarwal G. Ananthanarayanan, The emerging landscape of edge computing. GetMobile Mob. Comput. Commun. **23**(4) (2020). The Case for VM-based Cloudlets in Mobile Computing. IEEE Pervasive Computing, November, 2009
3. Qualcomm Developer Network, Vision AI Development Kit (2022) [online]. Available at: <https://developer.qualcomm.com/hardware/vision-ai-development-kit [Accessed 24 April 2022]
4. Samsung, Samsung S22 I S22+ 5G (2022) [online]. Available at: <https://www.samsung.com/sg/smartphones/galaxy-s22/ [Accessed 24 April 2022]
5. AWS, AWS DeepLens (2022) [online]. Available at: <https://aws.amazon.com/deeplens/ [Accessed 24 April 2022]
6. Y. Kang, J. Hauswald, C. Gao, A. Rovinski, T. Mudge, J. Mars, L. Tang, Neurosurgeon: collaborative intelligence between the cloud and mobile edge. ACM SIGARCH Comput. Archit. News **45**(1), 615–629 (2017)
7. M.-R. Ra, A. Sheth, L. Mummert, P. Pillai, D. Wetherall, R. Govindan, Odessa: enabling interactive perception applications on mobile devices, in *Proceedings of the 9th International Conference on Mobile Systems, Applications, and Services* (ACM, New York, 2011), pp. 43–56

8. S. Han, H. Shen, M. Philipose, S. Agarwal, A. Wolman, A. Krishnamurthy, MCDNN: an approximation-based execution framework for deep stream processing under resource constraints, in *Proceedings of the 14th Annual International Conference on Mobile Systems, Applications, and Services (MobiSys'16)* (2016), pp. 123–136

9. M. Xu, M. Zhu, Y. Liu, F.X. Lin, X. Liu, Deepcache: principled cache for mobile deep vision, in *Proceedings of the 24th Annual International Conference on Mobile Computing and Networking* (2018), pp. 129–144

10. Y. Li, A. Padmanabhan, P. Zhao, Y. Wang, G.H. Xu, R. Netravali, Reducto: on-camera filtering for resource-efficient real-time video analytics, in *Proceedings of the Annual Conference of the ACM Special Interest Group on Data Communication on the Applications, Technologies, Architectures, and Protocols for Computer Communication* (2020), pp. 359–376

11. T.Y.-H. Chen, L. Ravindranath, S. Deng, P. Bahl, H. Balakrishnan, Glimpse: continuous, real-time object recognition on mobile devices, in *Proceedings of the 13th ACM Conference on Embedded Networked Sensor Systems (SenSys'15)* (2015), pp. 155–168

12. A. Mathur, N.D. Lane, S. Bhattacharya, A. Boran, C. Forlivesi, F. Kawsar, DeepEye: resource efficient local execution of multiple deep vision models using wearable commodity hardware, in *Proceedings of the 15th Annual International Conference on Mobile Systems, Applications, and Services (MobiSys'17)* (2017), pp. 68–81

13. L.N. Huynh, Y. Lee, R.K. Balan, Deepmon: mobile gpu-based deep learning framework for continuous vision applications, in *Proceedings of the 15th Annual International Conference on Mobile Systems, Applications, and Services*(2017), pp. 82–95

14. G. Ananthanarayanan, V. Bahl, L. Cox, A. Crown, S. Nogbahi, S., Y. Shu, Video analytics-killer app for edge computing, in *Proceedings of the 17th Annual International Conference on Mobile Systems, Applications, and Services* (2019), pp. 695–696

15. S. Jain, G. Ananthanarayanan, J. Jiang, Y. Shu, J. Gonzalez, Scaling video analytics systems to large camera deployments, in *Proceedings of the 20th International Workshop on Mobile Computing Systems and Applications* (2019), pp. 9–14

16. D. Weerakoon, K. Jayarajah, R. Tandriansyah, A. Misra, Resilient collaborative intelligence for adversarial IoT environments, in *2019 22th International Conference on Information Fusion (FUSION)* (IEEE, 2019), pp. 1–8

17. P. Baqué, F. Fleuret, P. Fua, Deep occlusion reasoning for multi-camera multi-target detection, in *Proceedings of the IEEE International Conference on Computer Vision* (2017), pp. 271–279

18. T. Chavdarova, F. Fleuret, Deep multi-camera people detection, in *2017 16th IEEE international conference on machine learning and applications (ICMLA)* (IEEE, 2017), pp. 848–853

19. Y. Hou, L. Zheng, S. Gould, Multiview detection with feature perspective transformation, in *European Conference on Computer Vision* (Springer, Cham, 2020), pp. 1–18

20. K. Jayarajah, D. Wanniarachchige, T. Abdelzaher, A. Misra, Comai: enabling lightweight, collaborative intelligence by retrofitting vision dnns, in *IEEE INFOCOM 2022-IEEE Conference on Computer Communications* (IEEE, 2022), pp. 41–50

21. W. Liu, D. Anguelov, D. Erhan, C. Szegedy, S. Reed, C.Y. Fu, A.C. Berg, Ssd: single shot multibox detector, in *European Conference on Computer Vision* (Springer, Cham, 2016), pp. 21–37

22. J.R. Uijlings, K.E. Van De Sande, T. Gevers, A.W. Smeulders, Selective search for object recognition. Int. J. Comput. Vision **104**(2), 154–171 (2013)

23. J. Ferryman, A. Shahrokni, Pets2009: dataset and challenge, in *2009 Twelfth IEEE International Workshop on Performance Evaluation of Tracking and Surveillance* (IEEE, 2009), pp. 1–6

24. T. Chavdarova, P. Baqué, S. Bouquet, A. Maksai, C. Jose, T. Bagautdinov, L. Lettry, P. Fua, L. Van Gool, F. Fleuret, Wildtrack: a multi-camera hd dataset for dense unscripted pedestrian detection, in *Proceedings of the IEEE Conference on Computer Vision and Pattern Recognition* (2018), pp. 5030–5039

25. C. Canel, T. Kim, G. Zhou, C. Li, H. Lim, D.G. Andersen, M. Kaminsky, S.R. Dulloor, Scaling video analytics on constrained edge nodes, in *2nd SysML Conference* (2019)
26. K. Hsieh, G. Ananthanarayanan, P. Bodik, S. Venkataraman, P. Bahl, M. Philipose, P.B. Gibbons, O. Mutlu, Focus: querying large video datasets with low latency and low cost, in *13th USENIX Symposium on Operating Systems Design and Implementation (OSDI 18)* (USENIX Association, Carlsbad, 2018), pp. 269–286. https://www.usenix.org/conference/osdi18/presentation/hsieh
27. D. Kang, J. Emmons, F. Abuzaid, P. Bailis, M. Zaharia, NoScope: optimizing neural network queries over video at scale, in *VLDB* (2017)
28. S. Liu, S. Yao, X. Fu, R. Tabish, S. Yu, A. Bansal, H. Yun, S. Lui, T. Abdelzaher, On removing algorithmic priority inversion from mission-critical machine inference pipelines, in *2020 IEEE Real-Time Systems Symposium (RTSS)* (IEEE, 2020), pp. 319–332
29. W. Kang, S. Chung, J.Y. Kim, Y. Lee, K. Lee, J. Lee, K. G. Shin, H.S. Chwa, DNN-SAM: split-and-Merge DNN execution for real-time object detection, in *2022 IEEE 28th Real-Time and Embedded Technology and Applications Symposium (RTAS)* (IEEE, 2022), pp. 160–172
30. S. Jain, X. Zhang, Y. Zhou, G. Ananthanarayanan, J. Jiang, Y. Shu, P. Bahl, J. Gonzalez, Spatula: efficient cross-camera video analytics on large camera networks, in *2020 IEEE/ACM Symposium on Edge Computing (SEC)* (IEEE, 2020), pp. 110–124
31. E. Ristani, F. Solera, R. Zou, R. Cucchiara, C. Tomasi, Performance measures and a data set for multi-target, multi-camera tracking, in *ECCV Workshops* (2016)
32. H. Guo, S. Yao, Z. Yang, Q. Zhou, K. Nahrstedt, CrossRoI: cross-camera region of interest optimization for efficient real time video analytics at scale, in *Proceedings of the 12th ACM Multimedia Systems Conference* (2021), pp. 186–199
33. J. Jiang, G. Ananthanarayanan, P. Bodik, S. Sen, I. Stoica, Chameleon: scalable adaptation of video analytics, in *Proceedings of the 2018 Conference of the ACM Special Interest Group on Data Communication* (2018), pp. 253–266
34. S. Jiang, Z. Lin, Y. Li, Y. Shu, Y. Liu, Flexible high-resolution object detection on edge devices with tunable latency, in *Proceedings of the 27th Annual International Conference on Mobile Computing and Networking* (2021), pp. 559–572
35. J. Y. Wu, V. Subhasharan, T. Tran, A. Misra, MRIM: enabline mixed-resolution imaging for low-power prevasive vision tasks, in *2022 IEEE International Conference on Pervasive Computing and Communications* (PerCom) (pp. 44–53). IEEE.
36. Y.J. Li, X. Weng, Y. Xu, K.M. Kitani, Visio-temporal attention for multi-camera multi-target association, in *Proceedings of the IEEE/CVF International Conference on Computer Vision* (2021), pp. 9834–9844
37. C.T. Liu, C.W. Wu, Y.C.F. Wang, S.Y. Chien, Spatially and temporally efficient non-local attention network for video-based person re-identification. Preprint (2019), arXiv:1908.01683
38. A. Porrello, L. Bergamini, S. Calderara, Robust re-identification by multiple views knowledge distillation, in *European Conference on Computer Vision* (Springer, Cham, 2020), pp. 93–110
39. X. Gu, H. Chang, B. Ma, H. Zhang, X. Chen, Appearance-preserving 3d convolution for video-based person re-identification, in *European Conference on Computer Vision* (Springer, Cham, 2020), pp. 228–243
40. S. Liu, T. Wang, H. Gui, X. Fu, P. David, M. Wigness, A. Misra, T. Abdelzaher, Multi-view scheduling of onboard live video analytics to minimize frame processing latency, in *Proceedings of the 2022 IEEE 42nd International Conference on Distributed Computing Systems* (2022), (pp. 503–514).
41. Q. Hang, X. Liu, S. Rallapalli, A.J. Bency, R. Urgaonkar B.S. Manjunath, R. Govindan, KESTREL: video Analytics for Augmented multi-camera vehicle tracking, in *Proceedings of ACM/IEEE International Conference on Internet-of-Things Design and Implementation (IoTDI)* (2018)
42. J. Shi, W. Wang, X. Wang, H. Sun, X. Lan, J. Xin, N. Zheng, Leveraging spatio-temporal evidence and independent vision channel to improve multi-sensor fusion for vehicle environmental perception, in *2018 IEEE Intelligent Vehicles Symposium (IV)* (IEEE, 2018), pp. 591–596

43. Q. Chen, S. Tang, Q. Yang, S. Fu, Cooper: cooperative perception for connected autonomous vehicles based on 3d point clouds, in *2019 IEEE 39th International Conference on Distributed Computing Systems (ICDCS)* (IEEE, 2019), pp. 514–524
44. B. Wu, A. Wan, X. Yue, K. Keutzer, Squeezeseg: convolutional neural nets with recurrent crf for real-time road-object segmentation from 3d lidar point cloud, in *2018 IEEE International Conference on Robotics and Automation (ICRA)* (IEEE, 2018), pp. 1887–1893
45. Y. Zhou, O. Tuzel, Voxelnet: end-to-end learning for point cloud based 3d object detection, in *Proceedings of the IEEE Conference on Computer Vision and Pattern Recognition* (2018), pp. 4490–4499
46. S. Ren, K. He, R. Girshick, J. Sun, Faster r-cnn: towards real-time object detection with region proposal networks. Adv. Neural Inf. Process. Syst. **28** (2015)
47. A. Geiger, P. Lenz, C. Stiller, R. Urtasun, Vision meets robotics: the kitti dataset. Int. J. Robot. Res. **32**(11), 1231–1237 (2013)
48. J.B. Kenney, Dedicated short-range communications (DSRC) standards in the United States. Proc. IEEE **99**(7), 1162–1182 (2011)
49. H. Qiu, F. Ahmad, F. Bai, M. Gruteser, R. Govindan, Avr: augmented vehicular reality, in *Proceedings of the 16th Annual International Conference on Mobile Systems, Applications, and Services* (2018), pp. 81–95
50. R. Mur-Artal, J.M.M. Montiel, J.D. Tardos, ORB-SLAM: a versatile and accurate monocular SLAM system. IEEE Trans. Robot. **31**(5), 1147–1163 (2015)
51. X. Zhang, A. Zhang, J. Sun, X. Zhu, Y.E. Guo, F. Qian, Z.M. Mao, Emp: edge-assisted multi-vehicle perception, in *Proceedings of the 27th Annual International Conference on Mobile Computing and Networking* (2021), pp. 545–558
52. H. Qiu, P. Huang, N. Asavisanu, X. Liu, K. Psounis, R. Govindan, Autocast: scalable infrastructure-less cooperative perception for distributed collaborative driving. Preprint (2021). arXiv:2112.14947
53. Q. Chen, X. Ma, S. Tang, J. Guo, Q. Yang, S. Fu, F-cooper: feature based cooperative perception for autonomous vehicle edge computing system using 3D point clouds, in *Proceedings of the 4th ACM/IEEE Symposium on Edge Computing* (2019), pp. 88–100
54. I. Goodfellow, D. Warde-Farley, M. Mirza, A. Courville, Y. Bengio, Maxout networks, in *International Conference on Machine Learning*. PMLR (2013), pp. 1319–1327
55. S. Marwecki, M. Brehm, L. Wagner, L.P. Cheng, F.F. Mueller, P. Baudisch, Virtualspace-overloading physical space with multiple virtual reality users, in *Proceedings of the 2018 CHI Conference on Human Factors in Computing Systems* (2018), pp. 1–10
56. X. Liang, H. Chen, Y. Li, Y. Liu, Visual laser-SLAM in large-scale indoor environments, in *2016 IEEE International Conference on Robotics and Biomimetics (ROBIO)* (IEEE, 2016), pp. 19–24
57. Z. Zhu, S. Yang, H. Dai, F. Li, Loop detection and correction of 3d laser-based slam with visual information, in *Proceedings of the 31st International Conference on Computer Animation and Social Agents* (2018), pp. 53–58
58. J. Graeter, A. Wilczynski, M. Lauer, Limo: Lidar-monocular visual odometry, in *2018 IEEE/RSJ International Conference on Intelligent Robots and Systems (IROS)* (IEEE, 2018), pp. 7872–7879
59. V. De Silva, J. Roche, A. Kondoz, Fusion of LiDAR and camera sensor data for environment sensing in driverless vehicles. 2018. Preprint (2018), arXiv:1710.06230
60. P. Fankhauser, M. Bloesch, P. Krüsi, R. Diethelm, M. Wermelinger, T. Schneider, M. Dymczyk, M. Hutter, R. Siegwart, Collaborative navigation for flying and walking robots, in *2016 IEEE/RSJ International Conference on Intelligent Robots and Systems (IROS)* (IEEE, 2016), pp. 2859–2866
61. C. Gan, Y. Zhang, J. Wu, B. Gong, J.B. Tenenbaum, Look, listen, and act: Towards audio-visual embodied navigation, in *2020 IEEE International Conference on Robotics and Automation (ICRA)* (IEEE, 2020), pp. 9701–9707
62. C. Chen, U. Jain, C. Schissler, S.V.A. Gari, Z. Al-Halah, V.K. Ithapu, P. Robinson, K. Grauman, Soundspaces: audio-visual navigation in 3d environments, in *European Conference on Computer Vision* (Springer, Cham, 2020), pp. 17–36

63. J. Yi, Y. Lee, Heimdall: mobile gpu coordination platform for augmented reality applications, in *Proceedings of the 26th Annual International Conference on Mobile Computing and Networking* (2020), pp. 1–14
64. Y.M. Chou, Y.M. Chan, J.H. Lee, C.Y. Chiu, C.S. Chen, Unifying and merging well-trained deep neural networks for inference stage. Preprint (2018). arXiv:1805.04980

Chapter 9
Dynamic Placement of Services at the Edge

Shiqiang Wang and Ting He

Abstract To realize the promise of low-latency service access in edge computing, the services need to be placed in close proximity to users. As a user moves, its service may need to be migrated to a new location. In this chapter, we first formulate this migration decision-making problem as a Markov decision process (MDP). Then, by analyzing the characteristics of this MDP, we provide efficient ways of obtaining the near-optimal policy for service migration. Finally, we discuss some practical aspects and present results from trace-driven simulation.

1 Introduction

Many artificial intelligence (AI) driven applications involve understanding and interacting with the physical world. In a typical application scenario, client devices collect data from the physical environment, where the devices can include sensors, smartphones, on-board computers in vehicles, etc. Then, the data are processed by a model that has been trained using machine learning techniques. This model analyzes the input data and makes predictions as required by the application. For example, it may detect objects and their classes in an image, translate natural language into a machine-understandable form to facilitate automated decision making, and predict

---

©2019 IEEE. Reprinted, with permission, from S. Wang, R. Urgaonkar, M. Zafer, T. He, K. Chan and K. K. Leung, "Dynamic Service Migration in Mobile Edge Computing Based on Markov Decision Process," in IEEE/ACM Transactions on Networking, vol. 27, no. 3, pp. 1272–1288, June 2019, doi: https://doi.org/10.1109/TNET.2019.2916577.

S. Wang (✉)
IBM T. J. Watson Research Center, Yorktown Heights, NY, USA
e-mail: wangshiq@us.ibm.com

T. He
Pennsylvania State University, University Park, PA, USA
e-mail: tinghe@psu.edu

© The Author(s), under exclusive license to Springer Nature Switzerland AG 2023
M. Srivatsa et al. (eds.), *Artificial Intelligence for Edge Computing*,
https://doi.org/10.1007/978-3-031-40787-1_9

events that may happen in the near future. The output of these models can be used by a wide range of emerging applications, such as augmented/virtual reality and autonomous driving [1].

A challenge in this setup is that the client devices may collect a lot of data over time, but they usually have limited computational capability to process the data. This is particularly the case when the system involves AI and machine learning models, which often demand a high amount of computational resources and/or specialized hardware (e.g., GPU). An option to overcome this challenge is to leverage cloud resources, so that all the local data are sent to the cloud for further processing [2]. However, this approach incurs a substantial amount of bandwidth usage in the backbone Internet, and more importantly, querying the cloud may incur high latency that cannot be tolerated by many latency-sensitive applications.

To achieve low-latency access to a sufficient amount of computational resources, *edge computing* has emerged as an important paradigm in recent years [3–5]. The main idea of edge computing is to install computational components at the network edge, such as cellular base stations, access points, and road-side infrastructures. These components, referred to *edge servers* in the rest of this chapter, provide computational capability to a small group of users that are in close proximity to them. Being close, the users can access their edge servers with low latency. Still, they can offload resource-demanding computations to these edge servers, where such computations are infeasible to run on the local devices.

An important problem in this emerging edge computing setting is to determine where to run the computations for each user. We refer to these computations as a *service* in the following, since the ultimate goal of computing is to provide useful service to the user. We informally define the problem of *dynamic service placement* [6] as follows. Depending on the user location, the service needs to follow the user at least to some degree, in order to maintain low-latency access. However, because services may carry state information, which needs to be transferred if the service is moved (i.e., *migrated*) from one edge server to another edge server, migrating a service incurs additional cost. Our goal is to make service placement and migration decisions to minimize the overall cost, which includes the cost of additional latency when the user moves away from its closest edge server and the cost of migrating the service. Intuitively, when the user-server distance is not too large, it can be beneficial to run the service at the original edge server, incurring a slightly higher access cost but zero migration cost; when the user-server distance is very large, it can be beneficial to migrate the service to a new edge server that is closer to the user, because the long-term access cost would outweigh the additional migration cost in this case.

In this chapter, we present a rigorous mathematical framework for this dynamic service placement problem. Assuming a Markovian mobility model of users, we formulate the decision making problem of whether and where to migrate the services as a Markov decision process (MDP). We then discuss characteristics, efficient solutions, and insights from this MDP formulation. Our main results presented in this chapter are from [7].

A summary of the technical content in this chapter is as follows.

1. We introduce an MDP-based mathematical framework for designing optimal policies of service migration, while considering general cost models. The problem has a relatively large state space, and it can be computationally intensive to solve it. Therefore, we present a distance-based approximation to approximately solve the problem, where the approximation is exact for the uniform one-dimensional (1-D) mobility pattern. We also show several interesting properties of this approximated MDP, including a closed form solution to the long-term cost. These properties are then used for developing an algorithm to solve the MDP with a complexity that is lower by an order of magnitude compared to standard solution approaches.
2. We discuss how to use the above approximation to obtain an approximate solution to the MDP with two-dimensional (2-D) mobility patterns. The error of this approximation is upper-bounded by a constant when the 2-D mobility is uniform. We further illustrate the improvement in computational efficiency when using the approximation instead of solving the original problem directly.
3. We present trace-driven simulation results to show the improvement obtained by our MDP-based solution, compared to baseline methods such as myopic, always-migrate, and never-migrate decisions.

We outline the remaining sections of this chapter as follows. The problem formulation is described in Sect. 2. In Sect. 3, we describe the distance-based MDP model and its solution. Then, in Sect. 4, we discuss how to use this distance-based model to approximately solve 2-D problems. Some further details on applying our algorithms to real-world problems are discussed in Sect. 5. In Sect. 6, we present results from trace-driven simulations. Then, Sect. 7 gives a final summary.

2 Problem Formulation

For ease of presentation and analysis, we assume that our system is time-slotted and consider a single user. In this time-slotted system, the user location remains unchanged within each slot, and it may change between two adjacent slots according to a mobility model that follows a Markov process. We can consider this time-slotted system as a sampled variant of a system that operates continuously over time, where the sampling interval (i.e., length of each time slot) can be either equal or non-equal.

We further consider a partitioned 2-D area and let $\mathcal{L}$ denote the set of possible locations for both the user and the edge server. We assume that $\mathcal{L}$ is finite. Initially, we assume that each location $l \in \mathcal{L}$ represents an area that has an edge server. This server can serve users in the same area (location) and those in nearby locations. There is a distance metric $\|l_1 - l_2\|$ that represents the distance between two locations l_1 and l_2, which may or may not Euclidean distance. In an example of a cellular network where the edge servers are located at base stations and the user location is considered as the location of its current base station, we can denote the locations by 2-D vectors and compute the distance by the number of hops between the origin

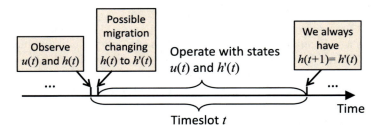

Fig. 9.1 Timing of the proposed service migration mechanism

cell and the destination cell. The locations of the user and the edge server hosting its service are denoted by $u(t)$ and $h(t)$, respectively, for time slot t, as shown in Fig. 9.1.

Our solution will be extended later in Sect. 5 to consider practical aspects, such as where multiple users co-exist in the system, edge servers are not available at all base stations, and there is a capacity limit at each edge server. In general, we assume that the users are independent of each other and each user only requests one service. Therefore, the notion of service in this chapter can also mean a specific instance of a type of service. For simplicity, we do not distinguish between a service and an instance of the service.

The main mathematical notations used throughout this chapter are summarized in Table 9.1.

2.1 Decisions and Costs

2.1.1 Migration

The system controller chooses to either migrate or not migrate at the beginning of each time slot t. These choices are explained in details as follows.

1. When deciding to migrate, the service (for the user under consideration) is moved from its previous location $h(t)$ to a new location $h'(t)$, where both $h(t), h'(t) \in \mathcal{L}$. This migration operation incurs a *migration cost* denoted by $b(x)$, which is assumed to be non-decreasing in x. Here, $x = \|h(t) - h'(t)\|$ is the distance between the locations $h(t)$ and $h'(t)$. After migration finishes, the system is in the new state of $(u(t), h'(t))$. Intuitively, the migration cost can capture the effect of service interruption and additional data transmission caused by migrating the service from one edge server to another edge server, such costs usually increase as the migration distance increases [8–10].
2. When deciding not to migrate, the migration cost is zero, i.e., $b(0) = 0$, and the service location remains the same, i.e., $h'(t) = h(t)$.

9 Dynamic Placement of Services at the Edge

Table 9.1 Summary of main notations

Notation	Description
:=	Is defined to be equal to
$\mathcal{L}$	Set of locations
l	Location
$\|l_1 - l_2\|$	Distance between locations l_1 and l_2
$u(t)$	User location at timeslot t
$h(t)$	Service location at timeslot t
$b(x)$	Migration cost
$c(y)$	Service access cost
$s(t)$	Initial state at slot t
$s'(t)$	Intermediate state at slot t
π	Decision policy
$a(s)\,(a^*(s))$	(Optimal) action taken when system is in state $s(t)$
$C_a(s)$	Sum of migration and access costs when taking action $a(s)$ in slot t
$V(s_0)$	Discounted sum cost when starting at state s_0
$P[s'_0, s_1]$	Transition probability from intermediate state s'_0 to the next initial state s_1 (in the next slot)
γ	Discount factor of the MDP
$d(t)$	User-service distance in slot t before possible migration (state in the distance-based MDP)
N	Number of states (excluding state zero) in the distance-based MDP
p_0, p, q	Transition probabilities of the distance-based MDP (see Fig. 9.2)
$\beta_c, \beta_l, \delta_c, \delta_l, \mu, \theta$	Parameters related to the constant-plus-exponential cost function (see (9.4) and (9.5))
A_k, B_k, D, H, m_1, m_2	Parameters related to the closed-form solution of the distance-based MDP (see (9.7)–(9.10))
$\{n_k : k \geq 0\}$	Series of migration states (i.e., all n_k such that $a(n_k) \neq n_k$)
r	Transition probability to one of the neighbors in the 2-D model
$e(t)$	Offset of the user from the service as a 2-D vector (state in the 2-D offset-based MDP)

2.1.2 Service Access

There is also a cost incurred for the user to connect to the edge server that currently hosts its service. This cost is referred to as the *access cost* in this chapter. The access cost $c(y)$ is also related to the distance between the user and the service, i.e., $y = \|u(t) - h'(t)\|$. If migration occurs at the beginning of the time slot t, the service location is regarded as the location of the edge server that hosts the service after migration. Intuitively, the access cost can capture the access latency and the communication bandwidth consumption for the user to interact with its service [3, 8, 11, 12].

We further assume that each time slot is long enough, so that the cost due to service interruption or access latency is incurred within a single time slot and it does not carry over to the next slot.

2.2 Control Objective

The overall goal of making control decisions of whether to migrate or not is to minimize the long-term cost, which includes both migration and access costs. For each time slot t, we denote the state of the system before possible migration as $s(t) = (u(t), h(t))$, which is referred to as the *initial state* of the time slot. A control policy π makes decision of whether or not to migrate based on the state $s(t)$. Let $a_\pi(s(t))$ denote the corresponding decision (action). After taking the action, the system transitions into an *intermediate state* of $s'(t) = a_\pi(s(t)) = (u(t), h'(t))$. Since we consider a deterministic policy π, the intermediate state $s'(t)$ has no randomness when $s(t)$ is given. Further, we let $C_{a_\pi}(s(t)) = b(\|h(t) - h'(t)\|) + c(\|u(t) - h'(t)\|)$ denote the sum of migration and access costs when taking action $a_\pi(s(t))$ in time slot t. We can then define the *value function* as the long-term expected discounted sum cost when starting from $s(0) = s_0$ as follows:

$$V_\pi(s_0) = \lim_{t \to \infty} \mathbb{E} \left\{ \sum_{\tau=0}^{t} \gamma^\tau C_{a_\pi}(s(\tau)) \middle| s(0) = s_0 \right\} \tag{9.1}$$

where $\gamma \in (0, 1)$ is a constant that is usually referred to as the discount factor in MDP literature. In the above expression, the expectation is over random user locations and the policy π is deterministic.

Our control objective is to minimize the above long-term cost for any initial state s_0. Formally, the problem is stated as follows:

$$V^*(s_0) = \min_\pi V_\pi(s_0) \quad \forall s_0. \tag{9.2}$$

From MDP literature [13], the optimal control policy to solve this problem can be obtained by finding the unique minimum of the Bellman's equation:

$$V^*(s_0) = \min_a \left\{ C_a(s_0) + \gamma \sum_{s_1 \in \mathcal{L} \times \mathcal{L}} P[a(s_0), s_1] \cdot V^*(s_1) \right\} \tag{9.3}$$

where $P[a(s_0), s_1]$ is the transition probability from state $s'(0) = s_0' = a(s_0)$ to $s(1) = s_1$. We note that the randomness here is due to the random user location in the next time slot $t = 1$ (assuming that we start at $t = 0$).

2.3 Properties of Optimal Policy

Before proceeding, we first present a basic property of the optimal policy π that solves (9.2). The proofs of this result and other results presented later in this chapter are given in [7].

Theorem 1 *For any state $s = (u, h)$ and the action $a^*(s) = (u, h')$ obtained from the optimal policy, we have $\|u - h'\| \leq \|u - h\|$.*

In essence, this theorem shows that the user-service distance after migration should not be larger than the distance before migration, as intuition expects.

2.4 Reducing the Search Space

The state space $\{s(t)\}$, which includes all possible user and service locations, can be large in general. To compute the optimal policy efficiently and also to obtain the action from the optimal policy using a very small amount of resource overhead during system operation, it is desirable to reduce the search space for optimal policies and their actions. From Theorem 1, we know that it is unnecessary to search for actions that migrate the service to a location that is farther away from the user. However, the state s is still a tuple that includes both user and service (edge server) locations. A further simplification can be to consider the distance, i.e., $d(t) = \|u(t) - h(t)\|$ between the user and its service. We consider such a distance-based MDP in the next section, and afterwards, we illustrate how such a simplification can be used to approximate more generic location patterns and mobility models.

In the following, we will consider slightly different variants of MDPs, and reuse $P, C_a(\cdot), V(\cdot)$, and $a(\cdot)$ to denote transition probabilities, instantaneous costs, long-term costs, and actions of such MDPs, respectively.

3 Distance-Based MDP and Its Optimal Policy

The distance-based MDP has the user-service distance as states. We define $d(t) = \|u(t) - h(t)\|$ as the distance before migration (if migration is performed) in time slot t. Then, the set of states of this MDP is $\{d(t)\}$. Let N denote a control parameter that specifies the maximum distance allowed by the system, so that migration is always performed if $d(t) \geq N$. The destination of migration in the case of $d(t) > N$ is configured to be the same as the destination of $d(t) = N$, so that our control algorithm described as follows only needs to make decisions when the system is in the states the states $\{d(t) : d(t) \in [0, N]\}$. We model the state transitions, i.e., change in distance, using a 1-D random walk Markovian model as shown in Fig. 9.2.

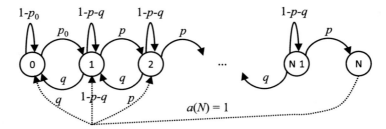

Fig. 9.2 Distance-based MDP with the distances $\{d(t)\}$ (before possible migration) as states. In this example, migration is only performed at state N, and only the possible action of $a(N) = 1$ is shown for compactness. The solid lines denote state transitions without migration

We will show later in Sect. 4.2 that this model can also approximate 2-D mobility patterns by properly choosing the parameters related to transition probabilities, namely p_0, p, and q. We further denote an intermediate state $d'(t) = a(d(t))$ that represents the distance after service migration but before the next user movement, similar to the notion of $h'(t)$ in Fig. 9.1. With this definition, the cost of a single timeslot following an action policy a is $C_a(d(t)) = b(|d(t) - d'(t)|) + c(d'(t))$, where $b(\cdot)$ and $c(\cdot)$ are the migration and access costs defined in Sect. 2.1.

3.1 Analytical Form of Cost Functions

For ease of analysis and in particular to obtain a closed-form expression of the value function of the MDP, we further specialize the definition of $b(x)$ and $c(y)$ in a specific analytical form that includes a constant and an exponential form whenever x or y is zero. These cost functions are defined as follows:

$$b(x) = \begin{cases} 0, & \text{when } x = 0 \\ \beta_c + \beta_l \mu^x, & \text{when } x > 0 \end{cases}, \qquad (9.4)$$

$$c(y) = \begin{cases} 0, & \text{when } y = 0 \\ \delta_c + \delta_l \theta^y, & \text{when } y > 0 \end{cases}, \qquad (9.5)$$

where the parameters β_c, β_l, δ_c, δ_l, μ, and θ are real values chosen in a way that the resulting $b(x)$ and $c(y)$ are both *non-decreasing* in x and y.

This definition of $b(x)$ and $c(y)$ can be used to approximate different shapes of actual cost functions, that can be either concave or convex in x or y for the case of $x > 0$ or $y > 0$. The step change from $x = 0$ or $y = 0$ to $x > 0$ or $y > 0$ can be used to model a constant portion of cost whenever migration occurs or whenever a non-local service access is present.

9 Dynamic Placement of Services at the Edge

3.2 Closed-Form Expression of Value Function

Based on (9.1), we have the following relation for the value function:

$$V(d) = C_a(d) + \gamma \sum_{\tilde{d}=a(d)-1}^{a(d)+1} P\left[a(d), \tilde{d}\right] \cdot V\left(\tilde{d}\right) \tag{9.6}$$

Here, we assume that the action map $a(\cdot)$ is defined by some policy π but we omit π in the above equation for simplicity. The notation d denotes the user-service distance at an arbitrary timeslot, and $\tilde{d}$ denotes the distance at the next timeslot after this.

We can see that (9.6) expresses the value function as a recurrence relation between two adjacent timeslots. Using techniques for solving difference equations, we can obtain the following result.

Theorem 2 *Let $\{n_k : k \geq 0\}$ be the set of all the states where migration happens, i.e., $a(n_k) \neq n_k$, as specified by a given policy π. We note that $0 \leq n_k \leq N$. The value function $V(d)$ for $d \in [n_{k-1}, n_k]$ (for convenience, we define $n_{-1} := 0$) is*

$$V(d) = A_k m_1^d + B_k m_2^d + D + \begin{cases} H \cdot \theta^d & \text{if } 1 - \frac{\phi_1}{\theta} - \phi_2\theta \neq 0 \\ Hd \cdot \theta^d & \text{if } 1 - \frac{\phi_1}{\theta} - \phi_2\theta = 0 \end{cases}. \tag{9.7}$$

Here, A_k and B_k are constants related to the interval $[n_{k-1}, n_k]$, and the coefficients m_1, m_2, D, and H are defined as

$$m_1 = \frac{1 + \sqrt{1 - 4\phi_1\phi_2}}{2\phi_2}, m_2 = \frac{1 - \sqrt{1 - 4\phi_1\phi_2}}{2\phi_2}, \tag{9.8}$$

$$D = \phi_3 / (1 - \phi_1 - \phi_2), \tag{9.9}$$

$$H = \begin{cases} \frac{\phi_4}{1 - \frac{\phi_1}{\theta} - \phi_2\theta} & \text{if } 1 - \frac{\phi_1}{\theta} - \phi_2\theta \neq 0 \\ \frac{\phi_4}{\frac{\phi_1}{\theta} - \phi_2\theta} & \text{if } 1 - \frac{\phi_1}{\theta} - \phi_2\theta = 0 \end{cases}, \tag{9.10}$$

where we further define $\phi_1 := \frac{\gamma q}{1 - \gamma(1-p-q)}$, $\phi_2 := \frac{\gamma p}{1 - \gamma(1-p-q)}$, $\phi_3 := \frac{\delta_c}{1 - \gamma(1-p-q)}$, and $\phi_4 := \frac{\delta_l}{1 - \gamma(1-p-q)}$.

For two different states d_1 and d_2, if the actions defined by the policy π satisfies $a(d_1) = d_2$ and $a(d_2) = d_2$, then we also have the following relation between the value functions computed at the states d_1 and d_2:

$$V(d_1) = b(|d_1 - d_2|) + V(d_2). \tag{9.11}$$

In the following, we discuss how to obtain the values of the parameters A_k and B_k in (9.7). We assume that $1 - \frac{\phi_1}{\theta} - \phi_2\theta \neq 0$ and $1 - \frac{\phi_2}{\theta} - \phi_1\theta \neq 0$. We can derive the results for other cases similarly and omit them for brevity.

Interval $[0, n_0]$ From (9.6), when $d = 0$, we have

$$V(0) = \gamma p_0 V(1) + \gamma(1 - p_0)V(0). \tag{9.12}$$

Substituting (9.7) into (9.12), we obtain

$$A_0(1 - \phi_0 m_1) + B_0(1 - \phi_0 m_2) = D(\phi_0 - 1) + H(\phi_0\theta - 1) \tag{9.13}$$

where $\phi_0 := \frac{\gamma p_0}{1 - \gamma(1 - p_0)}$. We also have the following relation by substituting (9.7) into (9.11):

$$A_0\left(m_1^{n_0} - m_1^{a(n_0)}\right) + B_0\left(m_2^{n_0} - m_2^{a(n_0)}\right) = \beta_c + \beta_l\mu^{n_0 - a(n_0)} - H\left(\theta^{n_0} - \theta^{a(n_0)}\right). \tag{9.14}$$

Now, using (9.13) and (9.14), we can solve for the values of A_0 and B_0.

Interval $[n_{k-1}, n_k]$ Assume that we know $A_{k'}$ and $B_{k'}$ for all $k \leq k - 1$, so that we have the exact expressions for $V(d)$ for all $d \leq n_{k-1}$. Let $d = n_{k-1}$ in (9.7), and we obtain

$$A_k m_1^{n_{k-1}} + B_k m_2^{n_{k-1}} = V(n_{k-1}) - D - H \cdot \theta^{n_{k-1}}. \tag{9.15}$$

Next, we consider two different cases. If $a(n_k) \leq n_{k-1}$, then

$$A_k m_1^{n_k} + B_k m_2^{n_k} = \beta_c + \beta_l\mu^{n_k - a(n_k)} + V(a(n_k)) - D - H \cdot \theta^{n_k}. \tag{9.16}$$

If $n_{k-1} < a(n_k) \leq n_k - 1$, then

$$A_k\left(m_1^{n_k} - m_1^{a(n_k)}\right) + B_k\left(m_2^{n_k} - m_2^{a(n_k)}\right) = \beta_c + \beta_l\mu^{n_k - a(n_k)} - H\left(\theta^{n_k} - \theta^{a(n_k)}\right). \tag{9.17}$$

Then, we can obtain the values of A_k and B_k from (9.15) combined with either (9.16) or (9.17).

According to the above discussion, the solutions of A_k and B_k can be obtained by solving a system of linear equations with two variables. Such solutions have closed-form expressions. Hence, (9.7) gives a closed-form expression for all $d \in [0, N]$. Numerically, we can also compute $V(d)$ using (9.7) in $O(N)$ time.

9 Dynamic Placement of Services at the Edge

3.3 Finding the Optimal Policy

Our result in (9.7) gives a closed-form expression of $V(d)$ for a given policy π that prescribes the state-to-action map $a(\cdot)$. In the following, we describe how to find the optimal policy itself, so that our overall objective (9.1) is minimized. There exist standard techniques for finding the optimal policy of MDPs, including value iteration and policy iteration [13, Chapter 6]. In general, if the exact value function $V(d)$ can be computed efficiently, policy iteration is usually more efficient than value iteration. The essence of policy iteration is to update the policy so that the objective function keeps decreasing, until such an update cannot be found anymore. Combining this policy searching procedure with our value function expression in (9.7), we obtain Algorithm 1 for finding the optimal policy.

Algorithm 1: Policy-iteration with efficient value function computation

---

1 Initialize $a(d) \leftarrow 0$ for all $d = 0, 1, 2, \ldots, N$;
2 Find constants $\phi_0, \phi_1, \phi_2, \phi_3, \phi_4, m_1, m_2, D$, and H;
3 **repeat**
4 $k \leftarrow 0$;
5   **for** $d = 1 \ldots N$ **do**
6   **if** $a(d) \neq d$ **then**
7 $n_k \leftarrow d, k \leftarrow k+1$;
8   **for** *all* n_k **do**
9   **if** $k = 0$ **then**
10   Solve for A_0 and B_0 from (9.13) and (9.14);
11   Find $V(d)$ with $0 \le d \le n_k$ from (9.7) with A_0 and B_0 found above;
12   **else if** $k > 0$ **then**
13   **if** $a(n_k) \le n_{k-1}$ **then**
14   Solve for A_k and B_k from (9.15) and (9.16);
15   **else**
16   Solve for A_k and B_k from (9.15) and (9.17);
17   Find $V(d)$ with $n_{k-1} < d \le n_k$ from (9.7) with A_k and B_k found above;
18   **for** $d = 1 \ldots N$ **do**
19 $a_{\text{prev}}(d) \leftarrow a(d)$;
20 $a(d) \leftarrow \arg\min_{a \le d} \left\{ C_a(d) + \gamma \sum_{j=a-1}^{a+1} P[a, j] \cdot V(j) \right\}$;
21 **until** $a_{\text{prev}}(d) = a(d)$ *for all* d;
22 **return** $a^*(d) \leftarrow a(d)$ for all d;

In this algorithm, Lines 4–7 compute the migration states n_k, Lines 8–17 compute $V(d)$ for the policy in the current iteration, and Lines 18–20 update the policy that has been optimal so far. We can see that the time complexity of each iteration in this algorithm is $O\left(N^2\right)$.

4 Using Distance-Based MDP to Approximate 2-D Mobility

Our distance-based MDP can be used to approximate more realistic mobility patterns such as where users move on a 2-D space. In this section, we describe such an extension and how to map the distance-based MDP to an "offset"-based MDP with 2-D mobility on a hexagonal cell structure. Other 2-D mobility models, e.g., Manhattan grid, can also be supported after modifications. For our hexagonal model, we assume that the user in a cell moves to one of its six neighboring cells with a probability of r at the beginning of each timeslot, and the user stays in the same cell with a probability of $1 - 6r$.

4.1 Offset-Based MDP

For the hexagonal mobility model, we define an MDP that captures the *offset* between the user and the service in its states. We define the offset as $e(t) = u(t) - h(t)$, where $e(t), u(t), h(t)$ are all 2-D vectors. In this hexagonal model, the distance $\|l_1 - l_2\|$ represents the smallest number of hops needed to move from a cell l_1 to another cell l_2.

We refer to those states with the same $\|e(t)\|$ as a *ring*. Each ring contains states where the user-service distance is the same. For ease of presentation, we express the states $\{e(t)\}$ with polar indices (i, j), where i denotes the ring index and k denotes the state index within the ring. See Fig. 9.3 for an example. With this notation, we have $\|e(t)\| = i$ for $e(t) = (i, j)$. When the user and its service are in the same cell, we have $u(t) = h(t)$, and thus $e(t) = (0, 0)$ and $\|e(t)\| = 0$.

Similar to the distance-based MDP, we always migrate in the 2-D offset-based MDP when $\|e(t)\| \geq N$. Here, N is a system design parameter that specifies how far away the user and service can be. With this design, we only need to consider

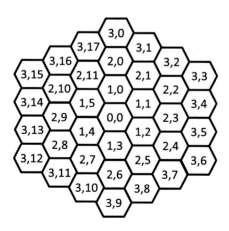

Fig. 9.3 Example of 2-D offset model on hexagon cells, where $N = 3$

9 Dynamic Placement of Services at the Edge 309

the state space of $\{e(t)\}$ with $\|e(t)\| \leq N$, because the user-service distance cannot exceed N.

After taking action $a(e(t))$, the system operates in the intermediate state $e'(t) = u(t) - h'(t) = a(e(t))$. The next state $e(t + 1)$ is random, following the transition probability of $P[e'(t), e(t + 1)]$. We have

$$P[e'(t), e(t + 1)] = \begin{cases} r, & \text{when } e(t + 1) \text{ is a neighbor of } e'(t) \\ 1 - 6r, & \text{when } e(t + 1) = e'(t) \\ 0, & \text{otherwise} \end{cases}.$$

The one-timeslot cost is $C_a(e(t)) = b(\|e(t) - e'(t)\|) + c(\|e'(t)\|)$, since $e(t) - e'(t) = h'(t) - h(t)$ by definition.

We note that this 2-D offset-based MDP has $M = 3N^2 + 3N + 1$ states, whereas there are only N states for the distance-based MDP. In the following, we discuss how to use the distance-based MDP to approximate the 2-D offset-based MDP, so that we can solve for the (near-)optimal policy more efficiently.

4.2 Approximation by Distance-Based MDP

To approximate the 2-D MDP using the distance-based MDP, we note that when starting at state $(i'_0, j'_0) = (0, 0)$ in the 2-D MDP, we transition to an arbitrary state in ring $i_1 = 1$ with a probability of $6r$ (see Fig. 9.3). Furthermore, when starting at any state of $(i'_0, j'_0) \neq (0, 0)$, we transition to a higher-indexed ring $i_1 = i'_0 + 1$ with a probability of either $2r$ or $3r$, and we transition to a lower-indexed ring $i_1 = i'_0 - 1$ with a probability of either r or $2r$. Based on this observation, we choose $p_0 = 6r$, $p = 2.5r$, and $q = 1.5r$ in the distance-based MDP (Fig. 9.2), where the values of p and q are set to the median values of the transition probabilities of the 2-D MDP.

With this approximation, we can solve for the optimal policy of the 2-D MDP by first finding the optimal policy of the distance-based MDP. Then, this optimal policy is mapped to a policy for the 2-D MDP. The mapping here notes the fact that, in the 2-D MDP, there always exists a path from a state in ring i to some (i.e., non-specific) state in ring i' such that the length of this path is $|i - i'|$. For example, when starting from state $(i, j) = (3, 2)$ and moving to an arbitrary state in ring $i' = 1$, a possible path of length $|i - i'| = 2$ is $\{(3, 2), (2, 1), (1, 0)\}$. In the distance-based MDP, the state index corresponds to the ring index i of the 2-D MDP. When taking an action of $a(i) = i'$ according to the distance-based MDP, we migrate the service from the original state (i, j) to some state in ring i' as seen in the 2-D MDP, while preserving a migration distance of $|i - i'|$ as explained above. If there are multiple migration paths with the same length, one of them is chosen arbitrarily. For example, if the distance-based MDP's policy gives $a(3) = 2$, then we can choose either $a(3, 2) = (2, 1)$ or $a(3, 2) = (2, 2)$ in the 2-D MDP.

This mapping of policies between the two MDPs ensures that the single timeslot costs $C_a(d(t))$ and $C_a(e(t))$ for the distance-based and 2-D MDPs, respectively, are the same. This is because: (1) the migration distances of both MDPs are the same, as explained above; (2) all the states in the same ring of the 2-D MDP satisfies $i' = \|e'(t)\| = d'(t)$, thus they have the same service access cost $c(\|e'(t)\|) = c(d'(t))$.

4.3   Bound on Approximation Error

Approximating the 2-D MDP using the distance-based MDP incurs some error, because the transition probabilities of these two MDPs are not exactly the same. As explained above, our choice of p and q as the median value gives a difference of at most $0.5r$. The following theorem gives an upper bound on the difference of the optimal value function when using the distance-based MDP to approximate 2-D MDP.

Theorem 3 *Let $V^*(e)$ denote the value function obtained by the true optimal policy of the 2-D MDP, and let $V_{\text{dist}}^*(e)$ denote the value function obtained by the optimal policy of the distance-based MDP and mapped to the 2-D MDP. For any state e in the 2-D MDP, we have*

$$V_{\text{dist}}^*(e) - V^*(e) \leq \frac{\gamma r \kappa}{1 - \gamma} \tag{9.18}$$

where $\kappa := \max_x \{b(x + 2) - b(x)\}$.

When the parameters γ, r, and κ are fixed, this error bound has a constant value. The bound increases in γ, but the value function defined in (9.1) also increases in γ due to the sum, so the relative error may remain similar.

5   Application to Practical Scenarios

In practice, the migration policy can be computed by a controller that is a logical entity located in either a cloud server or an edge server. The transition probabilities of the MDP can be estimated using statistics of user connectivity across different timeslots, where the timeslot here has a pre-specified physical time length. Usually, each edge server provides service to a specific geographical region, thus they are often co-located with cellular base stations or Wi-Fi access points. The distance can be measured in metrics such as the number of hops between different edge servers or the Euclidean distance after quantization. Finally, the parameters in the cost function β_c, β_l, μ, δ_c, δ_l, and θ, as well as the discounting factor γ, can be chosen depending on the application scenario and system characteristics. Their values may also vary based on the overall traffic load of the system, and the migration policy

9 Dynamic Placement of Services at the Edge

is recomputed at a certain given interval to accommodate the most up-to-date cost definition.

In addition to supporting the homogeneous mobility pattern as described in previous sections, the migration decisions can be extended (in a heuristic manner) to support scenarios where some locations may not have edge servers and some edge servers may have a fixed capacity limit. If the policy obtained by the MDP migrates a service to a location that either does not have an edge server or has an edge server that is already at its maximum load, a nearby location with an edge server that is feasible to run the service is chosen. The choice of the nearby location here is determined by a similar one-step optimization problem (9.3) subject to all the feasibility constraints.

6 Trace-Driven Simulation

To evaluate the empirical performance of the method described in this chapter, we ran a simulation with a real-world dataset of taxi locations in San Francisco [14, 15]. This dataset includes a total of 536 taxis operating on the day of May 31, 2008, where the exact number of active taxis is different at different times of the day. We consider each taxi as a user that requests a service, where the services of different users are assumed to be independent of each other.

We consider possible edge server locations in two separate settings. The first setting is a hexagonal grid as described in Sect. 4, and the second setting includes real cell tower locations obtained from a separate dataset.[1] These two settings are illustrated in Fig. 9.4. Among all these possible locations, we randomly choose 100 locations that have an edge server and each edge server can host a maximum of 50 services. Physically, these possible locations can represent cellular base stations that the users communicate with. Some base stations have edge servers attached to them while other base stations do not have an edge server. The distance in our model is computed based on the distance between these base stations, while assuming that a user always connects to its closest base station.

For each timeslot in our model, the physical time length is defined to be 60 seconds. We recompute the migration policy in every slot, using statistics computed over the most recent 60 slots to estimate the transition probabilities. We also choose $N = 10$ and $\gamma = 0.9$.

In the simulation, we compare the cost performance of our MDP-based method to several baseline policies, including policy that *always migrates* to the edge server that is closest to the user, *never migrates*, and one that is *myopic* and only tries to minimize the instantaneous cost.

---

[1] The cell tower locations were obtained from https://www.antennasearch.com.

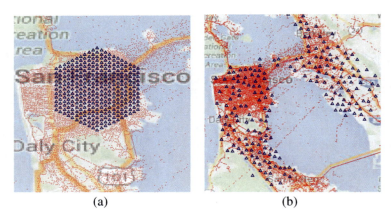

Fig. 9.4 BS and taxi locations: (**a**) Hexagonal BS placement, (**b**) Real BS placement. The blue triangles indicate the BS location and the red dots indicate possible taxi locations. There appears to be a small amount of erroneous/inaccurate data for taxi locations but the majority of them are correct. Map data: Google

Cost Definition We define the cost so that it is related to both the system load and the distance. An assumption we make is that the system load is proportional to the number of active taxis, i.e., those that are operating at the given time of the day. We note that the number of active taxis varies during the day that the data is collected. We further define $R_t > 1$ and $R_p > 1$. Intuitively, the values of R_t and R_p can represent the transmission and processing resource availability, respectively. We then define $G_t := 1 / \left(1 - \frac{m_{\text{cur}}}{R_t m_{\text{max}}}\right)$ and $G_p := 1 / \left(1 - \frac{m_{\text{cur}}}{R_p m_{\text{max}}}\right)$, where m_{cur} denotes the number of active taxis when computing the migration policy, and m_{max} denotes the maximum number of active taxis at any time in the dataset. These expressions of G_t and G_p are in the same form as the average queueing delay [16, 17].

In our simulation, we use the cost definitions in (9.4) and (9.5), and we choose $\mu = \theta = 0.8$. Then, we consider two types of costs: the first is a *non-constant cost* that keeps increasing with the distance, and the second is a *constant cost* that remains unchanged as long as the distance is non-zero. We also note that according to (9.4) and (9.5), when the distance is zero, the costs are also zero.

For the non-constant cost definition, we choose $\beta_c = G_p + G_t$, $\beta_l = -G_t$, $\delta_c = G_t$, and $\delta_l = -G_t$. With this definition, the migration cost includes components related to both processing and transmission, while the service access cost only includes the transmission component. For the constant cost, we choose $\beta_c = G_p$, $\delta_c = G_t$, and $\beta_l = \delta_l = 0$, where we ignore the transmission component in the migration cost, since the assumption in this case is that there is a fixed amount of cost whenever service migration or access to a distant service occurs.

Results We present the results in Fig. 9.5, showing both the instantaneous and average cost values for different policies. The instantaneous costs are also averaged

9 Dynamic Placement of Services at the Edge

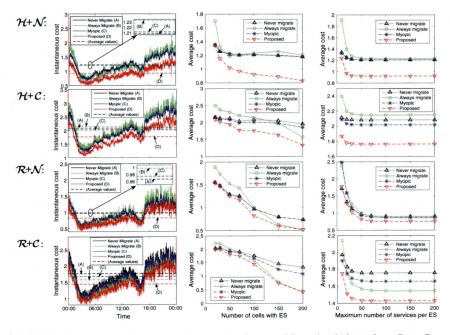

Fig. 9.5 Instantaneous and average costs over a day in trace-driven simulation, where $R_t = R_p = 1.5$, $\mathcal{H}$ and $\mathcal{R}$ respectively stand for $\mathcal{H}$exagonal and $\mathcal{R}$eal BS placement, $\mathcal{N}$ and $\mathcal{C}$ respectively stand for $\mathcal{N}$on-constant and $\mathcal{C}$onstant costs. In some plots, an enlarged plot of the circled area is shown. The arrows annotated with (A), (B), (C), and (D) point to the average values over the whole day of the corresponding policy

over all the users that a active in each single timeslot, while the average costs are averaged over the entire day. We can see that the MDP-based policy produces a lower cost than the baseline policies in most cases. The instantaneous cost fluctuates because of different number of active taxis, hence different system load, seen during the day. Such system variation also suggests that it is necessary to recompute the migration policy at a certain frequency over time, to accommodate the most updated system status. When the number of edge servers or the capacity per edge server is small, the average costs produced by the never-migrate and myopic policies get closer to the cost given by the MDP-based policy, because it is difficult to migrate the service to a better location in this resource-constrained scenario.

7 Summary

In this chapter, we have discussed the service placement problem in edge computing systems. Our primary focus has been on the decision of whether and where to migrate the service when the user moves to different locations over time. We have formulated this problem as an MDP, and then discussed how to reduce the state space of the MDP by approximating a 2-D MDP using a distance-based MDP. We have also presented theoretical results on the approximation error and the optimal solution to the distance-based MDP. Our results from trace-driven simulations provide further verification on the effectiveness of this approach.

References

1. W. Zhang, J. Chen, Y. Zhang, D. Raychaudhuri, Towards efficient edge cloud augmentation for virtual reality mmogs, in *Proc. of ACM/IEEE Symposium on Edge Computing (SEC)*, Oct. (2017)
2. H.T. Dinh, C. Lee, D. Niyato, P. Wang, A survey of mobile cloud computing: architecture, applications, and approaches. Wirel. Commun. Mob. Comput. **13**(18), 1587–1611 (2013)
3. M. Satyanarayanan, The emergence of edge computing. Computer **50**(1), 30–39 (2017)
4. P. Mach, Z. Becvar, Mobile edge computing: A survey on architecture and computation offloading. IEEE Commun. Surv. Tutorials **19**(3), 1628–1656 (2017)
5. Y. Mao, C. You, J. Zhang, K. Huang, K.B. Letaief, A survey on mobile edge computing: The communication perspective. IEEE Commun. Surv. Tutorials **19**(4), 2322–2358 (2017). https://doi.org/10.1109/COMST.2017.2745201
6. S. Wang, Dynamic service placement in mobile micro-clouds, Ph.D. dissertation, Imperial College London, 2015
7. S. Wang, R. Urgaonkar, M. Zafer, T. He, K. Chan, K.K. Leung, Dynamic service migration in mobile edge computing based on markov decision process. IEEE/ACM Trans. Netw. **27**(3), 1272–1288 (2019)
8. K. Ha, Y. Abe, Z. Chen, W. Hu, B. Amos, P. Pillai, M. Satyanarayanan, Adaptive VM handoff across cloudlets, Technical Report CMU-CS-15-113, CMU School of Computer Science, Tech. Rep., 2015
9. L. Ma, S. Yi, Q. Li, Efficient service handoff across edge servers via docker container migration, in *Proc. of ACM/IEEE Symposium on Edge Computing (SEC)*, Oct. (2017)
10. A. Machen, S. Wang, K.K. Leung, B.J. Ko, T. Salonidis, Live service migration in mobile edge clouds. IEEE Wirel. Commun. **PP**(99), 2–9 (2017)
11. A. Ceselli, M. Premoli, S. Secci, Mobile edge cloud network design optimization. IEEE/ACM Trans. Netw. **25**(3), (2017)
12. Z. Xu, W. Liang, W. Xu, M. Jia, S. Guo, Efficient algorithms for capacitated cloudlet placements. IEEE Trans. Parallel Distrib. Syst. **27**(10), 2866–2880 (2016)
13. M.L. Puterman, *Markov Decision Processes: Discrete Stochastic Dynamic Programming*, vol. 414 (John Wiley & Sons, 2009)
14. M. Piorkowski, N. Sarafijanovoc-Djukic, M. Grossglauser, A parsimonious model of mobile partitioned networks with clustering, in *Proc. of COMSNETS*, Jan. (2009)
15. M. Piorkowski, N. Sarafijanovic-Djukic, M. Grossglauser, CRAWDAD data set epfl/mobility (v. 2009-02-24), Downloaded from http://crawdad.org/epfl/mobility/, Feb. (2009)
16. L. Kleinrock, *Queuing Systems, Volume II: Computer Applications* (John Wiley and Sons, Hoboken, NJ, 1976)
17. J. Li, H. Kameda, Load balancing problems for multiclass jobs in distributed/parallel computer systems. IEEE Trans. Comput. **47**(3), 322–332 (1998)

Chapter 10
Joint Service Placement and Request Scheduling at the Edge

Ting He and Shiqiang Wang

Abstract This chapter focuses on the joint allocation of multiple types of resources when trying to run machine learning applications on edge computing platforms. To have the maximum applicability, the machine learning workloads will be simply modeled as demands for various types of resources (storage, communication, computation), and the resource allocation algorithms are designed to optimally satisfy these demands within the limited resource capacities of edge clouds. Different problem formulations differ in terms of the performance objective, the types of resources considered, and the forms of resource constraints, which all originate from the different application scenarios of interest. These differences in turn lead to differences in the problem complexities and the applicable solutions.

Admittedly, the above workload model makes the resource allocation problem largely application-agnostic, and heavily overlap with the generic resource allocation problems for arbitrary applications in *mobile edge computing*. However, there is a critical difference that machine learning applications are generally *data-intensive*, requiring both a large amount of data to be stored on the server and a non-negligible amount of data to be provided by the user. One typical example is augmented reality, where the server must store the trained machine learning model for object recognition and the object database for augmenting the recognized object with relevant information. Meanwhile, a user using this application also needs to provide a camera view from his/her mobile device, which will then be used as input for object recognition. We will thus start with an overview of existing works on generic resource allocation in mobile edge computing and then discuss in detail a

---

T. He (✉)
Pennsylvania State University, University Park, PA, USA
e-mail: tinghe@psu.edu

S. Wang
IBM T. J. Watson Research Center, Yorktown Heights, NY, USA
e-mail: wangshiq@us.ibm.com

© The Author(s), under exclusive license to Springer Nature Switzerland AG 2023
M. Srivatsa et al. (eds.), *Artificial Intelligence for Edge Computing*,
https://doi.org/10.1007/978-3-031-40787-1_10

state-of-the-art resource allocation framework for data-intensive workloads, which is suitable for machine learning applications.

1 Background on Resource Allocation at the Edge

In *mobile edge computing* [1], small servers or server clusters are deployed at the edge of wireless networks (e.g., base stations) to offload heavy-lifting workloads from mobile devices. These platforms are generally referred to as *edge clouds* [2], *cloudlets* [3], *fog* [4], *follow me cloud* [5], or *micro clouds* [6], and hereafter we will use the term "edge clouds" for consistency. The benefit of edge clouds is that they are located close to the mobile users, and hence can provide support for resource-intensive applications within the network infrastructure, without the large latency in accessing data centers located deep in the Internet. This property makes it a popular choice for resource-intensive and delay-sensitive applications, which includes but is not limited to many machine learning applications.

The small form factor of edge clouds leads to fierce resource contention among users. Thus, the problem of resource application in mobile edge computing has attracted significant attention in the research community that is summarized below:

Point-to-Point Offloading Early works assumed that users offload an entire workload in the form of a virtual machine (VM) or container, which is inherited from use cases in the traditional data center-based cloud. To minimize access delay, it is desirable to keep the offloaded VM/container on the edge cloud closest to the user [7], hence the terminology of "follow me cloud" [8]. In reality, however, the user is mobile, and there are various costs in migrating the VM/container to follow the user due to data transfer, service disruption, and other operations. Therefore, the research question is how to migrate the VM/container for each user to achieve a desirable tradeoff between the access delay and the operation cost [2, 8–10]. For example, under the assumption that the user mobility can be modeled as a Markov chain, the problem can be formulated as a *Markov decision process*, for which heuristic and optimal solutions have been found in several cases [2, 8–10].

Resource Pooling and Scheduling When some edge clouds are heavily loaded, it can be beneficial for users to run their workloads on non-local edge clouds. The benefits of utilizing non-local edge clouds have been verified for edge clouds located within the same metropolitan area network [11–13]. From a research perspective, allowing a user to access edge clouds multiple hops away substantially enlarges the solution space for resource allocation. In particular, the edge cloud system can now schedule user workloads onto a pool of resources provided by edge clouds within the same geographical region. This scheduling problem has been studied extensively in the literature, where different works differ in one or multiple of the following aspects:

1. *Objective:* Popular objectives are to minimize the operation cost [14] or the total completion time [15] in the under-loaded regime (i.e., all the workloads can be handled by the edge clouds), or to maximize the number of served requests [16, 17] in the overloaded regime (i.e., not all the workloads can be handled by the edge clouds).
2. *Workload model:* Existing workload models include the fluid model [14], the task model [15], and the multi-component application model [18], with an increasing level of generality and complexity.
3. *System architecture:* Most works assume a flat architecture where all the edge clouds are peer, but there are also studies that consider alternative architectures (e.g., a hierarchical architecture [19]).

We note that with the exception of [15–17], these works typically assume that each workload requires its own resource for execution, e.g., each unit of memory space or CPU time can only be used by one workload at a time, although the memory or CPU can be shared among workloads. Although this assumption is usually satisfied for computation and communication resources, it is too restrictive for storage resources, as one copy of data/code can be used to serve multiple user requests.

Content Placement at the Edge Another well-studied resource allocation problem is the content placement problem. In this context, the edge clouds only provide access to pre-stored data, and are hence reduced to "edge caches". The key resource of interest is the storage resources at the caches, and the allocation decides which contents to place at each cache such that the overall performance (e.g., hit ratio, access delay) is optimized. Depending on the knowledge of content popularities, existing works can be classified into *proactive content placement* [20, 21] and *reactive content placement* [22]. Further variations study the problem in more complicated settings, such as cooperative caching [23] and jointly optimized content placement and request routing [24]. However, works on the content placement problem mostly focus on the storage resource, while the other types of resources (e.g., CPU, bandwidth) are often ignored.

Joint Allocation of Multiple Types of Resources The trend of ongoing research is to jointly consider multiple types of resources (e.g., communication, computation, storage) and jointly optimize multiple types of resource allocation actions (e.g., service placement, request routing, request scheduling). Obviously, the resulting optimization problems will be harder than before, for which solutions with performance guarantees exist only in special cases. We briefly summarize the existing works in this space as follows:

1. A *dynamic service placement and workload scheduling* framework was proposed in [25] to jointly allocate storage and computation resources with the objective of cost minimization. An online algorithm based on Lyapunov optimization was proposed that had an asymptotically optimal cost, but to achieve the optimal cost, the worst-case delay would be unbounded, i.e., requests can be queued indefinitely.

2. An algorithm with performance guarantee was developed in [26] for *jointly placing a certain type of services (virtual network functions) and routing flows among the placed services* under service chaining constraints. However, there is no storage capacity constraint on the service placement. As we will show later, service placement under storage capacity constraint is the most difficult part of the joint resource allocation problem.

3. In our recent works [16, 17], we considered the problem of *joint service placement and request scheduling* to maximize the number of requests served at edge clouds under storage, computation, and communication capacity constraints. We fully characterized the complexity of the problem in all cases, and proposed optimal solutions when the problem is polynomial-time solvable, as well as general heuristics when the problem is NP-hard. The proposed heuristics are proved to achieve a guaranteed approximation ratio under certain conditions.

4. In a recent work closely related to ours [27], *joint service placement and request routing* was considered in a system where a user can be associated with multiple edge clouds (e.g., in the coverage areas of multiple base stations deployed with edge clouds). A bi-criteria algorithm was developed that achieved a guaranteed approximation ratio while incurring a bounded violation of the capacity constraints.

Remark We would like to point out two subtleties. First, although the formulations in [27] and [17] appear similar, they have a critical difference: [27] restricts all the resources consumed by a request (bandwidth, computation, storage) to a single edge cloud, while [17] allows them to be with two different edge clouds, where the bandwidth consumption is with the edge cloud directly covering the user (via wireless communications), and the computation and storage resources are with the edge cloud scheduled to process the request (which communicates with the directly covering edge cloud via backhaul links for input/output).

Moreover, the term "resource" has been used for different meanings. For example, in [28], an optimal algorithm was developed for joint resource placement and assignment in distributed networks, where a "resource" means a service, and a "type of resources" means a type of services. Here, we use "resource" to refer to a physical resource, e.g., bandwidth, storage space, or CPU cycles.

2 Resource Allocation for Data-Intensive Applications at the Edge

As reviewed in Sect. 1, existing formulations differ significantly in terms of the performance objective, the types of resources considered, and the forms of resource constraints. The fundamental reason for these differences is the difference in the application scenario of interest. As the applications of interest in this chapter are machine learning applications, we will now narrow down to formulations suitable for modeling the resource demands of these applications.

A notable feature of machine learning applications is data dependency, where running such an application at an edge cloud typically requires the edge cloud to pre-store a large amount of persistent data (e.g., trained machine learning models and databases) and the user to provide a non-negligible amount of transient data (e.g., the image to be analyzed). Moreover, different from caching, machine learning applications also require a non-negligible amount of computation to process each service request. The challenge in allocating resources for such applications is that serving each request consumes not only *dedicated* resources (bandwidth, CPU cycles) but also *amortized* resources (storage space) that are shared among requests of the same type.

To address this challenge, we formulate a set of coupled optimization problems to separately allocate different resources. The storage resources are allocated via *service placement*, and the communication/computation resources are allocated via *request scheduling*. In our terminology, a "service" is a bundle of code and data for providing one type of service to the users, e.g., classification based on a neural network trained on cars is one service, and classification based on a neural network trained on faces is another service. The code and data for one type of service can be replicated to multiple locations, each referred to as one *service replica*.

2.1 Solution Framework

As illustrated in Fig. 10.1, we consider a wireless edge network consisting of a set N of edge clouds, each accessible via a wireless access point or base station covering a specified area. We assume that all the edge clouds are connected by back-haul links that can be used for inter-cloud communications. There is a set L of services, of

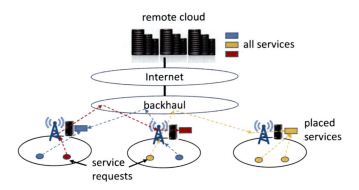

Fig. 10.1 System model

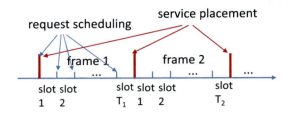

Fig. 10.2 Time scales of service placement and request scheduling

which a subset can be hosted by each edge cloud at a given point in time, subject to storage capacity constraints.

Services may be migrated/replicated between edge clouds, and/or from a remote cloud to an edge cloud. To access a certain service, a user will first send a request for this service to its local edge cloud, which may then route the request to another edge cloud for processing. Serving a request for service l submitted to edge cloud n at edge cloud m (possibly $m \neq n$) consumes communication resources for transferring input/output between the user and edge cloud n, and computation resources at edge cloud m. Additionally, edge cloud m must have a replica of service l. If $m \neq n$, communication resources are also consumed for transferring input/output between edge cloud n and edge cloud m, but as back-haul links usually have much higher bandwidth than access links, we will focus on the communication resources consumed at the access link in edge cloud n.

To ensure system stability while providing timely services, we adopt a two-time-scale framework as illustrated in Fig. 10.2, where service placement is performed once per frame, and request scheduling is performed once per slot. Furthermore, we impose a budget B to control the cost of migrating/replicating services in each frame. We refer to a request for service l that is submitted to edge cloud n as a "type-(l, n)" request. The average rate of type-(l, n) requests in frame f is denoted by λ_{ln}^f (unit: requests/slot), which is assumed to be predictable based on the request history. The actual number of type-(l, n) requests in slot t is denoted by λ_{ln}^t, which is only known at the beginning of slot[1] t.

Each edge cloud has limited communication, computation, and storage capacities. The capacities of different edge clouds may be different. Likewise the size of each service replica and the communication/computation resources required by each request may be different. There may be other constraints (e.g., latency, security) on whether a given edge cloud m is permitted to serve type-(l, n) requests, and we model that by an indicator a_{lnm} ('0': not permitted; '1': permitted). The main notations used in this paper are described in Table 10.1.

---

[1] This is feasible by considering all the requests received during slot $t - 1$ as being "submitted" in slot t.

10 Joint Service Placement and Request Scheduling at the Edge 321

Table 10.1 Table of notations

Notation	Meaning
N	Set of edge clouds
$N_+ = N \cup \{n_0\}$	Set of edge clouds plus the remote cloud n_0
L	Set of all possible services
R_n	Storage capacity of edge cloud n
W_n	Processing capacity of edge cloud n (per slot)
K_n	Communication capacity of edge cloud n (per slot)
r_l	Size per replica of service l
κ_l	Size of input/output data per request for service l
ω_l	Computation requirement per request for service l
$a_{lnm} \in \{0, 1\}$	Indicates whether edge cloud m is permitted to serve type-(l, n) requests
$\lambda_{ln}^t, \lambda_{ln}^f$	Actual number of type-(l, n) requests in slot t and average number of type-(l, n) requests per slot in frame f
$c_{ln'n}$	Cost of replicating or migrating service l from cloud n' to edge cloud n, where cloud n' can be either a remote cloud or an edge cloud
B	Maximum cost for service placement in one frame
$x_{ln}^f \in \{0, 1\}$	Placement variable for frame f, 1 if service l is placed on edge cloud n and 0 otherwise
$y_{lnm}^t, y_{lnm}^f \in [0, 1]$	Scheduling variable representing the probability that a type-(l, n) request is scheduled to edge cloud m in slot t (under soft resource constraints) or frame f
z_{lnm}^t	Scheduling variable representing the number of type-(l, n) requests that are scheduled to edge cloud m in slot t (under hard resource constraints)

2.2 Service Placement Problem

Problem Formulation To evaluate the service placement cost, we assume that the services always exist on the remote cloud n_0, i.e., $x_{ln_0}^f \equiv 1$, and deleting a service replica from an edge cloud incurs no cost. Furthermore, we always replicate a service from the nearest location hosting the service. That is, the cost of placing service l at edge cloud n in frame f is $c_{ln}^f = \min_{n' \in N_+ : x_{ln'}^{f-1} = 1} c_{ln'n}$, where $c_{lnn} \equiv 0$.

The optimization problem for service placement can be formulated as (10.1) [29]:

$$\max \sum_{l \in L} \sum_{n \in N} \lambda_{ln} \sum_{m \in N} y_{lnm} \tag{10.1a}$$

$$\text{s.t.} \sum_{m \in N} y_{lnm} \leq 1, \qquad \forall l \in L, n \in N, \tag{10.1b}$$

$$\sum_{l \in L} x_{lm} r_l \leq R_m, \qquad\qquad \forall m \in N, \qquad\qquad (10.1c)$$

$$\sum_{l \in L} \lambda_{ln} \kappa_l \sum_{m \in N} y_{lnm} \leq K_n, \qquad\qquad \forall n \in N, \qquad\qquad (10.1d)$$

$$\sum_{l \in L} \omega_l \sum_{n \in N} \lambda_{ln} y_{lnm} \leq W_m, \qquad\qquad \forall m \in N, \qquad\qquad (10.1e)$$

$$y_{lnm} \leq a_{lnm} x_{lm}, \qquad\qquad \forall l \in L, n \in N, m \in N, \qquad\qquad (10.1f)$$

$$\sum_{l \in L} \sum_{n \in N} x_{ln} c_{ln} \leq B, \qquad\qquad (10.1g)$$

$$x_{ln} \in \{0, 1\}, y_{lnm} \geq 0, \qquad\qquad \forall l \in L, n \in N, m \in N. \qquad\qquad (10.1h)$$

Objective (10.1a) maximizes the expected number of requests served per slot. Constraint (10.1b) guarantees that the scheduling variables are valid. Constraint (10.1c) ensures that each edge cloud n does not store more than its storage capacity R_n. Constraint (10.1d) guarantees that the total communication demand on an edge cloud n stays within its communication capacity K_n on the average. Constraint (10.1e) ensures that the total computation demand scheduled to an edge cloud m is within its computation capacity W_m on the average. Constraint (10.1f) states that an edge cloud can only serve a request if it contains the requested service and is a candidate server. Constraint (10.1g) ensures that the total service placement cost is within the budget. Constraint (10.1h) specifies valid ranges of the decision variables.

At the beginning of each frame f, we solve (10.1) with the predicted request rates $\lambda_{ln} = \lambda_{ln}^f$ and the placement costs $c_{ln} = c_{ln}^f$ for the service placement x_{ln}^f and the corresponding request scheduling y_{lnm}^f. Then only x_{ln}^f will be used (to place services). Although the scheduling variable y_{lnm}^f will not be used for actual scheduling, it is needed to evaluate the served request rate (10.1a) under a given service placement. For this reason, we refer to y_{lnm}^f as the *shadow scheduling variable*.

Complexity Analysis The optimization problem (10.1) is a *mixed integer linear program (MILP)*, which can be hard to solve. The first task in tackling this problem is therefore to understand its complexity. To this end, we note that while (10.1) is related to several known problems in the literature, e.g., the knapsack problem, the *data placement problem (DPP)* [30], the *generalized assignment problem (GAP)* [31], and the *distributed caching problem (DCP)* [32], it is fundamentally different. Specifically, the above known problems are all special cases of the *separable assignment problem (SAP)* [32]. SAP considers packing items into bins under general packing constraints that can model both dedicated and non-dedicated resources. For example, if items represent requests and bins represent edge clouds, then SAP can model service placement where requests for the same service can share a service replica, while each consuming a dedicated share of computation resource and bandwidth. However, SAP requires all the resources consumed for

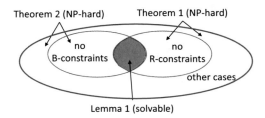

Fig. 10.3 Complexity of service placement (10.1)

serving a request to be with a single edge cloud, which is generally not true for the service placement problem.

To understand the complexity of (10.1), let us consider three special cases: (1) having B-constraint (10.1g) only, (2) having R-constraint (10.1c) only, and (3) removing both R- and B-constraints. The following has been shown in [16]:

Theorem 1 *The B-constraint alone makes the problem (10.1) NP-hard.*

Theorem 2 *The R-constraint alone makes the problem (10.1) NP-hard.*

Lemma 1 *Removing R- and B-constraints makes the problem (10.1) polynomial-time solvable.*

The insight from these results is that the service placement problem (10.1) is only polynomial-time solvable when there is neither a constraint on the budget (B-constraint) nor a constraint on the storage capacity (R-constraint), and it is NP-hard in all the other cases. Therefore, we have not only proved that this problem is NP-hard in general, but also characterized all the solvable special cases, which is highlighted in Fig. 10.3.

Approximation and Heuristic Algorithms The NP-hardndess of the problem in general signals the need of efficient suboptimal algorithms, preferably with performance guarantees. To this end, multiple approaches have been applied, including the greedy heuristic [16, 17], LP-relaxation with randomized rounding [27], and a simple solution based on local service popularities [16, 17]. Among these, the randomized rounding approach has the limitation that the solution may not be strictly feasible (i.e., it may exceed some of the capacities), and the popularity-based approach has the limitation that it does not jointly consider the scheduling of requests, which can distort the actual demand of a certain type of service at a given edge cloud (as some requests might be scheduled to nearby edge clouds). We will therefore detail the greedy heuristic as an exemplary solution.

To better understand this solution, we reformulate the problem (10.1) as a set function maximization. Let $S \subseteq L \times N$ denote the set of selected single-service placements, where $(l, n) \in S$ means to place a replica of service l at edge cloud n. Let $\Omega(S)$ denote the optimal objective value of (10.1) for a fixed **x** given by $x_{ln} = 1$ if and only if $(l, n) \in S$. This can be calculated by solving the following (shadow) request scheduling problem, where $\mathbb{1}_{l,m}$ is the indicator function:

$$\max \sum_{l \in L} \sum_{n \in N} \lambda_{ln} \sum_{m \in N} y_{lnm} \tag{10.2a}$$

$$\text{s.t. } (10.1b), (10.1d), (10.1e), \tag{10.2b}$$

$$y_{lnm} \leq a_{lnm} \mathbb{1}_{(l,m) \in S}, \qquad \forall l \in L, n \in N, m \in N, \tag{10.2c}$$

$$y_{lnm} \in [0, 1], \qquad \forall l \in L, n \in N, m \in N. \tag{10.2d}$$

After that, we can rewrite the service placement problem as:

$$\max \ \Omega(S) \tag{10.3a}$$

$$\text{s.t. } \sum_{l:(l,n) \in S} r_l \leq R_n, \qquad \forall n \in N, \tag{10.3b}$$

$$\sum_{(l,n) \in S} c_{ln} \leq B, \tag{10.3c}$$

$$S \subseteq L \times N, \tag{10.3d}$$

where $S_n \triangleq L \times \{n\}$ is the set of all possible service placements at edge cloud n.

The reformulation (10.3) brings out the gist of the service placement problem: *select a set of single service placements (denotes by (l, n)'s) such that the rate of served requests can be maximized under the optimal request scheduling.* Applying the generic greedy heuristic to this problem yields an algorithm called the *Greedy Service Placement based on Shadow Scheduling (GSP-SS)* that works as follows:

1. find a pair (l, n) such that placing a replica of service l at edge cloud n yields the maximum increase in $\Omega(S)$;
2. add (l, n) to S;
3. repeat steps 1–2 until no more service placement is allowed under (10.3b–10.3d).

While a greedy algorithm like GSP-SS generally does not have any performance guarantee, it has been shown in [16, 29] that under the following conditions, GSP-SS achieves a constant-factor approximation to the optimal solution.

Theorem 3 *If $\kappa_l \equiv \kappa$ ($\forall l \in L$), and*

1. *$\lfloor R_n/r_l \rfloor \leq 1$ for all $n \in N$ and $l \in L$, or*
2. *$W_m \geq \sum_{l \in L} \omega_l \sum_{n \in N} \lambda_{ln}$ for all $m \in N$,*

then GSP-SS yields a $1/(1 + p)$-approximation for (10.1), where $p = \left\lceil \frac{\max c_{ln}}{\min_{c_{ln} > 0} c_{ln}} \right\rceil + \left\lceil \frac{\max r_l}{\min_{l:r_l > 0} r_l} \right\rceil$.

The insight leading to the above result is that under these conditions, the objective function $\Omega(S)$ is a *monotone sub-modular* function [33], which is the discrete counterpart of monotone concave function. Moreover, constraints (10.3b)–(10.3d) form a *p-independence system*, which intuitively means that to add a new item

10 Joint Service Placement and Request Scheduling at the Edge 325

into S, we only need to take out at most p existing items. It is well-known that a monotone sub-modular function can be approximately maximized by the greedy heuristic with an approximation ratio of $1/(1 + p)$, if the set of feasible solutions forms a p-independence system [34].

2.3 Request Scheduling Problem

Problem Formulation Under Soft Resource Constraints Depending on whether requests submitted in a slot can be postponed till a later slot, the optimization problem for request scheduling differs slightly. If the requests can be postponed, then the average resource constraints (10.1d,10.1e) suffice, as temporary bursts in demands can be absorbed over time. In this case, at the beginning of each slot t within frame f, we solve (10.1) with the current demands $\lambda_{ln} = \lambda_{ln}^t$ and the previously determined service placement $x_{lm} = x_{lm}^f$ for the scheduling variable y_{lnm}^t, which is then used to schedule requests probabilistically in this slot.

Problem Formulation Under Hard Resource Constraints If the requests cannot be postponed, e.g., for real-time services with hard deadlines, then we must impose hard resource constraints such that all the requests scheduled to the edge clouds in slot t can be finished within the same slot (the unscheduled requests will be routed to the remote cloud for processing). The corresponding optimization problem can be formulated as (10.4) ($\mathbb{N}$: natural numbers):

$$\max \sum_{l \in L} \sum_{n \in N} \sum_{m \in N} z_{lnm} \tag{10.4a}$$

$$\text{s.t.} \sum_{m \in N} z_{lnm} \leq \lambda_{ln}, \qquad \forall l \in L, n \in N, \tag{10.4b}$$

$$\sum_{l \in L} \kappa_l \sum_{m \in N} z_{lnm} \leq K_n, \qquad \forall n \in N, \tag{10.4c}$$

$$\sum_{l \in L} \omega_l \sum_{n \in N} z_{lnm} \leq W_m, \qquad \forall m \in N, \tag{10.4d}$$

$$z_{lnm} \leq a_{lnm} x_{lm} \lambda_{ln}, \qquad \forall l \in L, n \in N, m \in N, \tag{10.4e}$$

$$z_{lnm} \in \mathbb{N}, \qquad \forall l \in L, n \in N, m \in N. \tag{10.4f}$$

Optimization (10.4) is similar to (10.1) under a fixed service placement x_{lm}, after replacing $\lambda_{ln} y_{lnm}$ by a new variable z_{lnm}. The only difference is that we now impose an integer constraint (10.4f), which means that instead of only specifying the expected number of type-(l, n) requests to schedule to edge cloud m (i.e., $\lambda_{ln} y_{lnm}$), we specify the exact number (i.e., z_{lnm}). At the beginning of each slot

t, we solve (10.4) with the demand $\lambda_{ln} = \lambda^t_{ln}$ and the service placement $x_{lm} = x^f_{lm}$ (f: the current frame) for the scheduling variable z^t_{lnm}, which is used to schedule requests deterministically in this slot. The deterministic scheduling ensures that instead of satisfying the communication/computation capacities on the average as in (10.1d,10.1e), we now satisfy them strictly, which ensures that all the scheduled requests can finish within the slot.

Complexity Analysis The request scheduling problem under soft constraints is a *linear program (LP)* and hence polynomial-time solvable using existing LP solvers. Under hard constraints, however, the problem (10.4) is an *integer linear program (ILP)*, which can be hard to solve. To understand its complexity, let us consider the following special cases: (1) the requests are heterogeneous in their resource demands (i.e., κ_l and ω_l can be different for different services), and (2) the requests are homogeneous in their resource demands.

In the case of heterogeneous requests, the following has been shown in [29]:[2]

Theorem 4 *The problem (10.4) is NP-hard even if $K_n \equiv K$ and $W_n \equiv W$ for all* $n \in N$.

Sketch of Proof The proof is to reduce the NP-complete *partition problem* to our problem. Given a set of positive integers $A = \{t_1, \ldots, t_m\}$, the partition problem decides whether A can be partitioned into two subsets A_1 and A_2 with equal sum. The idea is to construct two edge clouds corresponding to the two subsets, each having a computation capacity of $W := \frac{1}{2} \sum_{t_i \in A} t_i$ (and unlimited communication capacity), and one request for each integer $t_i \in A$ with computation requirement $\omega_i := t_i$. Scheduling request i to edge cloud j corresponds to partitioning integer t_i into subset A_j. Then it is easy to see that all the requests can be scheduled if and only if the partition problem is feasible. Note that this reduction holds even if there are only two edge clouds with $K_n \equiv \infty$ and $W_n \equiv W$.

In the case of homogeneous requests, the following has been shown in [29]:

Theorem 5 *In the special case of $\kappa_l \equiv \kappa$ and $\omega_l \equiv \omega$ for all $l \in L$, the problem (10.4) is polynomial-time solvable.*

Proof This statement is proved constructively by providing a polynomial-time optimal solution (see Theorem 6).

The insight from these results is that the request scheduling problem (10.4) is only polynomial-time solvable when the requests are homogeneous in their communication and computation demands, and it is NP-hard otherwise. Therefore, we have, again, not only proved the NP-hardness in general, but also characterized all the solvable special cases, as illustrated in Fig. 10.4.

---

[2] Although the theorem corresponding to Theorem 4 ([29, Theorem 3]) is for the general case of possibly heterogeneous K_n's and W_n's, the proof still holds in the case of homogeneous K_n's and W_n's.

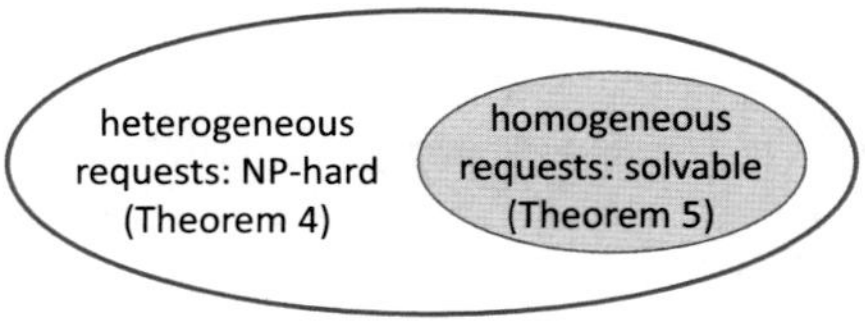

Fig. 10.4 Complexity of request scheduling under hard constraints (10.4)

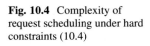

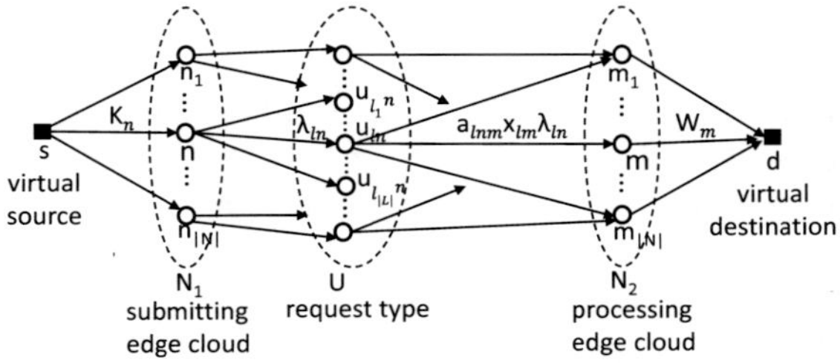

Fig. 10.5 Auxiliary graph $\mathcal{G}$ for request scheduling (assuming $\kappa_l \equiv \omega_l \equiv 1$, $K_n, W_n \in \mathbb{N}$)

Optimal and Heuristic Algorithms As mentioned above, the request scheduling problem under soft constraints is an LP and hence solvable by a standard LP solver, and hence the attention is focused on the version with hard constraints.

Let us start with the special case of homogeneous requests, i.e., $\kappa_l \equiv \kappa$ and $\omega_l \equiv \omega$ for all $l \in L$. By Theorem 5, we should be able to develop a polynomial-time algorithm that can solve the problem optimally. In fact, this theorem was proved constructively by explicitly developing such an algorithm. Without loss of generality, assume $\kappa_l \equiv 1$, $\omega_l \equiv 1$, and K_n and W_n are integers for all $n \in N$. The key observation is that in this special case, (10.4) can be converted to a *maximum flow problem in an auxiliary graph*, and is thus polynomial-time solvable by any of the existing maximum flow algorithms.

Specifically, consider an auxiliary graph $\mathcal{G}$ illustrated in Fig. 10.5. The nodes in $\mathcal{G}$ consist of a source s, a destination d, a set of nodes U in 1-1 correspondence with the types of requests $\{(l, n)\}_{l \in L, n \in N}$, and two sets of nodes N_1 and N_2, each in 1-1 correspondence with the edge clouds. Node s is connected to each node $n \in N_1$ by a directed link of capacity K_n, and each node $m \in N_2$ is connected to node d by a directed link of capacity W_m. Moreover, each node $n \in N_1$ is connected to each node $u_{ln} \in U$ (representing type-(l, n) requests) by a directed link of capacity λ_{ln}, and each node $u_{ln} \in U$ is connected to each node $m \in N_2$ by a directed link of capacity $a_{lnm}x_{lm}\lambda_{ln}$.

It has been shown in [29] that the problem of scheduling homogeneous requests under hard resource constraints is equivalent to the problem of finding the maximum flow from s to d in $\mathcal{G}$, as stated below.

Theorem 6 *For homogeneous requests, the optimal value of (10.4) equals the maximum flow from s to d in $\mathcal{G}$ (Fig. 10.5), and the optimal solution is to set z_{lnm} to the flow rate on link (u_{ln}, m) under the maximum integral flow from s to d.*

Sketch of Proof The idea is to use link capacities in the auxiliary graph $\mathcal{G}$ to represent the constraints in (10.4). Specifically, the capacity between $n \in N_1$ and $u_{ln} \in U$ ensures constraint (10.4b), the capacity between s and $n \in N_1$ ensures constraint (10.4c), the capacity between $m \in N_2$ and d ensures constraint (10.4d), and the capacity between $u_{ln} \in U$ and $m \in N_2$ ensures constraint (10.4e). The last constraint (10.4f) is ensured by requiring the s-to-d flow to be integral (i.e., having an integral rate on each link). This ensures that setting z_{lnm} to the flow rate on link (u_{ln}, m) under any integral s-to-d flow gives a feasible solution to (10.4). Similarly, each feasible solution to (10.4) corresponds to an integral flow from s to d in $\mathcal{G}$. Moreover, as all the link capacities in $\mathcal{G}$ are integers, the *integral flow theorem* [35] implies that there must be a maximum s-to-d flow that is integral. Hence, the scheduling solution corresponding to this maximum integral flow must be the optimal solution to (10.4).

Therefore, the optimal request scheduling algorithm for homogeneous requests under hard constraints works as follows:

1. construct an auxiliary graph $\mathcal{G}$ as in Fig. 10.5;
2. compute the maximum integral flow from s to d in $\mathcal{G}$;
3. set z_{lnm} to the flow rate on link (u_{ln}, m) in $\mathcal{G}$ for every $(l, n, m) \in L \times N \times N$.

Importantly, the Ford-Fulkerson algorithm [35] can be used to solve step 2, which is guaranteed to give an integral solution for a graph with integral link capacities as what we have here.

In the general case where requests for different services can have different communication/computation demands, only heuristics are known. For example, we can apply LP relaxation with rounding, or the greedy heuristic. We will discuss the performances of these heuristics in Sect. 3.

3 Performance Evaluations

We complete our discussions with a sample of evaluation results and refer to [16, 17, 29] for more results.

Consider the scenario where $|N| = 6$ and $|L| = 100$. For each service l, the resource demands κ_l, ω_l, and r_l are drawn uniformly from [0.5, 1]. For each edge cloud n, the resource capacities R_n, K_n and W_n are drawn uniformly from the intervals [24, 36], [16, 24] and [32, 48], respectively. Note that the resource capacities are relative to the resource demands, and thus the absolute units are not important. What this parameter setting means is that the communication capacity can serve 16–24 most I/O-intensive requests, the computation capacity can serve 32–48 most computation-intensive requests, and the storage capacity can hold 24–36 maximum-sized services. The above setting makes the communication bandwidth the most restrictive resource, followed by the storage and then the

Fig. 10.6 Performance of service placement algorithms

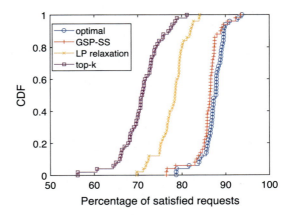

computation resources, but qualitatively similar observations have been made under other settings.

Assuming that the edge clouds are associated with hexagon cells arranged into two rows, we set the costs of replicating a service from an edge cloud k hops away or the remote cloud to $0.2k$ and 2, respectively, and the budget B for service placement to $0.2 \cdot |N| \cdot |L|$. We set a_{lnm} such that each request can only be served by edge clouds within 2 hops of the edge cloud it is submitted to. Each edge cloud n receives a total request rate λ_n that is randomly drawn from $[3, 5]$ (unit: requests/slot). Of these, p_{ln} fraction of requests are for service l, which follows the Zipf's distribution with skewness $\alpha = 0.5$. We initialize the system by randomly placing $|L|/8$ services and repeat the simulations for 50 Monte Carlo runs.

For service placement, we compare the optimal solution obtained by solving the MILP (10.1) (which is proved to be NP-hard) with GSP-SS [16], LP relaxation with rounding, and a baseline solution that ranks services by their local popularities and places services in the descending order of popularities within B- and R-constraints ('top-k'). Fig. 10.6 shows the cumulative distribution function (CDF) of the percentage of requests served by the edge clouds computed over the Monte Carlo runs. We see that GSP-SS, although being a heuristic in this case as the conditions in Theorem 3 do not hold, performs near-optimally, with a notable gap to other heuristics. Similar comparisons have been observed in other settings [16, 29]. The takeaway message is: *the design of service placement algorithm has a significant impact on the system performance*, and the greedy heuristic appears to perform very well.

For request scheduling, the nontrivial case is under hard resource constraints, as the problem is an LP under soft resource constraints. In the case of hard resource constraints with heterogeneous requests, only heuristics are known. We thus compare two popular heuristics for solving ILPs: LP relaxation with rounding, and the greedy heuristic. The CDFs of the percentage of served requests of these algorithms are compared against that of the optimal solution obtained by solving the ILP (10.4), as shown in Fig. 10.7. We see that in contrast to the clear gaps between

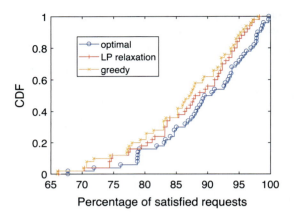

Fig. 10.7 Performance of request scheduling algorithms (under hard resource constraints)

service placement algorithms in Fig. 10.6, the two request scheduling algorithms are close to each other, both close to the optimal. Similar observations have been made in other settings [16, 29]. The takeaway message is: *the design of request scheduling algorithm does not have a significant impact on the system performance*, and common heuristics (e.g., LP relaxation, greedy) appear to be sufficient.

In practice, there are a number of further engineering decisions to make, e.g., how long is one slot (for request scheduling), how long is one frame (for service placement), and these parameters should relate to the delay tolerance and the level of dynamics. We refer to the case study based on real traces in [29] for details.

4 Summary

Due to the proximity to users and data sources, the network edge is an ideal location for many machine learning applications. However, its resource-constrained nature makes it crucial to employ efficient resource allocation. This chapter reviews the state of the art on resource allocation for machine learning at the edge based on given resource demands. At the high level, resource allocation for machine learning at the edge can be treated similarly to resource allocation for other workloads in mobile edge computing, both typically formulated as static optimization problems that are repeatedly solved at each decision epoch. Machine learning applications, however, have a distinctive requirement of the communication and storage of large amounts of data. To accommodate such requirement, we propose a resource allocation framework that employs different time scales for the decision of service data/code placement and the decision of request scheduling. We show that the corresponding optimization problems possess unique properties that lead to new challenges and discoveries, despite their conceptual similarity to many existing packing/assignment problems. In contrast to the complete characterization of their complexities, efficient solutions with performance guarantees are only known in special cases (e.g., under the conditions in Theorems 3 and 6). For many such problems, whether there exists an efficient solution that can achieve guaranteed performance in the general case remains open. As mobile edge computing is still an active research field, we expect that more models, frameworks, and algorithms will be developed in the near future that can be applied to manage resource allocation to better support machine learning applications at the network edge, where they are most needed.

References

1. P. Mach, Z. Becvar, Mobile edge computing: a survey on architecture and computation offloading. IEEE Commun. Surv. Tuts. **19**(3), 1628–1656 (2017)
2. S. Wang, R. Urgaonkar, M. Zafer, T. He, K. Chan, K.K. Leung, Dynamic service migration in mobile edge-clouds, in *IFIP Networking* (2015)
3. M. Satyanarayanan, G. Lewis, E. Morris, S. Simanta, J. Boleng, K. Ha, The role of cloudlets in hostile environments. IEEE Pervasive Comput. **12**(4), 40–49 (2013)
4. F. Bonomi, R. Milito, J. Zhu, S. Addepalli, Fog computing and its role in the Internet of Things, in *MCC'12: Proceedings of the First Edition of the MCC Workshop on Mobile Cloud Computing* (2012)
5. T. Taleb, A. Ksentini, Follow me cloud: interworking federated clouds and distributed mobile networks. IEEE Netw. **27**(5), 12–19 (2013)
6. S. Wang, R. Urgaonkar, T. He, K. Chan, M. Zafer, K.K. Leung, Dynamic service placement for mobile micro-clouds with predicted future costs. IEEE Trans. Parallel Distrib. Syst. **28**(4), 1002–1016 (2017)
7. K. Ha, Y. Abe, Z. Chen, W. He, B. Amos, P. Pillai, M. Satyanarayanan, Adaptive VM handoff across cloudlets. Technical Report CMU-CS-15-113 (2015). https://www.cs.cmu.edu/~satya/docdir/CMU-CS-15-113.pdf
8. T. Taleb, A. Ksentini, P. Frangoudis, Follow-me cloud: when cloud services follow mobile users. IEEE Trans. Cloud Comput. **7**, 369–382 (2016)
9. A. Ksentini, T. Taleb, M. Chen, A Markov decision process-based service migration procedure for Follow Me cloud, in *IEEE International Conference on Communications* (2014)
10. S. Wang, R. Urgaonkar, T. He, M. Zafer, K. Chan, K.K. Leung, Mobility-induced service migration in mobile micro-clouds, in *IEEE Military Communications Conference* (2014)
11. M. Jia, J. Cao, W. Liang, Optimal cloudlet placement and user to cloudlet allocation in wireless metropolitan area networks. IEEE Trans. Cloud Comput. **5**, 725–737 (2015)
12. Z. Xu, W. Liang, W. Xu, M. Jia, S. Guo, Efficient algorithms for capacitated cloudlet placements. IEEE Trans. Parallel Distrib. Syst. **27**(10), 2866–2880 (2016)
13. A. Ceselli, M. Premoli, S. Secci, Mobile edge cloud network design optimization. IEEE/ACM Trans. Netw. **25**(3), 1818–1831 (2017)
14. L. Wang, L. Jiao, J. Li, M. Muhlhauser, Online resource allocation for arbitrary user mobility in distributed edge clouds, in *2017 IEEE 37th International Conference on Distributed Computing Systems (ICDCS)* (2017)
15. H. Tan, Z. Han, X.-Y. Li, F. Lau, Online job dispatching and scheduling in edge-clouds, in *IEEE INFOCOM* (2017)
16. V. Farhadi, F. Mehmeti, T. He, T.L. Porta, H. Khamfroush, S. Wang, K.S. Chan, Service placement and request scheduling for data-intensive applications in edge clouds, in *IEEE INFOCOM* (2019)
17. T. He, H. Khamfroush, S. Wang, T.L. Porta, S. Stein, It's hard to share: joint service placement and request scheduling in edge clouds with sharable and non-sharable resources, in *2018 IEEE 38th International Conference on Distributed Computing Systems (ICDCS)* (2018)

18. S. Wang, M. Zafer, K.K. Leung, Online placement of multi-component applications in edge computing environments. IEEE Access **5**, 2514–2533 (2017)
19. L. Tong, Y. Li, W. Gao, A hierarchical edge cloud architecture for mobile computing, in *IEEE INFOCOM 2016* (2016)
20. G. Dán, N. Carlsson, Dynamic content allocation for cloud-assisted service of periodic workloads, in *IEEE INFOCOM* (2014)
21. S. Shukla, O. Bhardwaj, A.A. Abouzeid, T. Salonidis, T. He, Hold'em caching: proactive retention-aware caching with multi-path routing for wireless edge networks, in *ACM Mobihoc* (2017)
22. I.H. Hou, T. Zhao, S. Wang, K. Chan, Asymptotically optimal algorithm for online reconfiguration of edge-clouds, in *ACM MobiHoc* (2016)
23. S. Borst, V. Gupta, A. Walid, Distributed caching algorithms for content distribution networks, in *IEEE INFOCOM* (2010)
24. M. Dehghan, B. Jiang, A. Seetharam, T. He, T. Salonidis, J. Kurose, D. Towsley, R. Sitaraman, On the complexity of optimal request routing and content caching in heterogeneous cache networks. IEEE/ACM Trans. Netw. **25**(3), 1635–1648 (2017)
25. R. Urgaonkar, S. Wang, T. He, M. Zafer, K. Chan, K.K. Leung, Dynamic service migration and workload scheduling in edge-clouds. Perform. Eval. **91**, 205–228 (2015)
26. H. Feng, J. Llorca, A.M. Tulino, D. Raz, A.F. Molisch, Approximation algorithms for the NFV service distribution problem, in *IEEE INFOCOM* (2017)
27. K. Poularakis, J. Llorca, A.M. Tulino, I. Taylor, L. Tassiulas, Joint service placement and request routing in multi-cell mobile edge computing networks, in *IEEE INFOCOM* (2019)
28. Y. Rochman, H. Levy, E. Brosh, Resource placement and assignment in distributed network topologies, in *IEEE INFOCOM* (2013)
29. V. Farhadi, F. Mehmeti, T. He, T.F.L. Porta, H. Khamfroush, S. Wang, K.S. Chan, K. Poularakis, Service placement and request scheduling for data-intensive applications in edge clouds. IEEE/ACM Trans. Netw. **29**(20), 779–792 (2021)
30. I. Baev, R. Rajaraman, C. Swamy, Approximation algorithms for data placement problems. SIAM J. Comput. **38**(4), 1411–1429 (2008)
31. D.B. Shmoys, É. Tardos, An approximation algorithm for the generalized assignment problem. Math. Program. **62**(1–3), 461–474 (1993)
32. L. Fleischer, M.X. Goemans, V.S. Mirrokni, M. Sviridenko, Tight approximation algorithms for maximum general assignment problems, in *ACM-SIAM SODA* (2006)
33. M. Fisher, G. Nemhauser, L. Wolsey, An analysis of approximations for maximizing submodular set functions – II. Math. Prog. Study **8**, 73–87 (1978)
34. A. Gupta, A. Roth, G. Schoenebeck, K. Talwar, Constrained non-monotone submodular maximization: offline and secretary algorithms, in *International Workshop on Internet and Network Economics*. Lecture Notes in Computer Science (LNCS), vol. 6484, no. 12 (2010)
35. B. Korte, J. Vygen, Network flows, in *Combinatorial Optimization*, chap. 8 (Springer, Berlin, 2000), pp. 153–184

Part III
Cross-cutting Thoughts

Chapter 11
Criticality-Based Data Segmentation and Resource Allocation in Machine Inference Pipelines

Shengzhong Liu, Lui Sha, and Tarek Abdelzaher

Abstract This chapter introduces a criticality-aware data segmentation and resource allocation framework for real-time machine perception pipelines at the edge, for running DNN-based perception models in real time on resource-constraint edge platforms to process the sensing data stream (i.e., sequence of image frames). Mainstream machine inference frameworks commonly adopt a simple First-in-First-out (FIFO) policy to process the perceived images in a holistic manner without differentiating the data criticality, which results in a significant form of *algorithmic priority inversion* issue. Priority inversion happens when data of lower priority are processed ahead of or together with data of higher priority. The proposed framework first segments the input data into fine-grained subframe regions with different criticality, and processes them in a priority-based manner with differentiated deadlines and computation resource allocation. We design the general architecture in a modularized way and implement multiple alternative algorithms for data segmentation, prioritization, and resource allocation respectively for different edge scenarios. Experimental results on autonomous driving applications show that the framework is able to provide more timely responses to critical regions with only negligible degradation in overall perception quality. We also extend the idea into two generalized edge AI scenarios: *collaborative multi-camera surveillance* and *edge-assisted live video analytics*.

1 Introduction

This chapter focuses on optimizing the schedulability of mission-critical real-time perception tasks on resource-constrained edge devices to remove priority

---

S. Liu (✉)
Shanghai Jiao Tong University, Shanghai, China
e-mail: shengzhong@sjtu.edu.cn.

L. Sha · T. Abdelzaher
University of Illinois at Urbana Champaign, Urbana, IL, USA
e-mail: lrs@illinois.edu; zaher@illinois.edu

© The Author(s), under exclusive license to Springer Nature Switzerland AG 2023
M. Srivatsa et al. (eds.), *Artificial Intelligence for Edge Computing*,
https://doi.org/10.1007/978-3-031-40787-1_11

inversion issues. Perception is one of the key components that enable system autonomy. It is also a major efficiency bottleneck that accounts for a considerable fraction of resource consumption [1, 10]. Due to the resource constraints, the edge platforms are not able to process high-dimensional sensing streams with computation-intensive deep neural networks (DNN) in real time. In general, priority inversion occurs in real-time systems when computations that are less critical (or that have longer deadlines) are performed together with or ahead of those that are more critical (or that have shorter deadlines). Current neural-network-based machine intelligence software suffers from a significant form of priority inversion on the path from perception to decision-making, because current algorithms process input data sequentially, as opposed to processing important parts of a scene first. This limitation may result in inferior system responsiveness to critical events, or (equivalently) increased cost of the hardware to meet mission needs.

To understand the present gap, observe that current perception-related neural networks perform many layers of manipulation of large multidimensional matrices (called *tensors*). Yet, the current state of the art in designing the underlying neural network libraries (e.g., *TensorFlow*) is reminiscent of what used to be called the *cyclic executive* [2] in early operating system literature. Cyclic executives, in contrast to priority-based real-time scheduling [9], processed all pieces of incoming computation at the same priority and quality (e.g., as nested loops). Similarly, given incoming data frames (e.g., multi-color images or 3D LiDAR point clouds), modern neural network algorithms process all data rows and columns at the same priority and quality, with no regard to cues from the physical environment that impact time-constraints and criticality of different parts of the data scene.

This flat processing is in sharp contrast to the way *humans* process information. Humans have an innate ability to not only perceive their environment but also make critical and timely attention allocation decisions that help us expend limited cognitive resources where they are most needed in a critical dynamic situation. For example, given a complex scene, such as a freeway where one of the nearby vehicles appears to have temporarily lost control of steering, human drivers are good at understanding what to focus on to plan a valid path forward amidst the resulting confusion. The lack of prioritized allocation of processing resources to different parts of an input data stream (e.g., from a camera) creates what we henceforth call *algorithmic priority inversion* [15]. In the above example, all pixels of the entire freeway scene are processed by the same algorithm at the same priority, as opposed to giving the runaway vehicle more attention while possibly temporarily ignoring other less important elements of the scene (e.g., far-away objects).

To enable a similar "prioritized attention allocation" mechanism in machine perception pipelines, we propose a novel real-time architecture that is capable of segmenting the input data into subframe regions of different criticality and subsequently processing them in a priority-based manner with differentiated resource allocation. There are two major components: *data cueing* and *real-time scheduling*. First, the data cueing mechanism deals with how to extract the semantically-meaningful regions from the original sensing stream. It answers the question about where the DNN models should look at. Second, the real-time scheduling module deals with the prioritization, resource allocation, and task batching decisions for processing the extracted subframe regions, which answers the question about how should the DNN models look at the subframe regions. In the proposed architecture, the algorithm priority inversion can be greatly resolved because we can not only process the critical data regions with prioritized order but also with higher processing fidelity. In the meanwhile, the degradation of the overall perception quality is negligible. As we will introduce later, we design the general architecture in a modularized way and provide multiple alternative implementations for each functional component that are suitable for different edge AI scenarios.

We provide extensive evaluation results on NVIDIA Jetson platforms with real-world autonomous driving datasets to demonstrate the improvement of the proposed real-time architecture in responding to critical situations on edge platforms, which effectively resolves the priority inversion in the existing machine inference pipelines. Besides, we also show that this improvement comes at only negligible degradation in the overall machine perception quality. To further demonstrate the general applicability of the proposed architecture to machine perception at the edge, we extend the key idea into two different edge AI applications, that are *multi-camera surveillance* and *edge-assisted live video analytics*. Our general "segment-and-inspect" strategy achieves substantial resource savings with respect to GPU computation and network transmission respectively, by allocating more resources to where they are most needed.

2 Architecture Overview

Let us consider an intelligent edge system equipped with a camera that observes its physical environment, a deep neural network (DNN) that processes the observations, and a control unit that must react in real time. As mentioned earlier, we focus on scheduling perception tasks for timely response to critical data regions. For example, the neural network might identify the locations and types of objects (i.e., object detection) present in the field of view so that subsequent path planning can be done accordingly. Specifically, we mainly consider a visual machine perception system with object recognition/detection as the target DNN task, which we generally call as the *frame inspection* task.

Figure 11.1 contrasts the traditional design of machine inference pipelines in such systems to the proposed criticality-aware architecture. In the traditional design, input data frames captured by sensors are processed sequentially by the neural network. DNN execution is typically non-preemptive. It considers one frame at a time, producing an output on each frame before the next frame is handled. Unfortunately, the multi-dimensional data frames captured by modern sensors (e.g., colored camera images and 3D LiDAR point clouds) carry information of different degrees of criticality at different regions of a frame. Data of different degrees of criticality may require a different processing latency. For example, processing parts of the image that represent faraway objects does not need to happen every frame,

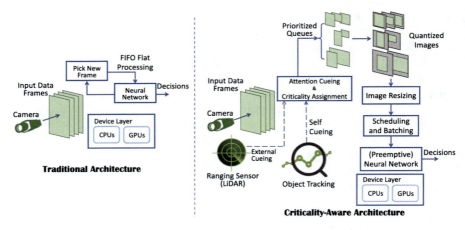

Fig. 11.1 The overview of the proposed criticality-aware real-time architecture

whereas processing nearby objects, such as a vehicle in front, needs to be done immediately because the nature of nearby objects (e.g., car versus pedestrian) has an impact on immediate path planning.

The above observation calls for two types of mechanisms. First, it is desired to efficiently identify subsets of input data frames that are likely to contain targets of interest. Second, it is desired to schedule data processing in a manner consistent with their importance and latency constraints. The two types of mechanisms respectively correspond to: (i) Data cueing:[1] *which subset of data* the aforementioned components will process and (ii) Real-time scheduling: *at what priority* to process the extracted regions with respect to both computation order and processing fidelity. The following two paragraphs briefly summarize solutions to such data segmentation and real-time scheduling problems.

Data Cueing Our first challenge is to quickly determine which parts of the frames to pass for inspection by the downstream DNN model (or other processing components), such that more critical data are processed first. This is a "chicken-and-eg" problem because we need to find important regions in the scene *before this scene is inspected by the DNN* and thus before some level of semantic understanding of the scene is achieved. Besides, the data cueing module need to run fast in real time to be used as a lightweight preprocessing before the actual DNN execution. We provide two different approaches: an external-cueing approach and a self-cueing approach. In the external-cueing method, the depth information from a synchronized range sensor is used to quickly identify the clusters of image pixels that potentially contain an object. Alternatively, in the self-cueing approach, we use the previously detected object locations (from the previous frame) and the optical flow estimated between the two consecutive frames to roughly project the object locations into the new

---

[1] We interchangeably use *data segmentation* and *data cueing* to denote the process of extracting regions of interest from a frame in this chapter.

frame. Both approaches are not expected to accurately estimate the object locations, but to be inclusive of the appeared objects where some background boundaries might be included in the extracted regions.

Real-Time Scheduling After extracting the list of subframe regions that potentially contain objects, we formulate their inspection tasks by downstream computational components (i.e., DNN model) as a real-time scheduling problem and solve it in a manner that meets data urgency and criticality constraints, while maximizing the resource utilization. We define the problem as *machine attention scheduling*. Such a problem was originally posed in [15], but has seen several extensions since [7, 11, 14]. Through this line of effort, we identified a multi-dimensional optimization space for machine attention scheduling, which includes: (1) **Prioritization**: It decides the processing order of the subframe regions. Intuitively, high-criticality regions (or with close deadlines) should be processed before the less-critical regions, but this can be violated in some cases for more batching opportunities (introduced next). (2) **Resource allocation**: We define two dimensions that can offer differentiated processing fidelities to regions with different criticality: *imprecise computation* and *image resizing*. With the imprecise computation model, the execution of DNN inspection on less-critical regions can be early exited in predefined exit points to save time for processing the critical regions. With the image resizing approach, we can downsize the less critical regions to lower resolutions to reduce their inspection latency. This is enabled by the advancements in recent block-wise and CNN-based neural network designs. (3) **Task batching**: Modern GPUs can process the image regions with the spatial size in parallel without causing significant extra latency, called *task batching*, which achieves higher processing throughput than the sequential processing. We maximally utilize the task batching opportunities to increase the volume of data that can be processed on the GPU before the deadlines. Ultimately, the scheduling problem needs to be formulated as one that balances the response efficiency to critical regions and the overall perception quality.

Generally, we believe the proposed criticality-based architecture opens up the space for fine-grained resource allocation optimization for machine perception pipelines on edge platforms. More details about each functional module will be introduced in the next two sections.

3 Data Cueing

We start with introducing the data cueing module, which should not only be able to extract the regions of interest in real time and reduce the overall areas to be inspected by the downstream visual DNN model, but also inclusively cover the appeared objects. As mentioned before, we propose two alternative cueing approaches: an external-cueing method that relies on the synchronized ranging sensor clusters to cue the regions of interest on the image plane, and a self-cueing approach that projects the previously detected object locations into the newly arriving frame

Fig. 11.2 An example of a camera frame segmented by the pixel distances measured by the synchronized LiDAR. Image extracted from Waymo open dataset [17]

with estimated optical flows. Between the two options, each method has its own advantages and limitations: The external cueing approach relies on input from a synchronized ranging sensor, which might not be available on all intelligent platforms. Reliance on multiple sensors increases cost and requires precise calibration and synchronization, where the degradation of either could cause downstream detection issues. However, the external cueing signals provide an estimate of locations or regions of interest to be inspected at every frame, thus avoiding the accumulation of estimation errors with time and simplifying the scheduling problem. With a self-cueing approach that works without dependency on external sensor inputs, we side-step the above fusion and synchronization challenges. However, in the absence of an external cue, a full frame needs to be processed occasionally to detect all objects of interest first. Uncertainty in object locations then keeps growing with the duration elapsed since the last time a full frame was processed. Essentially, no single approach comprehensively outperforms the other, but they can potentially be used in different application scenarios. Below, we summarize the technical details used in each approach and their associated evaluation performance.

3.1 External-Cueing Approach

In the external cueing approach [15], we assume that the observer is equipped with a *ranging* sensor, and its data can be projected into the image plane (as visualized in Fig. 11.2). For example, in autonomous driving systems, a LiDAR sensor measures distances between the vehicle and other objects. LiDAR point cloud based object localization techniques have been proposed [4] that provide a fast (i.e., over 200 Hz) and accurate ranging and object localization capability. The computed object locations can then be projected onto the image obtained from the camera, allowing the extraction of regions (subareas of the image) that represent these localized objects, sorted by distance from the observer. For simplicity, we restrict those subareas to rectangular regions, or *bounding boxes*. We define the priority (of bounding boxes) by time-to-collision, given the trajectory of the observer and the

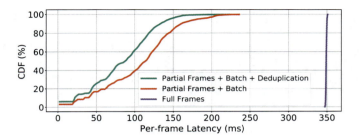

Fig. 11.3 The latency saving with the external-cueing approach

location of the object. Computing the time-to-collision is a well-studied topic and is not our contribution [16].

Besides, we also implement a deduplication function that automatically eliminates redundant bounding boxes. Since the same objects generally persist across many frames, the same bounding boxes will be identified in multiple frames. The set of bounding boxes pertaining to the same object in different frames is called a *tubelet*. Since the best information is usually the most recent, only the most recent bounding box in a tubelet needs to be acted on. The deduplication module identifies boxes with large overlap as redundant and stores the most recent box only. For efficiency reasons, we quantize the used bounding box sizes. The deduplication module uses the same box size for the same object throughout the entire tubelet. As we show in Fig. 11.3, when using ResNet-50 [6] as the DNN model, the external-cueing approach and a greedy batching on the extracted regions can reduce the processing latency on Jetson Xavier from 350 ms to below 200 ms, while the deduplication function can further reduce the latency for most frames.

3.2 Self-Cueing Approach

A self-cueing approach [11] builds on the assumption that strong temporal correlations exist in sensing data streams. For example, the consecutive frames in a video stream are highly similar. A self-cueing approach will process a full input frame at a lower frequency (e.g., once every few seconds) to localize and recognize all important targets, then use the observed target locations together with appropriately inferred motion vectors to approximately guess the locations of these same targets in the intermediate frames. Only the partial regions thought to contain the targets are processed in the intermediate frames. The remaining regions are skipped to save resources. The approach saves time by reducing the area that needs to be inspected by a deep neural network, or (in the case of data offloading) it saves by reducing the bandwidth consumption due to the data that need to be shared. To determine where a target is in intermediate frames, we use lightweight optical flow computations [8] to map the last-detected object locations onto the new frame (with

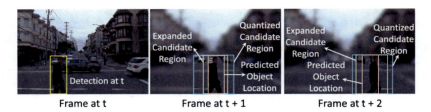

Fig. 11.4 Visualization of the self cueing process. The detection is shifted with the optical flow to get the predicted object location. It is then enlarged to the expanded candidate region according to the optical flow distribution, and further quantized to enable more batching opportunities

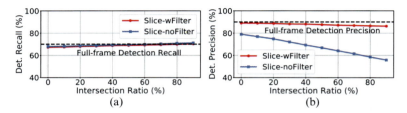

Fig. 11.5 The impact of the self-cueing approach with respect to the object detection quality. (**a**) Detection recall. (**b**) Detection precision

appropriate region expansion to account for uncertainty). The key idea is inspired by the encoding done by common video codecs, except that it directly incorporates application-level semantics (e.g., the purpose of the mission) into the selection of regions to focus on. For example, motion of clouds in the sky need not be tracked. Specifically, to identify the regions of interest on a new frame, we perform a three-step process starting from the last detected object bounding boxes, which is also visually explained in Fig. 11.4.

1. **Predicted Object Location:** It tightly bounds the most likely (predicted) object location from the optical flow map. We use this updated location as the best guess of the current object location in the absence of an actual object inspection.
2. **Expanded Candidate Region:** It expands the predicted location on account of uncertainty. This is the area that should be inspected by the detector if we want to localize the object again. It is a box whose area keeps expanding until an inspection of this region is scheduled.
3. **Quantized Candidate Region:** We pad the expanded candidate region to the nearest quantized size from a preset set. This is done to improve subsequent batching opportunities since same-size images can be processed in parallel (batched).

However, due to the visual overlaps among objects, there could be fragmented parts of objects detected in the extracted candidate regions. To resolve their impact, we propose a simple filtering approach to filter out the fragmented detections (details can be found in [11]). As we show in Fig. 11.5, the self-cueing approach

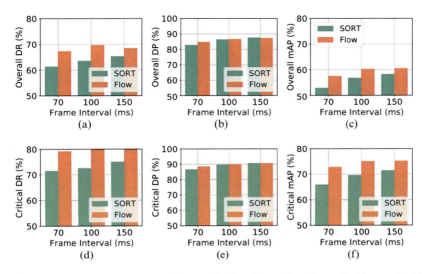

Fig. 11.6 The impact of tracking algorithms on the detection quality of overall objects and critical objects. DR means detection recall, DP means detection precision, and mAP means mean average precision. (**a**) Overall DR. (**b**) Overall DP. (**c**) Overall mAP. (**d**) Critical DR. (**e**) Critical DP. (**f**) Critical mAP

and corresponding partial-frame inspections, along with the filtering method, cause almost no degradation to the object detection accuracy, compared to inspecting the full image frames. We also justify the choice of tracking with the optical flow in Fig. 11.6, instead of the conventional tracking algorithms (e.g., SORT [3] with Kalman filter), by comparing the achieved object detection quality on extracted regions, where the optical flow-based tracking shows better performance in terms of detection recall (DR), detection precision (DP), and mean average precision (mAP) metrics on both overall objects and critical objects. The critical objects are defined according to their semantic classes and the approximated physical object distances according to the spatial bounding box sizes.

4 Real-Time Scheduling

We define a multi-dimensional scheduling space to simultaneously optimize the response efficiency to critical data regions and the overall machine perception quality, including *prioritization, resource allocation*, and *task batching*. Prioritization deals with the processing order of the extracted subframe regions. Resource allocation deals with how to allocate different degrees of attention (processing fidelity and latency) to regions of different criticality, such that more resources can be reserved for critical regions. Task batching is useful for increasing the overall data processing throughput, such that the overall perception quality can be improved.

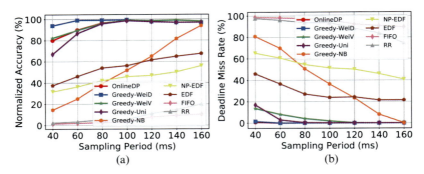

Fig. 11.7 The performance improvement of distance-based and relative-velocity-based prioritization with the external-cueing signal. (**a**) Normalized accuracy of critical objects. (**b**) Deadline miss rate of critical objects

However, it also means the priority among subframe regions can be violated within and across the batches, thus leading to a tradeoff between the two perspectives of system performance (responsiveness to critical regions vs. GPU utilization).

4.1 Prioritization

Prioritization decides the order of task processing, where mission-critical regions get a higher priority for a shorter response time. Different heuristics were developed to decide on priority assignments. In our architecture, we typically formulate the scheduling problem as a utility maximization problem, where the priority of each frame inspection task can be reflected in their corresponding weights of utility.

For example, when the sensing data stream comes with physical distance information, distance-based [15] or relative-velocity-based criticality [14] can be applied through different weighting mechanisms. If we consider close objects with a distance less than 10m as critical, the results in Fig. 11.7 indicate that using a distance-based prioritization algorithm (i.e., Greedy-WeiD) achieves higher accuracy and lower deadline miss rate than a relative velocity-based prioritization algorithm (i.e., Greedy-WeiV) and unweighted algorithm (i.e., Greedy-Uni). It is also noteworthy that all three variants use batching so their achieved performance is significantly better than the unbatched baselines.

Alternatively, with the self-cueing signal, we consider objects with higher uncertainty growth as more critical and prioritize their processing with higher inspection frequency over others [11], to bound the overall system uncertainty on tracked objects. When scheduling the object inspection frequency together with the task batching (introduce below) decisions, we also need to carefully align the inspection of objects with similar image sizes, so these tasks can be batched as much as possible. Therefore, a batched proportional balancing (BPB) algorithm is

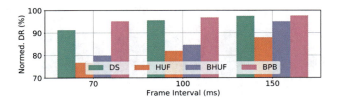

Fig. 11.8 The performance improvement of uncertainty-based prioritization with the self-cueing signal

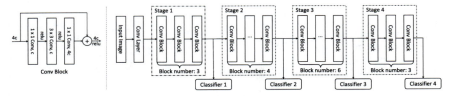

Fig. 11.9 ResNet [6] architecture with 4 stages and 50 layers. In the left part, we show the design of bottleneck block, which is the basic building block of ResNet. *c* represents the feature dimension. The classifier is simply the concatenation of a max pooling layer and a fully connected layer

correspondingly proposed in [11]. As we show in Fig. 11.8, it is able to achieve higher detection recall on physically close objects than directly downsizing (DS) the full image frame, especially when the frame sampling interval is low (i.e., 70 ms).

4.2 Resource Allocation

Resource allocation decides the amount of computation to be allocated to each task. This allocation depends on the task urgency requirement and its processing fidelity requirement. We developed two approaches to trade off fidelity and resource demand, both of which generally make the DNN execution more flexible. They are an imprecise computation model that divides the DNN execution into multiple stage-wise computations, and an image resizing mechanism that saves the computation time by downsizing the resolution of images.

In the first approach, instead of using a single-exit DNN, we proposed an imprecise computation model for DNNs [18], where multiple preset exit points are defined at different depths (see Fig. 11.9). Earlier exits save time by skipping the remaining DNN layers, but offer a lower fidelity (i.e., prediction accuracy). According to the profiled results with ResNet-50 on ImageNet in Fig. 11.10, as we execute more stages, the achieved overall model accuracy also increases. With the imprecise computation model, the number of stages executed for each task becomes a new dimension of scheduling decisions. The tasks with low criticality or those that already achieve high confidence after executing the first few stages can be early

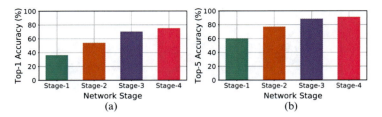

Fig. 11.10 ResNet stage accuracy change on ImageNet [5] dataset. (**a**) Top-1 accuracy. (**b**) Top-5 accuracy

Fig. 11.11 Average batch size comparison between imprecise computation (RTSS2020) and image downsizing (Proposed)

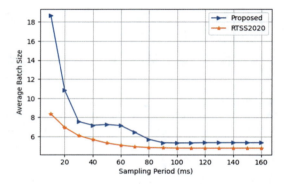

exited, to save the computation time for high-criticality regions and maximize the overall weighted system utility.

Alternatively, we can dynamically change the spatial resolution of the input data by resizing the frames (or frame regions) before feeding them into the computational components [7]. The ensuing execution latency is correlated with the input data dimensions. Lower-resolution images are faster to process but are lower in fidelity. This approach is easy to implement and does not require modification on the downstream DNN models. We downsize the extracted regions with low criticality and use high resolutions for critical regions, which turns out to offer more batching opportunities among tasks (see Fig. 11.11).

4.3 Task Batching

Task batching means we can combine multiple tasks that share the same processing kernel on the GPU. Batching typically requires that the tasks perform the same computations on inputs of the same dimension (e.g., same-size images). Batching exploits the advantages of parallel processing on modern GPUs; while batch response time increases slightly with batch size, it remains significantly lower compared to the latency of processing the tasks sequentially (as shown in Fig. 11.12). Since batching may entail putting tasks of different priorities into the

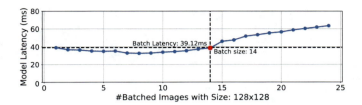

Fig. 11.12 The latency saving of task batching on GPU. Results profiled with YOLOv5 on Jetson Xavier SoC

same batch, it should be balanced carefully against prioritization to attain a good trade-off between the latency of high-priority tasks and overall GPU utilization.

Batching is regarded as the most straightforward way to exploit the parallel processing capacity on GPU with significantly higher data throughput, but it also complicates the scheduling problem significantly because of the conceptual contradiction between prioritization and batching. In addition, we think the GPU capacity is still utilized in a suboptimal way with batching because we have to include more regions in the quantization step, which is not needed from the perspective of downstream task accuracy. More intelligent data grouping strategies, like canvas-based grouping that arranges multiple small images into a large image template in an elegant way, can be proposed in the future for better efficiency.

Depending on the selected scheduling objectives, different scheduling algorithms were developed to optimize in the above multidimensional trade-off space, such as dynamic programming [15, 18], simple (batched) greedy heuristics [7, 14], and proportional planning algorithms [11]. Their technical details are skipped here but can be found in the individual papers. In summary, mission-informed real-time data management algorithms simultaneously improve the response efficiency to mission-critical data regions and the overall GPU utilization (and perception quality).

5 Generalized Applications

In this section, we extend the key design philosophy of criticality-based and fine-grained resource management into two generalized edge AI application scenarios: *multi-camera surveillance* and *edge-assisted live video analytics*. New application-specific optimization spaces are identified along with the general idea.

5.1 Collaborative Multi-Camera Surveillance

In this subsection, we extend our data segmentation and resource management architecture from the single-device scenario into a multi-camera collaboration scenario. Specifically, we consider DNN-based analytics of live video streams

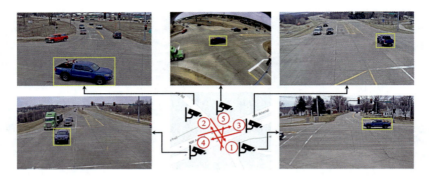

Fig. 11.13 An example of camera view overlaps. We use yellow boxes to highlight an object observable to every camera

associated with a multi-camera infrastructure deployment, where an overall region is collectively monitored by a set of cameras with partial spatial field-of-view (FoV) overlaps. DNN analytics happen locally on camera devices with limited computing capacities. One such example is given in Fig. 11.13.

We have the following two observations. First, with our cueing-and-inspection framework, the overall frame processing latency of an individual camera, proportional to the number of objects it currently tracks, shows significant variability across time. Second, due to the opportunistic, progressive nature of such in-the-wild infrastructure deployments, there is often significant heterogeneity in the processing capability across cameras–e.g., the memory capacity and GPU cores available. As a result, an identical object-level workload may generate very different processing latency on distinct devices. Meanwhile, objects that appear in the overlapped regions do not need to be repetitively tracked by multiple cameras.

We propose to reduce the DNN execution latency by eliminating the redundant DNN inspections, on multiple cameras, in such overlapping regions. The main idea [12] we proposed is a finer-grained *object-level, workload-aware, latency-balancing* approach, where the responsibility for tracking distinct objects in the overall shared region, is dynamically and periodically redistributed among the set of collaborating cameras. It simultaneously handles the computing power heterogeneity among the cameras, and the dynamic camera workload distribution at different times.

We design a batch-aware latency-balanced (BALB) scheduling algorithms to solve this problem. The BALB algorithm employs a hybrid centralized-cum-distributed scheduling mechanism to reduce and distribute the object-tracking workload. First, all cameras communicate with a *central scheduler*, which can be an edge node. After a full frame inspection, each camera uploads its list of detected objects to the central scheduler. The central scheduler first associates the detected bounding boxes and identifies the common objects across cameras, and then runs a *central stage* of the BALB algorithm to derive an initial object-to-camera assignment. The individual computing capacities and inspection workload

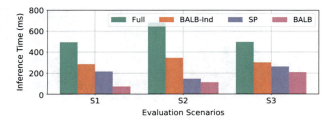

Fig. 11.14 Comparisons on per-frame YOLO inference latency for multi-camera collaboration

of cameras, as well as the task batching opportunities are all considered in the optimization.

To additionally tackle the unforeseen object dynamics between such periodic assignments (such as a new object entering the monitored region), in between full-frame inspections, each camera independently runs a *distributed stage* of the BALB algorithm to decide whether to track a new object or not. To minimize the need for frequent, per-frame communication, this distributed scheduler employs spatial partitioning to update the assignment for new objects and apportions the responsibility for tracking such object dynamics.

The experiment results (as shown in Fig. 11.14) on a testbed consisting of 5 NVIDIA Jetson platforms (of different models) with a realworld traffic surveillance dataset show that our system substantially improves the video processing speed, attaining multiplicative speedups of 2.45× to 6.85×, and consistently outperforms the competitive static region partitioning strategy.

5.2 Edge-Assisted Live Video Analytics

In this subsection, we show that the data cueing idea is not only applicable to the onboard processing tasks, but can also be extended to the edge-assisted DNN offloading scenario, and propose a novel machine-centric video streaming algorithm [13]. To set the context, video streams are offloaded from an end device to an edge server for remote video analytics, under limited and dynamic network bandwidth conditions. Note computation latency at the server side is no longer considered an efficiency bottleneck. We still consider DNN-based object detection as the downstream processing task. The objective of machine-centric video streaming is to maximize the downstream DNN inference accuracy, instead of the human viewing experience, in the presence of network dynamics (i.e., fluctuating bandwidth) (Fig. 11.15).

We observe that, by simply masking out the task-unrelated regions with black pixels, we can save the bandwidth of the offloaded video streams by 20–65%, while the degradation on the downstream object detection accuracy is negligible ($< 1\%$). This is because both the inter-frame correlation and intra-frame correlation among

Fig. 11.15 Original image vs. masked image (Original image from Waymo dataset [17]).

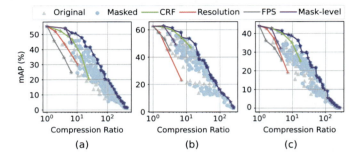

Fig. 11.16 The overall Pareto boundaries between video compression ratio and the downstream accuracy (mAP) in offloading

the video frame pixels are significantly enhanced, such that the video codecs like H.264 can store the video with fewer bits. In addition, the locations of objects of interest can be roughly extrapolated to future frames (from their last detected locations) [11] to inform masking. We further define a hierarchy of *frame masking levels* that mask out different portions of the images, because the static or slowly-moving objects can be masked out more often than those containing fast-moving objects. Masking, therefore, constitutes a novel task-driven control knob that we integrate with conventional H.264 control knobs (e.g., resolution, frame rate) to reshape the configuration space for video compression. A *configuration* denotes an instantiation of all control knobs, which can be changed dynamically at runtime to achieve different video bitrates. As we show in Fig. 11.16, the introduction of frame masking into the configuration space indeed improves the overall achievable Pareto boundary between the video compression ratio and the downstream model accuracy. The results are evaluated on three different video applications: dash cameras (Waymo), video surveillance cameras (AIC21), and drone cameras (VisDrone).

We further propose a runtime adaptation framework, called AdaMask, to automatically identify the Pareto-optimal configuration that maximizes the accuracy under bandwidth constraints. To design AdaMask, we first perform extensive profiling to study the impact of different control knobs on bandwidth consumption and inference accuracy over decoded frames. At runtime, to deal with the new scenarios, we further base on the online video content to customize the offline profiles for accuracy and bandwidth consumption estimation. We optimize the efficiency of the

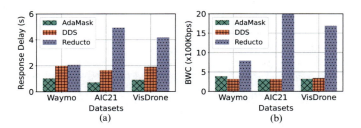

Fig. 11.17 End-to-end response delay and bandwidth consumption of AdaMask under real network connections. (**a**) Response delay. (**b**) Bandwidth

estimation algorithms with lookup tables, so the adaptation algorithm can efficiently find the Pareto-optimal configuration in real time. According to the experimental results with real WiFi connections in Fig. 11.17, AdaMask significantly reduces the response delays and the bandwidth consumption of the live video streams, without degrading the downstream model accuracy.

6 Conclusion

We introduced a criticality-based data segmentation and resource allocation architecture for DNN-based machine perception pipelines at the edge, whose advantage has been demonstrated in a wide range of edge AI application scenarios, including machine perception for autonomous systems, multi-camera collaborative surveillance, and edge-assisted live video analytics. Instead of processing the high-dimension image frames in a holistic manner, its key idea is to first use a lightweight cueing function to extract a set of subframe regions of interest from the original frame, and then schedule the processing of these regions in a criticality-aware manner with differentiated priority and processing fidelity. Multiple cueing functions have been introduced, and a multi-dimensional scheduling space is formulated through the combination of prioritization, imprecise computation, image resizing, and task batching mechanisms. We intend to open up the space of fine-grained machine attention scheduling problems for the readers but believe the current scheduling results can be further optimized with more intelligent design of future scheduling algorithms with new insights.

Acknowledgments Research reported in this chapter was sponsored in part by the Army Research Laboratory under Cooperative Agreement W911NF-17-20196, NSF CNS 20-38817, the IBM-Illinois Discovery Acceleration Institute, and the Boeing Company.

References

1. M. Alcon, H. Tabani, L. Kosmidis, E. Mezzetti, J. Abella, F.J. Cazorla, Timing of autonomous driving software: problem analysis and prospects for future solutions, in *2020 IEEE Real-Time and Embedded Technology and Applications Symposium (RTAS)* (IEEE, Piscataway, 2020), pp. 267–280
2. T.P. Baker, A. Shaw, The cyclic executive model and ada. Real-Time Syst. **1**(1), 7–25 (1989)
3. A. Bewley, Z. Ge, L. Ott, F. Ramos, B. Upcroft, Simple online and realtime tracking, in *2016 IEEE International Conference on Image Processing (ICIP)* (IEEE, Piscataway, 2016), pp. 3464–3468
4. I. Bogoslavskyi, C. Stachniss, Fast range image-based segmentation of sparse 3d laser scans for online operation, in *2016 IEEE/RSJ International Conference on Intelligent Robots and Systems (IROS)* (IEEE, Piscataway, 2016), pp. 163–169
5. J. Deng, W. Dong, R. Socher, L.J. Li, K. Li, L. Fei-Fei, Imagenet: a large-scale hierarchical image database, in *IEEE Conference on Computer Vision and Pattern Recognition, 2009. CVPR 2009* (IEEE, Piscataway, 2009), pp. 248–255
6. K. He, X. Zhang, S. Ren, J. Sun, Deep residual learning for image recognition, in *Proceedings of the IEEE Conference on Computer Vision and Pattern Recognition* (2016), pp. 770–778
7. Y. Hu, S. Liu, T. Abdelzaher, M. Wigness, P. David, On exploring image resizing for optimizing criticality-based machine perception, in *IEEE International Conference on Embedded and Real-Time Computing Systems and Applications (RTCSA)* (2021)
8. T. Kroeger, R. Timofte, D. Dai, L. Van Gool, Fast optical flow using dense inverse search, in *European Conference on Computer Vision* (Springer, Berlin, 2016), pp. 471–488
9. J. Lehoczky, L. Sha, Y. Ding, The rate monotonic scheduling algorithm: exact characterization and average case behavior, in *Real-Time Systems Symposium*, vol. 89 (1989), pp. 166–171
10. S.C. Lin, Y. Zhang, C.H. Hsu, M. Skach, M.E. Haque, L. Tang, J. Mars, The architectural implications of autonomous driving: constraints and acceleration, in *Proceedings of the Twenty-Third International Conference on Architectural Support for Programming Languages and Operating Systems* (2018), pp. 751–766
11. S. Liu, X. Fu, M. Wigness, P. David, S. Yao, L. Sha, T. Abdelzaher, Self-cueing real-time attention scheduling in criticality-aware visual machine perception, in *Proceedings of the 28th IEEE Real-Time and Embedded Technology and Applications Symposium (RTAS)* (2022)
12. S. Liu, T. Wang, H. Guo, X. Fu, P. David, M. Wigness, A. Misra, T. Abdelzaher, Multi-view scheduling of onboard live video analytics to minimize frame processing latency, in *2022 IEEE 42nd International Conference on Distributed Computing Systems (ICDCS)* (IEEE, Piscataway, 2022), pp. 503–514
13. S. Liu, T. Wang, J. Li, D. Sun, M. Srivastava, T. Abdelzaher, Adamask: enabling machine-centric video streaming with adaptive frame masking for DNN inference offloading, in *Proceedings of the 30th ACM International Conference on Multimedia* (2022), pp. 3035–3044
14. S. Liu, S. Yao, X. Fu, H. Shao, R. Tabish, S. Yu, A. Bansal, H. Yun, L. Sha, T. Abdelzaher, Real-time task scheduling for machine perception in intelligent cyber-physical systems. IEEE Trans. Comput. **71**, 1770–1783 (2021)
15. S. Liu, S. Yao, X. Fu, R. Tabish, S. Yu, H. Yun, L. Sha, T. Abdelzaher, On removing algorithmic priority inversion from mission-critical machine inference pipelines, in *Proceedings of IEEE Real-Time Systems Symposium (RTSS)* (2020)
16. M.M. Minderhoud, P.H. Bovy, Extended time-to-collision measures for road traffic safety assessment. Accident Anal. Prevent. **33**(1), 89–97 (2001)
17. P. Sun, H. Kretzschmar, X. Dotiwalla, A. Chouard, V. Patnaik, P. Tsui, J. Guo, Y. Zhou, Y. Chai, B. Caine, et al., Scalability in perception for autonomous driving: Waymo open dataset, in *Proceedings of the IEEE/CVF Conference on Computer Vision and Pattern Recognition* (2020), pp. 2446–2454
18. S. Yao, Y. Hao, Y. Zhao, H. Shao, D. Liu, S. Liu, T. Wang, J. Li, T. Abdelzaher, Scheduling real-time deep learning services as imprecise computations, in *2020 IEEE 26th International Conference on Embedded and Real-Time Computing Systems and Applications (RTCSA)* (IEEE, Piscataway, 2020)

Chapter 12
Model Operationalization at Edge Devices

Shikhar Kwatra, Utpal Mangla, and Mudhakar Srivatsa

Abstract This Chapter covers the core aspects related to Model Operationalization (ML-Ops). *MLOps* accelerates the journey of seamlessly building, training, validating and deploying the optimal models in production. This allows for the organization to accelerate their end to end pipeline from capturing the data from different sources and applying governance policies on the data to continuously monitoring the best candidate models deployed in production environment. This chapter further articulates the intersection of ML-Ops at edge devices and optimizations associated with efficiently creating ML-Ops pipeline across said edge devices.

1 Model Operationalization at a Glance

Data Engineers, Data scientists or Business Analysts, at scale have been experimenting with a variety of deep learning and machine learning models. Furthermore, in most enterprises, these personas have been involved in different stages of model creation, be it at model validation stage or at the step of presenting detailed insights to the third parties or respective customers.

According to a research conducted in 2016 (https://visit.figure-eight.com/rs/416-ZBE-142/images/CrowdFlower_DataScienceReport_2016.pdf), it is stated that 80% of the models created by other data scientists are side projects that are not put into production. Furthermore, there is widespread consensus among ML practitioners that data preparation accounts for approximately 80% of the time spent in developing a viable ML model (https://www.forbes.com/sites/gilpress/2016/03/23/data-preparation-most-time-consuming-least-enjoyable-data-science-task-survey-says/?sh=2fb540636f63).

S. Kwatra (✉)
AWS, Santa Clara, CA, USA
e-mail: sk4094@columbia.edu; shkwatra@amazon.com

U. Mangla · M. Srivatsa
IBM Thomas J. Watson Research Center, New York, NY, USA
e-mail: utpal@ibm.com; mudhakar@ibm.com

© The Author(s), under exclusive license to Springer Nature Switzerland AG 2023
M. Srivatsa et al. (eds.), *Artificial Intelligence for Edge Computing*,
https://doi.org/10.1007/978-3-031-40787-1_12

If we examine the statistics pertaining to different models being used in various industries, such as the financial, retail, and healthcare sectors, we find that the majority of them are rule-based models.

ML operationalization is the process of putting machine learning models into use in the real world so that businesses may benefit from them. The DevOps paradigm is layered on the model lifecycle management process through ML operationalization (CRISP-DM). One aspect for this process is monitoring not only prediction quality, but also system resources. This is important to correctly provide the necessary infrastructure, either using a fully-managed cloud platform or a local solution [1].

ML models can be compared to the engine of a car from an analogies standpoint. But when you examine a car, you don't just look at the engine; you also consider the chassis, sideview mirrors, transmission, entertainment system, etc., which round out the overall spec and infrastructure of the vehicle. Similar to this, the idea of operationalizing machine learning models includes the use of ML models along with independent validation, deployment, supporting a scalable servicing infrastructure, and having a Continuous Integration/Continuous Deployment (CI/CD) pipeline.

2 Pain Points

Operationalization of machine learning or deep learning models is difficult due to the following pain points:

- Time taken from model *conceptualization* to *productionization*—It can take weeks or even months to push or *productionize* a particular model post conceptualization due to several components involved in creating a ML-Ops pipeline.
- Serving millions of requests for creating a scalable infrastructure
- Effort in pushing from black-box to glass-box approach leading to more transparency
- The ability to explain to auditors why a particular model is predicting in a certain way
- Need to get Explanation for every case predicted by a specific model
- Need to have an optimized Infrastructure to support large number of Data Scientists and respective models

3 ML-Ops Key Metrics

In order to solve the above problems, Operationalization of models either at the cloud or edge needs to be solved, taking into consideration the following key metrics:

- Continuous Training

- Automated Validation and Deployment
- Insight Infusion at Scale
- Ensuring Transparency
- Removing Bias
- Business KPI Mapping
- Data and Model Governance
- Model Risk Management

4 Features of Model Operationalizaition Framework

In terms of inspecting the MLOps frameworks, the following features need to be considered for successful environment setup:

Features	Description
Flexibility or Customizability	How flexible is the platform in integrating and/or customizing new frameworks for AI model development.
Ease of Use	How easy is it to leverage these tools and proposed techniques from setup to application.
Integrations	How well does the platform integrate with Git or other model versioning and source control tools, catalogs (for governance and discoverability) or various data sources.
Governance	How well does the solution support governance and discoverability of assets (data assets, models, notebooks etc.)
Platform	Support for various platforms (public cloud, on-prem, hybrid cloud), and compute types (CPU/GPU) for training and scoring (or inference) AI models
Monitoring	How well does the solution support monitoring AI models for performance, explainability, fairness etc.
Scalability	How scalable is the platform in supporting various Data and AI users in different roles to explore, develop, and deploy AI models.
Openness	How well does the platform support open-source technologies which has become a key differentiator for platform providers.
Security	How well does the platform support enterprise-grade security access to the platform in terms of authorization and authentication.
Support for 5Cs	Support for continuous training, validation, deployment, integration and monitoring

5 Popular ML-Ops Frameworks

There are various frameworks and platforms available for ML Operationalization with varied degree of support for required Functional and Non Functional capabilities as demonstrated in the image below (Fig. 12.1).

6 Key Capabilities Needed Across the Steps of ML-Ops Pipeline

The flow shown below depicts different steps across the workflow that indicate the capabilities needed to build a successful ML-Ops life-cycle. The personas involved across each stage, for instance, Data Steward, Data Engineer may change based on the users within a respective organization who are involved in the end-to-end lifecycle of different machine learning projects.

In contrast to Application DevOps, ML-Ops not only takes care of Continuous Integration and Continuous Deployment but also covers Continuous Training, Continuous Validation and Continuous Monitoring (Fig. 12.2). [3] covers the various tooling that have emerged in automating MLOps pipeline.

6.1 Data Provisioning with Governance

This primary stage involves data Ingestion from various types of Data Sources – File Servers, Object Stores, Relational Databases, NoSQL, Hadoop, Rest Based Data

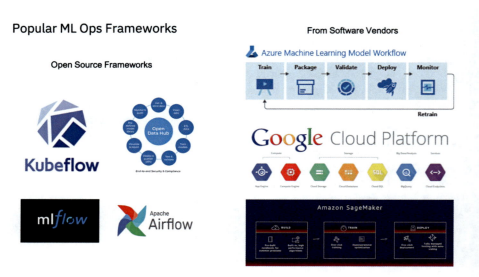

Fig. 12.1 Enterprise ML-Ops frameworks

12 Model Operationalization at Edge Devices

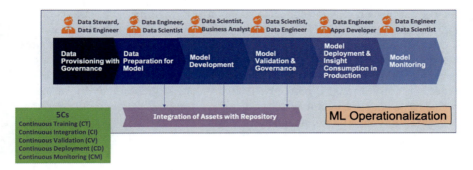

Fig. 12.2 ML-Ops pipeline

Services etc. This also implies metadata discovery, forming the data lineage, data quality control, overlaying canonical models, and enforcing of data security and data policies, along with virtualization of data including caching and joining between varied relational and non-relational data stores.

Data Lineage within the data provisioning layer aids in understanding, recording, and visualizing the data as it flows from a plethora of data sources to consumption. This can also include several data transformations that were executed and captured on the data itself. Data lineage can be further corroborated with tags for marking the data appropriately across the transformation lifecycle. Another form of data lineage may include parsing which relies on automatically reading logic such as SQL-based solutioning, extract transform load operations, etc. in order to process the data.

Akin to Data Lineage, Model Lineage assists in tracking data scientists and model builders with :

- Keeping a running history of model discovery experiments.
- Establishing model governance by tracking model lineage artifacts for auditing and compliance verification.

In medium-to-large scale enterprises, data lineage is infused with data classification as part of data governance as it plays an important part of an information security and compliance program. It can help in understanding and encapsulating sensitive data or critical information as data is parsed across a cross collaboration matrix of data engineers, data stewards, data scientists and data admin.

Hence, when it comes to data governance and classification, data protection plays a vital role. It involves many layers of protection including data loss prevention that aids in inspecting data in motion, or at rest, regardless of its location at a cloud or edge device. With data masking and encryption, encapsulating or obfuscating sensitive data like Social Security information, Date of birth or other PII (Personal Identifiable information) including health records can be prevented from being shared across multiple unauthenticated parties or get stolen.

Similarly, additional layers of protection Identity access management which can pertain to monitoring data access and activities of privileged users, database firewall for blocking Sql injection and common vulnerabilities, and alerting module to proactively and reactively alert the user of any suspicious activity with the data.

Data lineage assists users in ensuring their data is coming from a trusted source, has been transformed correctly, and loaded to the specified location. Hence, data becomes very difficult to verify or very costly and time-consuming, if such data processes are not tracked correctly. Even in Data Migrations, having a clearly defined data lineage can provide the tracking of information across multiple sources and access to such information quickly and easily.

6.2 Data Preparation for Models

As part of data selection, the user should be able to query the data from a variety of data sources, either via importing JDBC/ODBC driver or connecting via credentials secured by the authorized persona. Data scientists may not need access to all the data sources or data connections, and hence, data steward or data engineer may provision access to certain relevant data sources pertaining to the use case under consideration.

This stage of data preparation dives deeper into data selection, cleansing, exploration, and visualization of data. Feature engineering or data transformation can involve impute missing data with mean or medium, one hot encoding, masking sensitive or confidential information like names or addresses and time-series specific transformers to accelerate the preparation of time series data for ML.

Such feature engineering can be done either using UI based tools provided by different organizations like Data Refinery (IBM), Amazon Sagemaker Data Wrangler (AWS) etc. and via support for various Programming Paradigms inclusive of but not limited to Python, R, Scala, SQL etc. Simultaneously, support for industry standard IDEs including Rstudio, Jupyterlab etc. need to be handled with data profiling and visualization capabilities.

This should also involve handling large volumes of data via distributed computing and pushing different processes to data lake or data stores wherever needed. Data Preparation step should identify potential errors and extreme values with a set of robust pre-configured visualization templates including histograms, scatter plots, box and whisker plots, line plots, and bar charts etc. There are statistical techniques to further detect potential bias during data preparation.

Once your data is prepared, fully automated ML workflows can be built and pushed to the model building and development lifecycle.

6.3 Model Development

This stage involves building different machine learning models post data transformation. Further feature engineering may be done by the data scientist prior to building different machine learning models using various Machine Learning frameworks and libraries including TensorFlow, Keras, PyTorch, Scikit-learn, Spark etc.

For instance, optimization problems where linear programming and genetic algorithms are involved can be solved via Decision Optimization tools. Data Scientists or ML engineers use various industry standard IDEs including Rstudio, Jupyter Notebooks, Jupyter Labs etc. for the purposes of automated model development by applying candidate Algorithms, Hyper Parameter Optimization, Mathematical Feature Engineering, Code Generation and API support.

Best practices as part of Model Development include Experiment tracking and consistent model versioning, which helps to keep track of changes and/or revert back to previous working versions in cases where the model starts under-performing to incoming data.

It is crucial to track the code for the algorithm used, hyperparameters used, any changes to the dataset and the kind of results obtained for a certain set of hyperparameters. Model development is essentially an iterative process where multiple models are tested until the model fits the desired criteria suitable for staging the model or pushing the optimal model in production.

Various UI based automated machine learning tools exist inclusive of but not limited to AutoML, Sagemaker Autopilot, IBM AutoAI that eliminate the heavy lifting of building various candidate machine learning models and tuning the best ML Models based on the data, while maintaining full control and visibility, either via UI or API based access to the data scientists authorized to work on the use case. Some of these tools including Autopilot and AutoAI can generate model candidate notebooks with Model Explainability, showcasing how the model pipelines created feature engineering and model development steps. It also provides a feature importance score to showcase the relative importance of each input feature affecting the target label. Providing these notebooks can aid data scientists to accelerate the model building journey by directly performing custom operations in the generated notebooks and modifying the same as part of model development iterations.

Hence, we can constantly track the performance requirements of the models and deploy the model to production or iterate on the recommended solutions to further improve the model quality, thereby substantially reducing the likelihood of model drift as these optimal models are pushed into the production environment to handle live unseen traffic.

6.4 Model Validation and Governance

Independent model validation becomes an integral component of model risk management. Open Risk Manual defines Model Governance as:

An internal framework of a firm or organization that controls the processes for model development, validation and usage, assigns responsibilities and roles, etc.

Model Governance framework is needed to understand:

- Ownership of one or multiple models by the authorized persona
- Model compliance with relevant regulations and guidelines
- Model Artifacts information including training and validation data
- Performance tracking thresholds or approvals required at various stages of model development
- Post-deployment model monitoring means

Hence, this stage in the pipeline involves validating the developed machine learning and deep learning models with blind test data for Accuracy, Fairness, Potential Drift (data and model drift) and local feature importance.

Model validation should occur with proper model artifact versioning and model governance through workflow, while simultaneously, maintaining a proper model lineage. ML-Ops engineers/Model Validators are able to move approved model to production stage for continuous monitoring.

Model factsheets can also be created to showcase performance insights and model metadata. A factsheet is considered a collection of relevant information (facts) about the creation and deployment of an AI model or service. Facts could range from information about the purpose and criticality of the model, measured characteristics of the dataset, model, or service, or actions taken during the creation and deployment process of the model or service.

These model factsheets developed by IBM are tailored to the particular AI model or service being documented, and thus can vary in content. They are tailored to the needs of their target audience or consumer, and thus can vary in content and format, even for the same model or service. They capture model or service facts from the entire AI lifecycle; and said factsheets are compiled with inputs from multiple roles in this lifecycle as they perform their actions to increase the accuracy of these facts.

6.5 Model Deployment and Insight Consumption in Production

ML-Ops engineers or respective similar ML Model deployer personas in the industry are responsible for packaging and deploying the machine learning and deep learning models using various UI and API-based deployment for models.

Large enterprises working on hundreds of models at scale, with cross-collaboration of data scientists, data stewards, administrators, and business analysts are pushing dozens of optimal models in production pertaining to a use case under consideration. The purpose of deploying your model ensures you can make the predictions from a trained ML model available to others, whether that be users, management, or other systems.

There are several methods to deploy ML Models:

- **Batch or Offline Deployment**—Allows to take batches or sub-samples of data at a particular time, and get the batch predictions in offline or in other words,

12 Model Operationalization at Edge Devices 361

not requiring in real-time. Support for batch and online processing needs to be in place depending on the use case, along with support for high throughput Scoring requests from Consuming Applications with Scalability, Elasticity and Failover.

- **Real-Time Deployment**—There are certain cases where the users may want the prediction in real-time, for instance, determining the likelihood of the customer to churn or if the transaction is fraudulent or not. In such a case, there needs to be support for the REST interface which can be invoked via the endpoint to access the model for getting the prediction.

6.6 Model Monitoring

This stage supports continuous monitoring of models for online and offline inference with key performance metrics including accuracy, drift, fairness (to ensure trustworthiness in the models) along with Model Explainability to get reasoning behind the prediction as to why the model predicted a certain output.

Model monitoring can include:

- Data Quality monitoring: Monitoring the data draft. As more data or real time traffic is incoming to invoke the model endpoint, there might be drift in feature values that can degrade or modify the performance metrics of the model.
- Model Quality Monitoring: Model performance metrics, for instance, F1 Score or accuracy can be affected due to drift in model quality.
- Bias of Models in production—Due to imbalance in the data or skewed data for a use case, model's predictions may result in biased outcomes.

Tools like IBM AutoAI or Amazon Sagemaker can assist in determining Bias, model quality or data drift to create a fair model version. Logs can be captured to further send notifications to the authorized persona indicating the drift in model, data or bias being highlighted in the results.

Adding Human in the loop or having an aggregate can provide measures to ameliorate the results or model performance through balancing the data, generating synthetic data or taking care of bias in the data by taking into account fairness measures accordingly.

6.7 Integration with Source Code Repository

Adding automation to your workflow doesn't end with deploying code into production. Users have to keep track of new errors before they significantly impact them.

This stage involves efficient integration of model and data artifacts along with key metadata information to be tracked and stored in respective data and model repositories, for instance, GitHub, GitLab, BitBucket or other private container repositories.

The ML-Ops engineers should be able to leverage a UI based approach using existing Integration tools or API based approach.

6.8 Continuous Training, Integration, Validation Deployment and Monitoring

This stage is a universally running stage across the entire ML-Ops pipeline, wherein the ML-Ops engineer or persona responsible for pipeline workflow, should perform integration with external workflow tools. Said external tools include Jenkins, Proprietary Tools using Scripts and/or Command Line Interface (CLI) or proprietary internal workflow tools in order to maintain the 5Cs as part of best practices for operationalization of machine learning models. In [2], the team built a functional ML platform with DevOps capability from existing continuous integration (CI) or continuous delivery (CD) tools and Kubeflow, constructed and ran ML pipelines to train models with different layers and hyperparameters while time and computing resources consumed were recorded. This provides a valuable reference for ML pipeline platform construction in practice.

GitLab has Continuous Integration/Continuous Delivery (CI/CD) and DevOps workflows built-in. GitHub lets you work with the CI/CD tools of your choice, will need to be manually integrated by the users. GitHub users typically work with such third-party CI programs such as Jenkins, CircleCI, or TravisCI.

7 ML-Ops Intersection with Edge Computing

As articulated in previous sections, currently machine learning models are trained in cloud on servers with high compute and storage capacity. In order to get the inference from these models in the production environment, live data was processed and sent to the cloud via the internet gateway, invoke the rest endpoint (in case that model is packaged and deployed in production and can be invoked with REST API), get the results back from the cloud to the edge device and then perform the respective output as per the inference.

The aforementioned sequence for ML-Ops is applicable to edge computing as we are able to push models to the edge platforms and devices with efficient packaging of models and optimizing the sequence of operations performed to make them compatible to be executed on such edge devices.

Model frameworks such as Tensorflow, MXNet, PyTorch, XGBoost etc. can be optimized to be easily deployed across a fleet of devices independent of firmware and application updates. Automated ML-Ops pipelines can be created for edge servers, smart cameras and Internet of Things (IoT) enabled sensors.

12 Model Operationalization at Edge Devices

For instance, a computer vision emotion classifier model with the ability to recognize the emotions of the user based on user's expressions, first transfers the user's features at a given point of time to be classified by the emotion classifier to the edge device, further processed and inference regarding the emotion is delivered with near real time latency. In order to eliminate the latency, the machine learning models can now be optimized and deployed on the edge device without the need to process the data in the cloud. This leads to the rise of Operationalization of complex models at edge devices using the power of edge computing, IoT and ML-Ops.

As witnessed in the pain points, in ML-Ops in larger environments, from a general perspective, there are different sets of challenges incurred with respect to model deployments when targeted towards IoT enabled/edge devices.

- Since the models need to be deployed on different types of target platforms, there may be a lack of standardization or in other words, the models need to be highly device centric and hence, customized for each target platform along with taking security constraints, environment-based packaging and performance metrics into consideration.
- Such IoT based machine learning models need to be continually trained in a frequent manner as they tend to degrade faster due to rapidly drifting IoT data being ingested by such models.
- Inferences should be able to in online as well as offline mode since some IoT enabled Edge solutions may require offline operations with rapidly refreshing the model versions to adapt to such IoT solutions.

As different models are trained in the cloud, they need to be edge-ready. Hence, the models need to transformed to low-level framework agnostic operational understanding.

8 Edge Devices with MLOps Pipeline

Tools and frameworks like Amazon Sagemaker Edge can be used to operate ML Models on a fleet of devices. In a particular MLOps compute on edge scenario, we set an edge agent deployed on the edge device to sync with the edge manager. Edge agent is configured to capture the data from said edge devices to the cloud. We programmatically set triggers and sampling frequency to upload sampled data securely from any device. The agent runs on the CPU of your devices.

The agent runs inference on the framework and hardware of the target device you specified during the compilation job. For example, if you compiled your model for the Jetson Nano, the agent supports the GPU in the provided Deep Learning Runtime (DLR).

Once the edge agent has captured the sampled data, it stores the data to a data lake, such as cloud repository bucket like S3 for analysis and visualization.

Then, model can be built and trained using UI or custom modeling notebooks running in the cloud. The updated model satisfying the performance criteria goes through the optimization stage to create an optimized model version leveraging available hardware for edge inference. The optimized model is then packaged and signed to be device compatible. The model package if further deployed to fleet of edge devices. Monitoring agent can be deployed on the fleet of edge devices to continuously monitor all the models. Once the monitoring of models on edge devices is turned on, multiple models can seamlessly run on the edge devices.

9 Model Deployment on Edge Devices

In collaboration with Intel, Amazon developed DeepLens—a wireless camera with AI inferencing capabilities. This allows ML models to be trained on SageMaker with support for different ML/DL frameworks and then to be deployed on DeepLens in an integrated fashion. Optimizing models to deploy on the edge compresses the model size so it runs quickly.

Google has also entered the field with the underlying hardware—Google's Edge Tensor Processing Units (TPUs). Coral provides the full toolkit to train TensorFlow models and deploy them on different platforms using the Coral USB Accelerator.

As another example, IBM Edge Application Manager (IEAM) deploys machine learning (ML) models. In an edge solution, there could be many models created and deployed. These models need to be managed, and that's where the model management system (MMS) comes in to play.

IBM leverages MMS can be used to deploy, manage, and synchronize models across the edge tiers. It will facilitate the storage, delivery, and security of models/data and other metadata packages needed by edge and cloud services. edge nodes can send and receive models and metadata to and from the hybrid cloud.

For deployment on such edge devices, it is important to convert the framework-specific functions and operations into a framework-agnostic intermediate representation. Then, it is required to perform a series of optimizations for which you need to know the hardware architecture, instruction set, memory access patterns, and input data shapes etc.

Then, we generate binary code for the optimized operations, write them to a shared object library, and save the model definition and parameters into separate files. Then, the compiled model can be loaded and executed on the target edge device.

Hence, several methods including designing power-efficient ML algorithms, developing better and more specialized hardware, and inventing new distributed-learning algorithms where all IoT devices communicate and share data can be leveraged to use said complex ML models at edge devices. The last approach is limited by the network bandwidth, therefore future 5G networks, which provide ultra-reliable, low-latency communication services, will help immensely in the area of edge computing.

10 Conclusion

We covered the core aspects of Model Operationalization (ML-Ops) and its popular frameworks. We walked through the various components involved within the ML-Ops pipeline and elucidated the best practices to optimize the journey of machine learning and deep learning models. Further, we discussed the intersection of ML-Ops at edge devices and optimizations associated with efficiently creating ML-Ops pipeline across edge devices.

References

1. L. Cardoso Silva et al., Benchmarking machine learning solutions in production, in *2020 19th IEEE International Conference on Machine Learning and Applications (ICMLA)* (2020)
2. Y. Zhou, Y. Yu, B. Ding, Towards MLOps: a case study of ml pipeline platform, in *2020 International Conference on Artificial Intelligence and Computer Engineering (ICAICE)* (2020), pp. 494-500. https://doi.org/10.1109/ICAICE51518.2020.00102
3. P. Ruf, et al., Demystifying MLOps and presenting a recipe for the selection of open-source tools. Appl. Sci. **11**(19), 8861 (2021). https://doi.org/10.3390/app11198861